Xamarin iOS
移动开发实战

刘媛媛 编著

清华大学出版社
北　京

内 容 简 介

本书是国内第一本 Xamarin iOS 开发图书。本书由浅入深，全面系统地讲解了 Xamarin 开发 iOS 应用程序的各项技术。其内容贴近实际应用，涵盖开发的每个环节。在讲解的时候，大量地采用了实例的形式，帮助读者更快掌握相关技术。

本书共 15 章，分为 3 篇。第 1 篇为界面构建篇，主要介绍了 Xamarin 发展、Xcode 开发环境的搭建、第一个 iOS 应用程序的编写、真机测试、视图、控制器等内容。第 2 篇为资源使用篇，主要介绍了数据管理、数据显示、网络服务、多媒体资源、内置应用程序、与外部设备交互、位置服务和地图等内容。第 3 篇为高级应用篇，主要介绍了图形和动画、多任务处理、本地化、发布应用程序，以及高级功能等内容。

本书涉及面广，从基本界面构建到资源使用，再到高级技术，几乎涉及 iOS 应用程序开发的所有重要知识。本书不仅适合使用 Xamarin 开发 iOS 应用的专业人员，也适合有 C#语言基础的程序员及大中专院校的学生。对于经常使用 C#做开发的人员，本书更是一本不可多得的案头必备参考书。

本书封面贴有清华大学出版社防伪标签，无标签者不得销售。
版权所有，侵权必究。侵权举报电话：010-62782989　13701121933

图书在版编目（CIP）数据

Xamarin iOS 移动开发实战 / 刘媛媛编著. —北京：清华大学出版社，2015
ISBN 978-7-302-39573-7

Ⅰ. ①X… Ⅱ. ①刘… Ⅲ. ①移动终端–应用程序–程序设计 Ⅳ. ①TN929.53

中国版本图书馆 CIP 数据核字（2015）第 046543 号

责任编辑：杨如林
封面设计：欧振旭
责任校对：徐俊伟
责任印制：宋　林

出版发行：清华大学出版社
　　网　　址：http://www.tup.com.cn, http://www.wqbook.com
　　地　　址：北京清华大学学研大厦 A 座　　邮　　编：100084
　　社 总 机：010-62770175　　邮　　购：010-62786544
　　投稿与读者服务：010-62776969，c-service@tup.tsinghua.edu.cn
　　质 量 反 馈：010-62772015，zhiliang@tup.tsinghua.edu.cn

印　刷　者：清华大学印刷厂
装　订　者：三河市新茂装订有限公司
经　　　销：全国新华书店
开　　　本：185mm×260mm　　印　张：31　　字　数：774 千字
版　　　次：2015 年 5 月第 1 版　　印　次：2015 年 5 月第 1 次印刷
印　　　数：1～3500
定　　　价：99.80 元

产品编号：063106-01

前　　言

手机应用软件是软件开发的重要领域。根据基于的操作系统的不同，手机应用软件分为 iOS、Android 和 Windows Phone 三大类。由于系统所属的厂商不同，三大类软件所采用的开发语言和平台也不相同。这种局面给开发者造成了很大困扰，Xamarin 便因此而产生。

Xamarin 创始于 2011 年，到目前为止已有 3 年的历史了。在这期间，Xamarin 简化了针对多种平台的应用开发，包括 iOS、Android、Windows Phone 和 Mac。开发人员在 Xamarin 开发环境中，只要使用 C#语言就可开发出 iOS、Android 与 Windows 等平台的应用程序。

目前，国内图书市场上还没有一本 Xamarin 类图书，所以笔者结合自己多年的 C#开发经验和 iOS 开发经验，以及心得体会，花费了大量时间写作了本书。希望各位读者能在本书的引领下跨入 Xamarin 的 iOS 开发大门，并成为一名开发高手。

本书全面、系统、深入地介绍了 Xamarin iOS 应用程序的各项开发技术，并以大量实例贯穿于全书的讲解之中，最后还详细介绍了 iOS 应用程序的发布。学习完本书后，读者应该可以具备独立进行项目开发的能力。

本书特色

1．内容贴近实际开发

本书内容充分考虑开发者的需求。内容不仅包括环境的搭建、开发者账号申请和真机测试，还深入讲解了实际开发中的 Web 服务请求、多任务处理、本地化和发布程序等内容。

2．内容全面、系统、深入

本书介绍了使用 Xamarin 开发 iOS 应用程序的基础知识、界面构建，以及资源使用等。内容覆盖 iOS 开发各个方面。

3．贯穿大量的开发实例和技巧，迅速提升开发水平

本书在讲解知识点时贯穿了大量短小精悍的典型实例，并给出了大量的开发技巧，以便让读者更好地理解各种概念和开发技术，体验实际编程，迅速提高开发水平。

4．避免购买相应设备，降低学习成本

由于 iOS 开发必须基于苹果操作系统进行，因此读者进行开发时往往需要购买相应的苹果计算机。本书另辟蹊径，讲解如何在虚拟机中搭建相应的开发环境，给读者节省大量的设备购买费用。

本书内容及体系结构

第1篇　界面构建篇（第1~3章）

本篇主要内容包括 Xamarin 概述、Xcode 开发环境的搭建、使用 C#编写第一个 iOS 应用程序、真机测试、视图和控制器等内容。通过本篇的学习，读者可以具备基本的 iOS 开发能力，为后面的学习打下基础。

第2篇　资源使用篇（第4~10章）

本篇主要内容包括数据管理、数据显示、网络服务、多媒体资源、内置应用程序、与外部设备交互、位置服务和地图等内容。通过本篇的学习，读者可以掌握 iOS 各种资源的使用方法。

第3篇　高级应用篇（第11~15章）

本篇主要内容包括图形和动画、多任务处理、本地化、发布应用程序，以及高级功能等内容。通过本篇的学习，读者可以使应用程序更完善并且学会发布。

本书读者对象

- iOS 应用开发人员；
- 移动开发爱好者；
- 有 C#基础，想从事 iOS 开发的人员；
- C#开发爱好者；
- 大中专院校的学生；
- 社会培训班学员。

本书配套资源获取方式

本书涉及的源程序及开发环境需要读者自行下载。读者可以在 www.wanjuanchina.net 的相关版块上下载这些资源，也可以在清华大学出版社网站（www.tup.com.cn）上搜索到本书页面，然后按照提示下载。

本书售后服务方式

本书提供了完善的学习交流和沟通方式。主要有以下几种方式：

- 提供了技术论坛 http://www.wanjuanchina.net，读者可以将学习过程中遇到的问题发布到论坛上以获得帮助。
- 提供了 QQ 交流群 336212690，读者申请加入该群后便可以和作者及广大读者交流学习心得，解决学习中遇到的各种问题。

- 提供了 book@wanjuanchina.net 和 bookservice2008@163.com 服务邮箱，读者可以将自己的疑问发电子邮件以获取帮助。

本书作者

本书主要由刘媛媛编写。其他参与编写的人员有陈小云、陈晓梅、陈欣波、陈智敏、崔杰、戴晟晖、邓福金、董改香、董加强、杜磊、杜友丽、范祥、方家娣、房健、付青、傅志辉、高德明、高雁翔、宫虎波、古超、桂颖、郭刚、郭立峰、郭秋滟、韩德、韩花。

阅读本书的过程中若有任何疑问，都可以发邮件或者在论坛和 QQ 群里提问，会有专人为您解答。最后顺祝各位读者读书快乐！

编者

目　　录

第 1 篇　界面构建篇

第 1 章　使用 C#编写第一个 iOS 应用程序 ... 2
1.1　初识 Xamarin ... 2
1.1.1　Xamarin 发展 ... 2
1.1.2　Xamarin 特点 ... 2
1.1.3　Xamarin 版本 ... 3
1.1.4　工具需求 ... 4
1.2　搭建开发环境 ... 4
1.2.1　开发者账号 ... 4
1.2.2　下载和安装 Xamarin ... 7
1.2.3　下载和安装 Xcode ... 12
1.3　编写第一个应用程序 ... 15
1.3.1　创建工程 ... 15
1.3.2　编辑、连接、运行 ... 16
1.3.3　iOS Simulator ... 18
1.3.4　Interface Builder ... 21
1.3.5　编写代码 ... 23
1.3.6　调试程序 ... 24
1.3.7　文件简述 ... 25
1.4　使用真机测试应用程序 ... 26
1.4.1　申请付费开发者账号 ... 26
1.4.2　申请和下载证书 ... 28
1.4.3　实现真机测试 ... 36

第 2 章　用户界面——视图 ... 38
2.1　视图 ... 38
2.2　添加和定制视图 ... 38
2.2.1　使用 Interface Builder 添加视图 ... 38
2.2.2　使用代码添加视图 ... 40
2.2.3　删除视图 ... 42

2.2.4 视图的位置和大小 43
2.3 使用按钮接受用户输入 44
2.3.1 使用代码添加按钮 44
2.3.2 按钮的格式化设置 45
2.3.3 按钮的响应 49
2.4 显示图像 51
2.4.1 为视图显示图像 51
2.4.2 定制特殊的图像 54
2.5 显示和编辑文本 56
2.5.1 标签视图 56
2.5.2 文本框视图 59
2.5.3 文本视图 62
2.6 使用键盘 64
2.6.1 定制键盘的输入类型 65
2.6.2 显示键盘时改变输入视图的位置 67
2.6.3 为键盘添加工具栏 70
2.7 进度条 71
2.8 滚动视图 74
2.9 页面控件 77
2.10 警告视图 81
2.10.1 为主视图添加警告视图 81
2.10.2 常用的警告视图样式 82
2.10.3 响应警告视图 86
2.11 自定义视图 88
2.12 一次性修改相同的视图 91

第3章 用户界面——控制器 93
3.1 使用视图控制器加载视图 93
3.2 导航不同的视图控制器 100
3.2.1 导航控制器的基本组成 100
3.2.2 添加导航控制器 101
3.2.3 通过导航控制器实现视图的切换 102
3.2.4 管理导航栏上的按钮 108
3.3 在标签栏中提供控制器 111
3.3.1 添加标签栏控制器 111
3.3.2 标签栏控制器的常用属性 114
3.3.3 标签栏控制器的响应 115
3.4 模型视图控制器 118
3.5 创建自定义视图控制器 121
3.6 利用视图控制器的有效性 123

3.7 iPad 视图控制器 ······125
3.8 使用故事面板设计 UI ······129
3.9 故事面板中的 Unwind Segue ······135

第 2 篇 资源使用篇

第 4 章 数据管理 ······140
  4.1 文件管理 ······140
    4.1.1 创建文件 ······140
    4.1.2 写入/读取内容 ······143
    4.1.3 删除文件 ······145
  4.2 使用 SQLite 数据库 ······147
    4.2.1 创建数据库 ······147
    4.2.2 插入数据 ······152
    4.2.3 读取数据 ······152
    4.2.4 查看数据库 ······153
  4.3 使用 iCloud ······155
    4.3.1 启动 iCloud 服务 ······155
    4.3.2 在 iCloud 中存储键/值数据 ······156

第 5 章 显示数据 ······159
  5.1 选择列表 ······159
    5.1.1 日期选择器 ······159
    5.1.2 自定义选择器 ······162
  5.2 在表中显示数据 ······165
    5.2.1 表中内容的显示 ······166
    5.2.2 设置表 ······168
    5.2.3 设置表单元格 ······172
  5.3 编辑表 ······178
    5.3.1 选取行 ······178
    5.3.2 删除行 ······179
    5.3.3 插入行 ······181
    5.3.4 移动行 ······185
    5.3.5 缩进 ······187
  5.4 索引表 ······189
  5.5 数据的查找 ······191
  5.6 创建简单的网页浏览器 ······195
    5.6.1 加载网页视图的内容 ······195
    5.6.2 设置网页视图 ······199

5.6.3 网页视图常用事件 ·············· 201
5.7 在网格中显示数据 ················· 205
5.7.1 网格中内容的显示 ············· 205
5.7.2 自定义网格 ·················· 208
5.7.3 网格的响应 ·················· 213

第 6 章 网络服务 ······················ 214
6.1 使用 Web 服务 ···················· 214
6.1.1 构建一个 Web 服务 ············· 214
6.1.2 Web 服务的使用 ··············· 219
6.2 使用 REST 服务 ··················· 222
6.3 使用原生的 API 进行通信 ············ 225

第 7 章 多媒体资源 ···················· 228
7.1 选择图像和视频 ··················· 228
7.1.1 选择图像 ···················· 228
7.1.2 向模拟器中添加图像 ············ 230
7.1.3 设置图像显示来源 ············· 233
7.1.4 选择视频 ···················· 235
7.2 使用相机捕获媒体 ················· 238
7.2.1 打开相机 ···················· 238
7.2.2 设置相机 ···················· 239
7.2.3 捕获媒体 ···················· 241
7.2.4 自定义相机 ·················· 244
7.3 播放视频 ························· 248
7.3.1 播放视频文件 ················· 248
7.3.2 设置视频控制器 ··············· 250
7.3.3 视频播放控制器常用的监听事件 ·· 253
7.4 播放音频 ························· 256
7.4.1 播放较短的音频文件 ············ 256
7.4.2 播放较长的音频文件 ············ 259
7.4.3 访问音乐库 ·················· 264
7.5 使用麦克风录音 ··················· 268
7.6 直接管理相册 ····················· 271
7.6.1 获取相册中内容的路径 ·········· 271
7.6.2 读取相册中 EXIF 数据 ·········· 273
7.6.3 获取相册中的实际的照片 ········ 274

第 8 章 内置应用程序 ·················· 278
8.1 打电话 ·························· 278
8.2 使用 Safari ······················· 280

8.3 发送短信和电子邮件 ························· 283
   8.3.1 发送短信 ···································· 283
   8.3.2 发送电子邮件 ···························· 285
8.4 在应用程序中使用短信 ····················· 288
8.5 在应用程序中使用电子邮件 ··············· 291
8.6 管理地址簿 ·· 296
   8.6.1 访问地址簿 ································ 296
   8.6.2 打开地址簿 ································ 298
   8.6.3 添加联系人 ································ 299
   8.6.4 显示联系人信息 ························ 301
8.7 管理日历 ·· 305
   8.7.1 访问日历 ···································· 305
   8.7.2 打开日历事件界面 ···················· 307
   8.7.3 添加日历事件 ···························· 308

第 9 章 与外部设备交互 ······················· 316
9.1 检测设备的方向 ································ 316
9.2 调整 UI 的方向 ································· 317
9.3 近距离传感器 ···································· 319
9.4 获取电池信息 ···································· 321
9.5 处理运动事件 ···································· 323
9.6 处理触摸事件 ···································· 327
9.7 手势识别器 ·· 329
   9.7.1 轻拍 ·· 329
   9.7.2 捏 ·· 330
   9.7.3 滑动 ·· 332
   9.7.4 旋转 ·· 334
   9.7.5 移动 ·· 335
   9.7.6 长按 ·· 336
9.8 自定义手势 ·· 338
9.9 使用加速计 ·· 341
9.10 使用陀螺仪 ······································ 343

第 10 章 位置服务和地图 ······················ 346
10.1 确定位置 ·· 346
10.2 确定方向 ·· 349
10.3 使用区域监测 ·································· 352
10.4 使用 significant-change 位置服务 ····· 355
10.5 在后台运行位置服务 ······················ 357
10.6 使用地图 ·· 361
   10.6.1 显示地图 ·································· 361

10.6.2	改变地图的类型	361
10.6.3	在地图上显示当前位置	363
10.6.4	指定位置	365
10.6.5	添加标记	367
10.6.6	添加标注	369
10.6.7	限制地图的显示范围	372
10.6.8	添加覆盖图	374
10.7 地理编码		376

第3篇 高级应用篇

第11章 图形和动画 … 380
- 11.1 视图动画 … 380
  - 11.1.1 动画块 … 380
  - 11.1.2 修改动画块 … 382
  - 11.1.3 动画属性 … 383
  - 11.1.4 基于块的视图动画 … 383
- 11.2 视图的过渡动画 … 385
  - 11.2.1 旋转动画 … 386
  - 11.2.2 卷页动画 … 388
- 11.3 转换视图 … 391
- 11.4 计时器动画 … 393
- 11.5 图像动画 … 396
- 11.6 图层动画 … 398
- 11.7 图层的过渡动画 … 400
  - 11.7.1 公开的过渡动画 … 400
  - 11.7.2 非公开的过渡动画 … 402
- 11.8 绘制路径 … 406
  - 11.8.1 绘制线段 … 406
  - 11.8.2 绘制水平线 … 408
  - 11.8.3 绘制折线 … 409
  - 11.8.4 绘制曲线 … 410
- 11.9 绘制形状 … 412
- 11.10 绘制位图 … 414
  - 11.10.1 绘制单个位图 … 414
  - 11.10.2 绘制多个位图 … 415
- 11.11 绘制文字 … 416
- 11.12 创建一个简单的绘制应用程序——画板 … 418
- 11.13 创建位图图形上下文 … 419

第 12 章　多任务处理 …… 424
12.1　检测应用程序的状态 …… 424
12.2　接收应用程序状态的通知 …… 426
12.3　在后台运行代码 …… 428
12.4　在后台播放音频 …… 430
12.5　在后台更新数据 …… 433
12.6　禁用后台模式 …… 436

第 13 章　本地化 …… 438
13.1　创建一个具有多种语言的应用程序 …… 438
13.2　本地化资源 …… 442
13.3　区域格式 …… 444

第 14 章　发布应用程序 …… 447
14.1　申请发布证书 …… 447
14.1.1　申请证书 …… 447
14.1.2　申请证书对应的配置文件（Provision File） …… 449
14.2　准备提交应用程序 …… 451
14.2.1　创建应用及基本信息 …… 451
14.2.2　工程的相关设置 …… 454
14.3　提交应用程序到 App Store 上 …… 455
14.4　常见审核不通过的原因 …… 461

第 15 章　高级功能 …… 462
15.1　卷页效果 …… 462
15.2　粒子系统 …… 465
15.3　内容共享 …… 467
15.4　动作表单 …… 470
15.5　实现自定义过渡动画 …… 472
15.6　在 UI 元素中使用物理引擎 …… 477
15.7　实现文本到语言的功能 …… 479

第1篇　界面构建篇

▶▶　第1章　使用C#编写第一个iOS应用程序

▶▶　第2章　用户界面——视图

▶▶　第3章　用户界面——控制器

第1章 使用C#编写第一个iOS应用程序

C#原本是用来编写Windows以及Windows Phone的应用程序。自从Xamarin问世后，C#的作用就发生了很大的变化。它不仅可以编写关于Windows以及Windows Phone的应用程序，还可以编写iOS、Android的应用程序。本章将讲解如何使用C#编写一个简单的iOS应用程序。

1.1 初识Xamarin

Xamarin是一个跨平台的开发框架。Xamarin的产品简化了针对多种平台的应用开发，包括iOS、Android、Windows Phone和Mac App。本节将讲解Xamarin的发展、Xamarin的特点、Xamarin的版本，以及工具需求。

1.1.1 Xamarin发展

Xamarin创始于2011年，到现在为止已经有4年时间了。Xamarin自创建后到现在的发展如表1-1所示。

表1-1 Xamarin发展史

时间	事件
2011年	Xamarin被创建
2013年2月21日	Xamarin 2.0推出
2014年5月8日	Xamarin 3推出

1.1.2 Xamarin特点

Xamarin在短短的4年时间内，发展到现在众所周知，并且成为国内知名社区CSDN力推的开发框架，想必自有它的过人之处。以下就是Xamarin的几个重要特点。

1. 跨平台

Xamarin可以使用C#语言来编写iOS、Android、Mac，以及Windows应用程序。

2. 智能输入

开发应用程序时，开发工具会侦测开发者输入的部分字母，寻找对应的API并自动完成输入，俗称自动补全代码。

3．代码共享

现在由 Objective-C 所开发的 iOS 应用程序以及由 Java 所开发的 Android 应用程序，无法达到程序代码共享。Xamarin 则是采用 C#编写，因此透过良好的设计，可以在不同平台间共享商业逻辑以及数据存取等程序，无需重新编写。除了省下重新编写的时间成本外，对于版本维护及一致性也有相当大的帮助。

4．与Visual Studio整合

Xamarin 提供了 Visual Studio 2010/2012 的 plug-in，让原本就熟悉 Visual Studio 的开发者不用再学习其他的开发工具。在建立项目时，可以直接建立 iOS 及 Android 的项目模板。开发团队也可以将 iOS 及 Android 的程序代码纳入到 ALM，使用 Team Foundation Server 进行版本及建置的管理。

5．UI设计

目前在 Visual Studio 中已整合 Android 的 UI 设计功能，iOS 的 UI 编辑目前仍需依赖 Xcode。但是在 Xamarin Studio 中已经整合的 UI 的设计功能。为了方便开发者的学习，我们会使用 Xamarin Studio 进行 iOS 应用程序的开发，Xamarin Studio 是 Xamarin 开发框架的一部分。

6．确保第一时间更新

Xamarin 对于 iOS 及 Android 的版本更新不遗余力。在 iOS SDK 5.0、6.0 以及 6.1，都与 Apple 在同一天发表对应的 Framework 版本。

1.1.3 Xamarin 版本

Xamarin 提供了免费版和付费版。免费版本包含 Xamarin Studio 服务。付费版本分为普通版 299 美元/年、商业版 999 美元/年和企业版 1899 美元/年。开发者可以根据自身需要进行购买，如图 1.1 所示。

图 1.1　Xamarin 各个版本

> 注意：针对学生及研究人员，Xamarin 提供以 99 美元/年的价格购买商业版授权。

1.1.4 工具需求

要使用 C#编写 iOS 应用程序，需要使用如下 3 个工具：
- Mac 或者 Mac 虚拟机；
- Xamarin 开发框架；
- Xcode 开发工具。

1.2 搭建开发环境

每一种应用程序的开发都有自己特定的开发环境。所谓开发环境就是为了支持系统软件和应用软件工程化开发和维护的一组软件。它通常简称为 SDE。本节将讲解 C#开发 iOS 应用程序的开发环境搭建，其中包括 Xamarin 的下载和安装，以及 Xcode 的下载和安装。

1.2.1 开发者账号

只有注册了苹果开发者账号的成员才可以直接使用苹果公司的 iOS SDK。所谓 iOS SDK 包就是软件开发包，所以在 iOS 开发中我们要注册开发者账号。在苹果公司注册 iOS 开发者账号的成员一共可以分为 4 种，如表 1-2 所示。

表 1-2 iPhone开发者账号的成员

成 员 类 型	成　　本
在线开发成员	免费
标准 iPhone 开发成员	$99/年
企业 iPhone 开发成员	$299/年
大学 iPhone 开发成员	免费

下面，我们来讲解免费的苹果开发者账户的注册过程，具体步骤如下所述。

（1）在 Dock（Dock 一般指的是苹果操作系统中的停靠栏）中，找到 Safari，如图 1.2 所示。

图 1.2 操作步骤 1

（2）单击图标打开 Safrai，如图 1.3 所示。

（3）在搜索栏中输入网址（https://developer.apple.com/devcenter/ios/index.action），然

后按回车键,进入 iOS Dev Center-App Developer 网页,如图 1.4 所示。

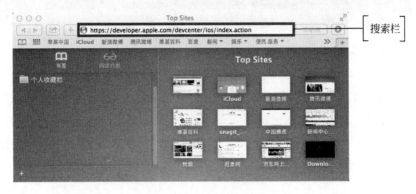

图 1.3 操作步骤 2

图 1.4 操作步骤 3

(4)选择 register for free 选项,进入 Apple Developer Registration-Apple Developer 网页,如图 1.5 所示。

图 1.5 操作步骤 4

(5)单击 Create Apple ID 按钮,进入 Apple-My Apple ID 网页,如图 1.6 所示。

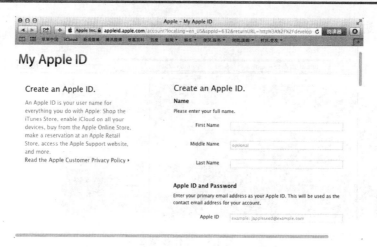

图 1.6　操作步骤 5

（6）在网页中输入一些内容后，单击网页最下方的 Create Apple ID 按钮，进入确定邮件地址的网页，如图 1.7 所示。

图 1.7　操作步骤 6

（7）单击 Continue 按钮，进入确定邮件地址的另一个网页。单击此网页中的 Send Verfication Email 按钮，进入确定邮件地址的另一个网页。

（8）打开需要确定的邮件地址，会看到 Apple 发来的一封确定邮件地址的邮件。打开该邮件，如图 1.8 所示。

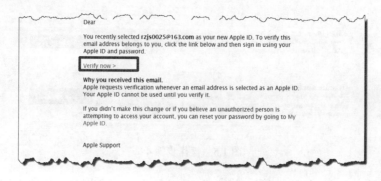

图 1.8　操作步骤 7

（9）单击 Verify now 链接，进入 Apple-My Apple ID-Email Verfication 网页，如图 1.9 所示。

图 1.9　操作步骤 8

（10）输入需要验证的邮箱以及地址，然后单击 Verify Address 按钮，进入下一个网页，此网页会提示开发者注册的 Apple ID 现在已经可以使用了。

1.2.2　下载和安装 Xamarin

以下是在 Mac（或者 Mac 虚拟机）上下载和安装 Xamarin 的具体步骤。

1. 下载Xamarin安装包

（1）在 Dock 中，找到 Safari，单击打开，在搜索框中输入网址（http://xamarin.com/download/）。然后按回车键，进入 DownLoad Xamarin for free to start building amazing native mobile apps-Xamarin 网页，如图 1.10 所示。

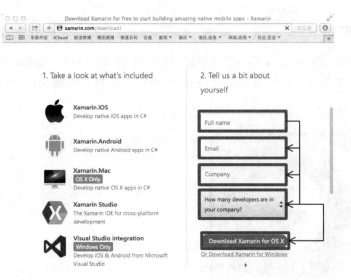

图 1.10　操作步骤 1

（2）输入名称、邮箱、公司名称，以及公司开发人员数目后，单击 DownLoad Xamarin for OS X 按钮，进入 Thanks for downloading Xamarin-Xamarin 网页，如图 1.11 所示。

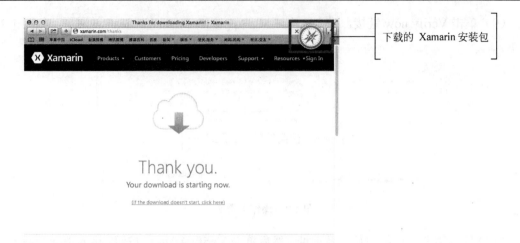

图 1.11　操作步骤 2

（3）一段时间后，在此网页中就会出现 Xamarin 安装包，此安装包将会移动到下载中去，进行下载，如图 1.12 所示。

图 1.12　操作步骤 3

2．安装 Xamarin

（1）XamarinInstaller.dmg 文件下载完毕后，双击，弹出"正在打开'XamarinInstaller.dmg'…"对话框。打开此文件后，弹出 Xamarin Installer 安装对话框，如图 1.13 所示。

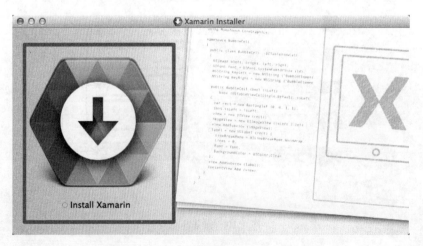

图 1.13　操作步骤 4

（2）双击 Install Xamarin 图标，弹性"'Install Xamarin'是从互联网下载的'应用程

序'。您确定要打开吗?"对话框,如图 1.14 所示。

图 1.14 操作步骤 5

(3) 单击"打开"按钮,弹出 Please review and accept the license in order to procceed 对话框,如图 1.15 所示。

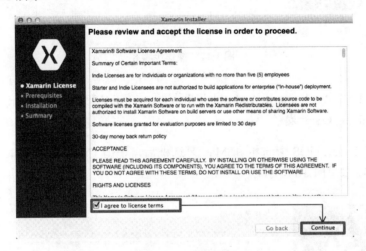

图 1.15 操作步骤 6

(4) 选择 I agree to license terms 复选框,然后单击 Continue 按钮,弹出 Welcome to the Xamarin Installer 对话框,如图 1.16 所示。

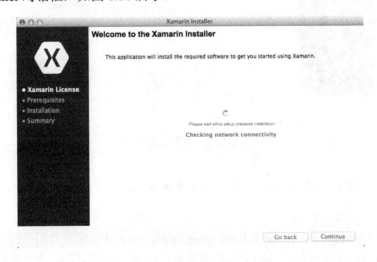

图 1.16 操作步骤 7

（5）经过一段时间后，会弹出 Please select products to install 对话框，如图 1.17 所示。

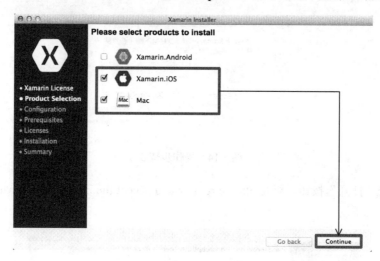

图 1.17　操作步骤 8

🔔注意：在图 1.17 中，Xamarin.Android 是用来做 Android 开发的，Xamarin.iOS 以及 Mac 是用来做 iOS 开发的。

（6）选择 Xamarin.iOS 复选框和 Mac 复选框后，单击 Continue 按钮，弹出 Xcode is Missing 对话框，如图 1.18 所示。

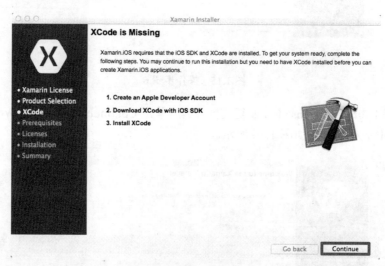

图 1.18　操作步骤 9

🔔注意：这时在 Mac 或者 Mac 虚拟机中还没有安装 Xcode。

（7）单击 Continue 按钮，弹出 Prerequisites 对话框，如图 1.19 所示。
（8）单击 Continue 按钮，弹出 Installation in progress 对话框，如图 1.20 所示。
（9）在此对话框中进行软件的下载以及安装，需要一段的时间，在这段时间中会时不时地出现"'Install Xamarin'想要进行更改。键入您的密码以允许执行此操作",如图 1.21 所示。

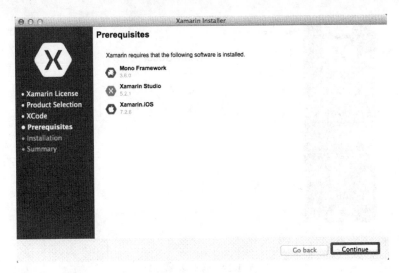

图 1.19 操作步骤 10

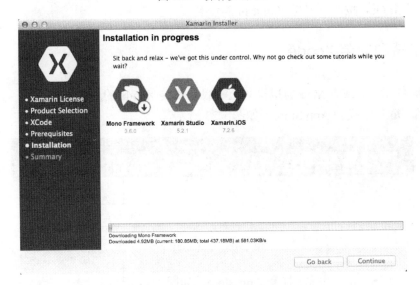

图 1.20 操作步骤 11

图 1.21 操作步骤 12

（10）输入密码，然后单击"好"按钮，才可以进行下一步的操作。在软件下载完毕后，就会弹出 Installation Finished 对话框，如图 1.22 所示。

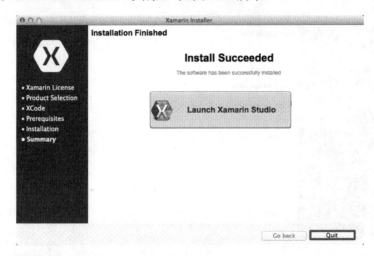

图 1.22　操作步骤 13

（11）单击 Quit 按钮，退出 Xamarin 的安装。

1.2.3　下载和安装 Xcode

以下是在 Mac（或者 Mac 虚拟机）上下载和安装 Xcode 的具体步骤。
（1）在 Dock 中找到 App Store，如图 1.23 所示。

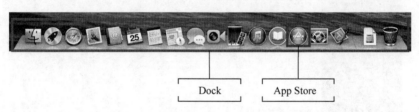

图 1.23　操作步骤 1

（2）单击 App Store 图标，打开 App Store，如图 1.24 所示。

图 1.24　操作步骤 2

（3）在搜索栏中输入要搜索的内容，即 Xcode，然后按回车键，进行搜索，如图 1.25 所示。

图 1.25　操作步骤 3

（4）单击 Xcode 右下方的"免费"按钮，此时"免费"按钮变为了"安装 APP"按钮，如图 1.26 所示。

图 1.26　操作步骤 4

（5）单击"安装 APP"按钮，弹出"登录 App Store 来下载"对话框，如图 1.27 所示。

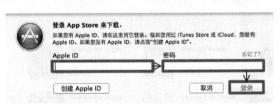

图 1.27　操作步骤 5

（6）输入 Apple ID 以及密码后，单击"登录"按钮。此时"安装 APP"按钮变为了"安装"按钮，如图 1.28 所示。并且 Xcode 会在 Launchpad 中进行下载和安装，如图 1.29 所示。

图 1.28　操作步骤 6

图 1.29　操作步骤 7

（7）一般在 Launchpad 中下载的软件，都可以在应用程序中找到。选择"前往"|"应用程序"命令打开应用程序，如图 1.30 所示。

图 1.30　操作步骤 8

（8）双击 Xcode 图标，弹出 Xcode and iOS SDK License Agreement 对话框，如图 1.31 所示。

（9）单击 Agree 按钮，弹出"'Xcode'想要进行更改。键入您的密码以允许执行此操作"，如图 1.32 所示。

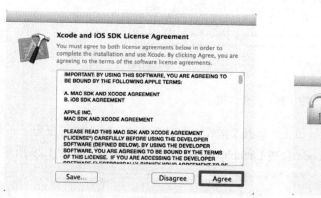

图 1.31　操作步骤 9

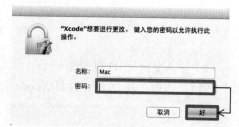

图 1.32　操作步骤 10

（10）输入密码，然后单击"好"按钮，进行组件的安装，组件安装完成后，就会弹出 Welcome to Xcode 对话框，如图 1.33 所示。

（11）单击左上角的"关闭"按钮，关闭 Xcode。

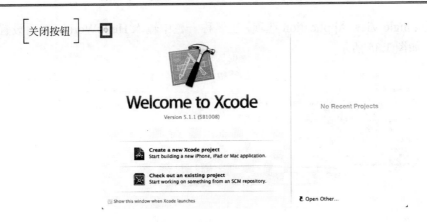

图 1.33　操作步骤 11

1.3　编写第一个应用程序

在 Xcode 以及 Xamarin 安装好后，就可以在 Xamarin Studio 中编写程序了。本节将主要讲解在 Xamarin Studio 中如何进行工程的创建，以及编写代码等内容。

1.3.1　创建工程

很多开发工具在编写代码之前，都必须创建一个工程，如 Visual Studio、Xcode 等。可以很好地将 iOS 应用程序开发中使用的文件都保存在这个工程中。在 Xamarin Studio 中该如何创建一个功能呢？以下就是它的步骤。

（1）单击 Xamarin Studio，弹出 Xamarin Studio 对话框，如图 1.34 所示。

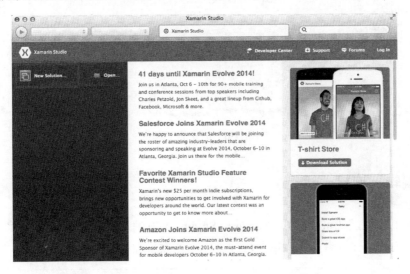

图 1.34　操作步骤 1

（2）选择 New Solution 选项，弹出"新建解决方案"对话框。选择 iOS 下的 iPhone

项目中的 Single View Application 选项,在名称一栏中输入 HelloWorld,然后设置工程的保存位置,如图 1.35 所示。

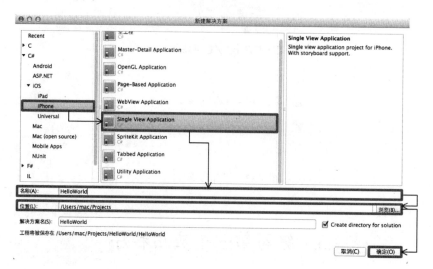

图 1.35　操作步骤 2

⚠ 注意:在输入名称时,不可以出现空格,只可以使用字符、数字、'.'、'_'。

(3)最后单击"确定"按钮,这样就创建好一个工程名为"HelloWorld"的工程了。在此工程中可以开发 iPhone 的应用程序,如图 1.36 所示。

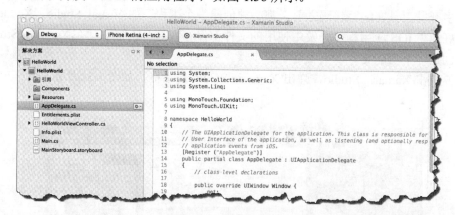

图 1.36　创建的工程

1.3.2　编辑、连接、运行

创建好工程后,就可以单击 Xamarin Studio 上方的运行按钮,如图 1.37 所示,对 HelloWorld 项目进行编辑、连接以及运行了。运行效果如图 1.38 所示。

图 1.37　运行按钮

由于在此 HelloWorld 的功能中没有做任何的事情,所以,运行结果是不会产生任何效果的。

> 注意:如果是第一次运行 Xamarin Studio 的程序,可能会出现如下的错误:

```
Error: A valid Xcode installation could not be found. If your copy of Xcode
is installed to a non-standard prefix, please specify the location in Xamarin
Studio's Preferences under 'SDK Locations'.
```

此错误是没有找到安装的 Xcode,解决此错误的步骤如下。

(1)选择菜单栏中的 Xamarin|Preferences 命令,弹出"选项"对话框,如图 1.39 所示。

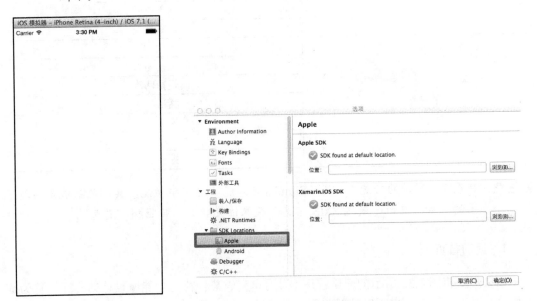

图 1.38　运行效果　　　　　　　　图 1.39　操作步骤 1

(2)选择 SDK Locations 下的 Apple 选项,在 Apple SDK 中单击"浏览"按钮,找到 Xcode 的位置,或者直接输入 Xcode 的位置,如图 1.40 所示。

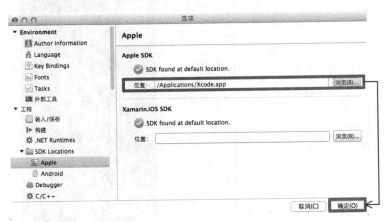

图 1.40　操作步骤 2

(3)单击"确定"按钮,就可以看到此错误就没有了。

1.3.3 iOS Simulator

在图 1.38 所示的运行效果中，所见到的类似于手机的模型就是 iOS Simulator。在没有 iPhone 或 iPad 设备时，可以使用 iOS Simulator 对程序进行检测。iOS Simulator 可以模仿真实的 iPhone 或 iPad 等设备的各种功能，如表 1-3 所示。

表 1-3 iOS Simulator

方面	功能
旋转屏幕	向上旋转
	向下旋转
	向右向左
手势支持	轻拍
	触摸与按下
	轻拍两次
	猛击
	轻弹
	拖动
	捏

> 注意：在表 1-3 所示的功能中，iOS Simulator 只能实现这些功能，其他的功能是实现不了的，如打电话、发送 SMS 信息、获取位置数据、照照相、麦克风等。

1. 返回首页

如果想要将图 1.38 所示的应用程序（为用户完成某种特定功能所设计的程序被称为应用程序）退出，该怎么办呢？这时就需要选择菜单栏中的"硬件"|"首页"命令，退出应用程序后的效果，如图 1.41 所示。

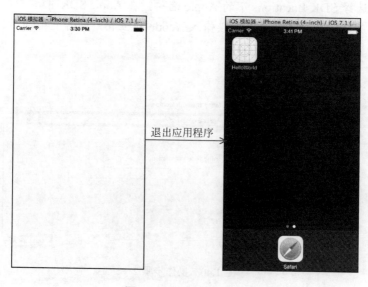

图 1.41 退出应用程序

2. 设置应用程序的图标

在图 1.41 中可以看到，退出应用程序后，此应用程序的图标是一个网格状的白色图标，它是 iOS 默认的图标，一般开发者都不会使用此图标，而是使用自己设置的图标。更改默认图标的具体步骤如下所述。

（1）回到 HelloWorld 工程中，双击打开 Info.plist 文件，如图 1.42 所示。

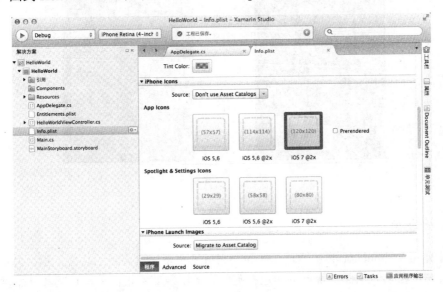

图 1.42　操作步骤 1

（2）找到 App Icons 下的 iOS 7 @2x，在 iOS 7 @2x 上方有一个空白区域，选择它，弹出选择文件对话框，选择某一图像文件，单击 Open 按钮，此时选择的图像就显示在这个空白区域上了，如图 1.43 所示。

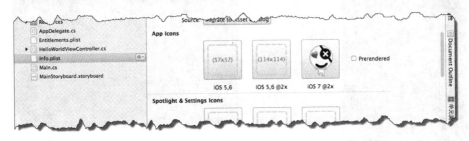

图 1.43　操作步骤 2

注意：在 App Icons 中，有 3 个空白区域。每一个空白区域对应的 iOS 都是不一样的，并且显示的内容大小也是不一样的。显示的内容必需符合空白区域的要求。例如，iOS 7 @2x 对应的空白区域要求显示 120×120 像素的内容，所选择的图像就必须是 120×120 像素的。

单击运行按钮，在出现运行效果后，退出应用程序，就可以看到更改后的图标了，如图 1.44 所示。

第1篇 界面构建篇

更改图标前　　　　　　　　　更改图标后

图1.44　运行效果

3．旋转屏幕

在讲解 iOS Simulator 的功能时，提到了它可以实现屏幕旋转的功能，那么怎样将 iOS Simulator 进行旋转，从而实现屏幕的选择呢？它的实现其实很简单，开发者只需要同时按住 Command 键和某一个方向键就可以了，如图1.45所示。

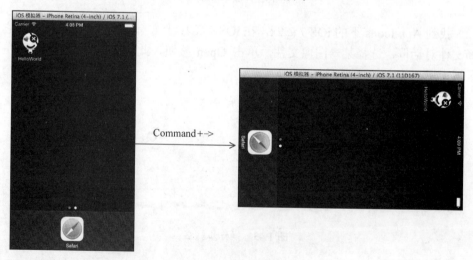

图1.45　旋转

4．删除应用程序

当开发者想要删除 iOS Simulator 上的某一应用程序，可以采用以下步骤。

（1）长按要删除的应用程序的图标，一段时间后，在要删除的应用程序的左上方会出现一个删除图标，如图1.46所示。

第 1 章　使用 C#编写第一个 iOS 应用程序

［删除图标］

图 1.46　操作步骤 1

（2）单击删除图标，会弹出一个提醒对话框，如图 1.47 所示。
（3）单击 Delete 按钮，删除此应用程序，此时 iOS Simulator 的效果如图 1.48 所示。

图 1.47　操作步骤 2　　　　　　　　　图 1.48　操作步骤 3

1.3.4　Interface Builder

　　Interface Builder 被称为编辑界面，它是一个虚拟的图形化设计工具，用来为 iOS 应用程序创建图形界面，单击打开 MainStoryboard.storyboard 就打开编辑界面了，在现在的 Xamarin Studio 5.2.1 中，编辑界面直接使用的是 Storyboard 故事面板，以前则使用的是 XIB（对于 XIB 的使用会在后面的章节中讲解）。它们之间最大的不同在于，Storyboard 故事面板可以对多个主视图进行设置，而 XIB 只可以对一个主视图进行设置。以下是对编辑界面的介绍。

1. 界面构成

单击 MainStoryboard.storyboard 文件打开编辑界面后,可以看到编辑界面由 4 部分组成,如图 1.49 所示。

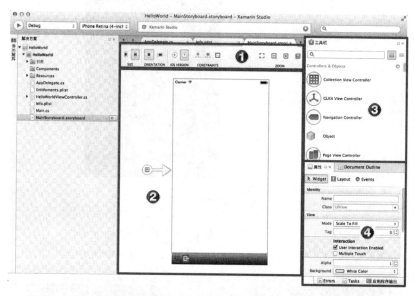

图 1.49　界面构成

其中,编号为 1 的部分为主视图的属性设置,例如可以设置主视图的尺寸大小、iOS 的版本等。编号为 2 的部分为画布,用于设计用户界面,在画布中用箭头指向的区域就是设计界面;在画布中可以有多个设计界面,一般将设计界面称为场景或者主视图。编号为 3 的部分为工具栏,在此工具栏中存放了很多视图对象。编号为 4 的部分为属性设置窗口,在其中可以对视图对象的属性进行设置。

2. 设计主视图

如果想要 iOS Simulator 上显示一个标签,就要对编辑界面进行设置。选择工具栏中的 Label 对象,将其拖动到画布的主视图中,如图 1.50 所示。

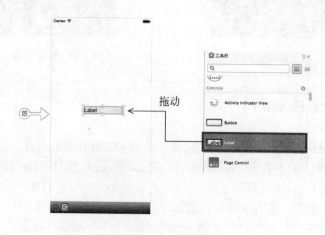

图 1.50　操作步骤

在属性中对 Label 标签对象的字体颜色以及对齐方式进行设置，如图 1.51 所示。设置后主视图的效果如图 1.52 所示。

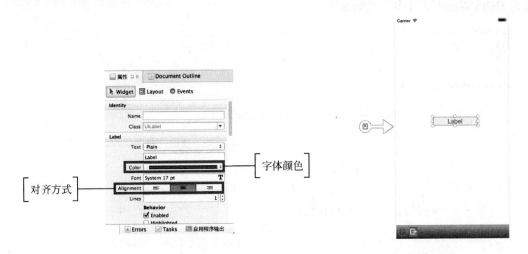

图 1.51　设置属性　　　　　　图 1.52　主视图的效果

运行效果如图 1.53 所示。

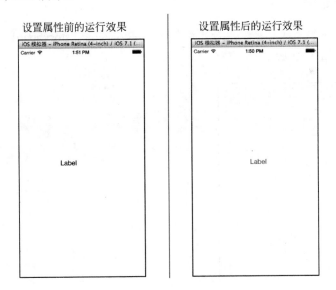

图 1.53　运行效果

1.3.5　编写代码

除了可以使用 Interface Builder 对主视图进行设置外，还可以使用代码进行设置。代码就是用来实现某一特定的功能，而用计算机语言编写的命令序列的集合。现在就来实现通过代码将标签中显示的内容设置为 Hello World 字符串，操作步骤如下所述。

（1）回到 MainStoryboard.storyboard 文件，选择主视图上的 Label 对象，然后在属性的 Identity 下将 Name 设置为 mylabel。Name 属性相当于为 Label 对象起了一个别名，开发者可以通过在 Name 中设置的内容来控制 Label 标签，其他的视图也一样。

（2）打开 HelloWorldViewController.cs 文件，编写代码，实现将标签中显示的内容设置为 Hello World 字符串。代码如下：

```
using System;
using System.Drawing;
using MonoTouch.Foundation;
using MonoTouch.UIKit;
namespace HelloWorld
{
    public partial class HelloWorldViewController : UIViewController
    {
        ……                         //这里省略了视图控制器的构造方法和析构方法
        #region View lifecycle
        public override void ViewDidLoad ()
        {
            base.ViewDidLoad ();
            mylabel.Text="Hello World";    //设置标签中显示的内容
        }
        ……                         //这里省略了视图加载和卸载前后的一些方法
        #endregion
    }
}
```

运行效果如图 1.54 所示。

💡注意：使用代码编写的部分，也会更新到 MainStoryboard.storyboard 文件的主视图中，效果如图 1.55 所示。

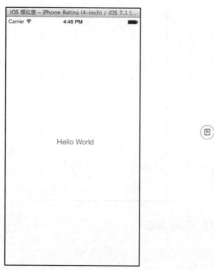

图 1.54　运行效果　　　　　　　图 1.55　主视图的效果

1.3.6　调试程序

调试又被称为排错，是发现和减少程序错误的一个过程。在 Xamarin Studio 中进行调试的需要实现以下几个步骤。

1．添加断点

在进行程序调试之前，首先需要为程序添加断点，断点是指可以暂停调试器中程序的运行，并可以让开发者查看程序的地方。将光标移动到要添加断点的地方，按住 Command+\ 键或者选择菜单栏中的"运行"|"切换断点"命令进行断点的添加，之后会在添加断点代码的最左边看到一个红色的小圆圈，并且这一行代码也被涂成了红色，如图 1.56 所示。

2．运行程序

单击运行按钮后，程序就会运行。这时运行的程序会停留在断点所在的位置，此代码最左边的圆圈中会出现黑色的箭头，并且代码行被涂成了黄色，表示现在程序运行到的位置，如图 1.57 所示。不仅如此，iOS Simulator 也会显示，但是没有内容。

图 1.56　添加断点　　　　　　　　　　图 1.57　执行断点

3．断点导航

在程序停留下来后，会出现断点导航，如图 1.58 所示。开发者可以使用此导航来控制程序的执行。

图 1.58　断点导航

1.3.7　文件简述

创建好工程后，会看到一些文件以及文件夹，以下针对几个重要并且常用的文件进行讲解。

- ❑ Resources 文件夹：存放了应用程序所需的资源，如图像、音乐文件等。
- ❑ AppDelegate.cs：主要的应用程序类别（class），并接听来自作业系统的事件及相对应的事件处理。
- ❑ Entitlements.plist：设置服务是否开启等内容。
- ❑ HelloWorldViewController.cs：负责视图控制器的生命周期，也就是 MVC 分层的 Controller。
- ❑ HelloWorldViewController.designer.cs：包含界面中物件的定义及动作（Action）的宣告。

- Info.plist：应用程序的资讯档，如名称、版本、图示等。
- Main.cs：应用程序的进入点。
- MainStoryboard.storyboard：可以用来设置应用程序的图形界面。

1.4 使用真机测试应用程序

在讲解 iOS Simulator 时，已经提到了虽然 iOS Simulator 可以模仿真实的设备，但是还是有很多的缺陷，如打电话、发送 SMS 信息、获取位置数据等。如果想要实现 iOS Simulator 实现不了的功能，就需要使用真机对应用程序进行测试。本节将讲解如何使用真机对应用程序进行测试。

1.4.1 申请付费开发者账号

使用真机测试，需要申请和下载证书。对于证书的申请和下载必须先成为一个付费的开发者成员，即标准 iPhone 开发成员或者企业 iPhone 开发成员。以下就是如何成为一名标准 iPhone 开发成员的具体步骤。

（1）在 Safari 中输入网址（https://developer.apple.com/programs/），然后按回车键，如图 1.59 所示。

图 1.59　操作步骤 1

（2）选择 iOS Developer Program 选项，进入 iOS Developer Program-Apple Developer 网页，如图 1.60 所示。

图 1.60　操作步骤 2

（3）单击 EnrollNow 按钮，进入 Enrolling in Apple Developer Programs-Apple Developer

网页，如图 1.61 所示。

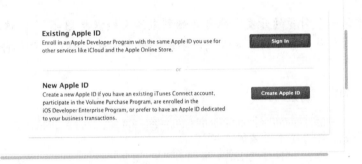

图 1.61　操作步骤 3

（4）选择 Continue 按钮，进入 Sign in or create an Apple ID-Apple Developer Program Enrollment 网页，如图 1.62 所示。

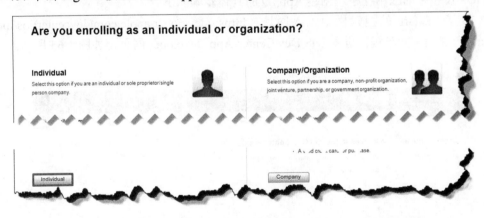

图 1.62　操作步骤 4

（5）单击 Sign In 按钮，进入 Apple Developer Program Enrollment 网页，如图 1.63 所示。

图 1.63　操作步骤 5

（6）单击 Individual 按钮后，进入 Sign in with your Apple ID-Apple Developer 网页，如图 1.64 所示。

（7）输入 Apple ID 以及密码后，单击 Sign In 按钮，进入 Apple Developer Program Enrollment-Update Information 网页，完善自己的信息，然后单击 Continue 按钮。以上这几步是申请付费开发者账号的重要步骤，剩下的步骤就需要根据开发者的需求自行填写了。

此处就不再做介绍了。

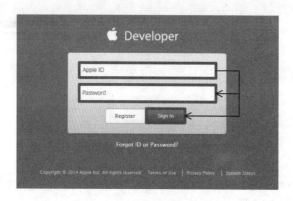

图 1.64　操作步骤 6

⚠注意：从申请一个付费的开发者账号开始到激活大概需要 3~5 天，这段时间需要开发者留心与苹果账号关联的邮箱，苹果公司会为此邮箱发一些邮件。

1.4.2　申请和下载证书

申请和下载证书的具体步骤如下所述。

1．创建App ID

在申请和下载证书之前，首先要创建一个 App ID。App ID 是一系列字符，用于唯一标识 iOS 设备中的应用程序。创建 App ID 的具体步骤如下。

（1）在 Safari 的搜索栏中输入网址（https://developer.apple.com/devcenter/ios/index.action），然后按回车键，进入 iOS Dev Center-App Developer 网页，如图 1.65 所示。

图 1.65　操作步骤 1

（2）单击 Log in 按钮，进入 Sign in with your Apple ID-Apple Developer 网页，在此网页中需要开发者输入 App ID 以及密码，然后单击 Sign In 按钮，此时会再次进入 iOS Dev Center-App Developer 网页，如图 1.66 所示。

⚠注意：图 1.66 所示的页面只有申请付费开发者账号后，才可以看到。

（3）选择 Certificates,Identifiers&Profiles 选项，进入 Certificates,Identifiers &Profiles-App Developer 网页，如图 1.67 所示。

第 1 章　使用 C#编写第一个 iOS 应用程序

图 1.66　操作步骤 2

图 1.67　操作步骤 3

（4）选择 Indentifiers 选项，进入 iOS App IDs-Apple Developer 网页，在此网页中，选择蓝色的 Register your App ID 字符串，进入 Register-iOS App IDs-Apple Developer 网页，在此网页中填入一些相关的内容。这些内容分为 4 部分，分别为 App ID Description、App ID Prefix、App ID Suffix、App Services。在填写 App ID Suffix 这部分内容时需要特别注意，如图 1.68 所示。

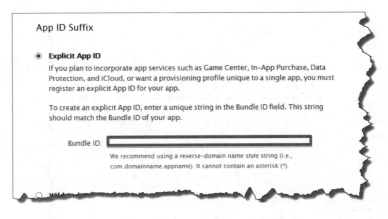

图 1.68　操作步骤 4

图 1.68 中，在 Bundle ID 中输入的内容是标识符，它会在第 4 章中使用到。

（5）单击 Continue 按钮，进入 Add-iOS App IDs-Apple Developer 网页，然后单击 Submit 按钮，之后再单击 Done 按钮。这样一个 App ID 就创建好了。

2. 获取设备的UDID

将设备连接到 Mac（或者 Mac 虚拟机）上，启动 Xcode。在菜单栏中选择 Window|Organizer 命令，弹出 Organizer-Devices 对话框，如图 1.69 所示。在对话框中显示的就是开发者的设备信息，其中 Identifier 就是 UDID。

图 1.69　操作步骤

3. 注册设备

如果开发者的设备是连接 Mac（或者 Mac 虚拟机）上的，回到 Certificates, Identifiers & Profiles-App Developer 网页，选择 Devices 选项，或者是如果开发者还处于创建 App ID 的网页，可以选择此网页右侧的 Devices 下的 All 选项，都会进入 iOS Devices-Apple Developer 网页，并会看到连接在 Mac（或者 Mac 虚拟机）上的设备已经被注册好了，如图 1.70 所示。

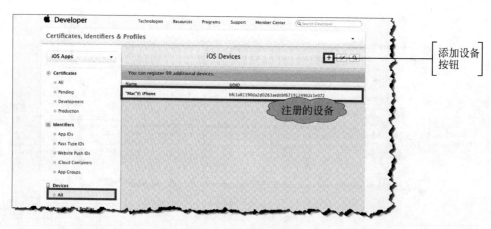

图 1.70　操作步骤

注意：如果开发者还需要注册其他的设备，可以单击添加设备的按钮，对设备进行添加，如图 1.71 所示。

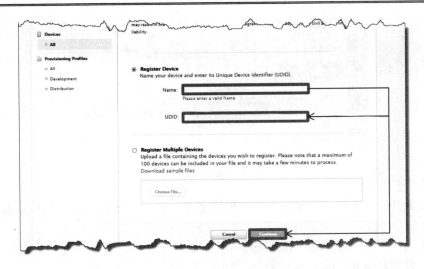

图 1.71　注册新的设备

在此图中，开发者只需要在 Name 文本框中输入设备的名称，在 UDID 文本框中输入设备的标识符就可以了，然后单击 Continue 按钮，进入对设备的检测和登记网页，然后单击 Register 按钮，进入登记设备成功的网页，最后单击 Done 按钮，一个新的设备就注册成功了。

4．生成证书签名申请

为了从 Apple 公司申请开发证书，需要生成一个证书签名申请。生成一个证书签名申请的具体步骤如下。

（1）选择菜单栏中的"前往"|"实用工具"命令，打开"实用工具"文件夹，如图 1.72 所示。

图 1.72　操作步骤 1

（2）找到"钥匙串访问"应用程序，双击图标将其打开，选择菜单栏上的钥匙串访问，如图 1.73 所示。

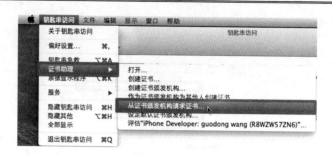

图 1.73　操作步骤 2

（3）选择"证书助理"|"从证书发布机构请求证书…"命令，弹出"证书助理"对话框，如图 1.74 所示。

（4）输入用户电子邮件地址，并选择存储到磁盘复选框，然后单击"继续"按钮，弹出"存储位置"对话框如图 1.75 所示。

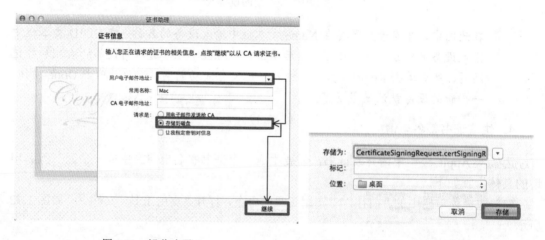

图 1.74　操作步骤 3　　　　　　　　图 1.75　操作步骤 4

注意：在"存储位置"对话框中，"存储为"以及"位置"下拉列表都有默认的选项。

（5）设置"位置"为桌面，单击"存储"按钮，就在桌面生成了一个证书签名申请，并回到"证书助理"对话框，告诉开发者证书请求已经在磁盘上创建了，然后单击"完成"按钮即可。

5．生成证书

以上这些准备工作都做好后，便可以生成证书了，它包括证书的申请和下载。具体的操作步骤如下。

（1）如果开发者还处于注册设备的网页，可以选择此网页右侧的 Certificates|Development 选项，进入 iOS Certificates (Development)-Apple Developer 网页，如图 1.76 所示。

（2）选择 iOS App Development 复选框，单击 Continue 按钮，进入到 Request 选项卡的网页中。在此网页中，单击 Continue 按钮，进入 Generate 选项卡的网页中，如图 1.77 所示。

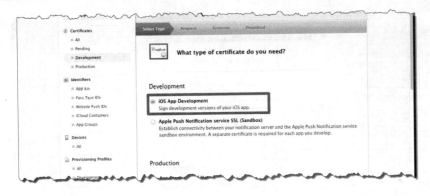

图 1.76　操作步骤 1

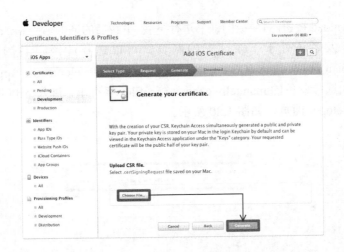

图 1.77　操作步骤 2

（3）单击 Choose File...按钮后，弹出选择文件对话框，如图 1.78 所示。

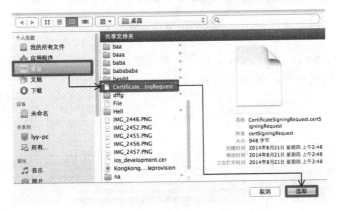

图 1.78　操作步骤 3

（4）选择在桌面的 CertificateSigningRequest.certSigningRequest 文件，此文件就是生成的证书签名申请，单击"选取"按钮。然后单击 Generate 按钮，进入 Download 选项卡的网页中，如图 1.79 所示。

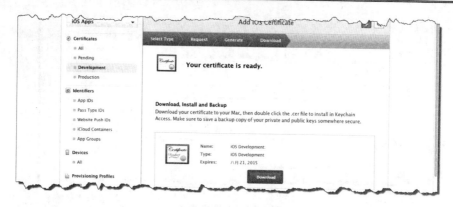

图 1.79　操作步骤 4

（5）单击 Download 按钮，对生成的证书进行下载。下载后的证书名为 ios_development.cer。

（6）如果开发者还处于下载证书的网页，可以选择此网页右侧的 Provisioning Profiles|Development 选项，进入 iOS Provisioning Profiles (Development)-Apple Developer 网页。在此网页中，选择蓝色的 manually generate profiles 字符串，进入 Add-iOS Provisioning Profile-Apple Developer 网页，如图 1.80 所示。

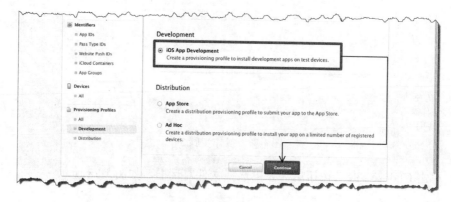

图 1.80　操作步骤 5

（7）选择 iOS App Development 复选框，然后单击 Continue 按钮，进入 Configure 选项卡的选择 App ID 的网页中，如图 1.81 所示。

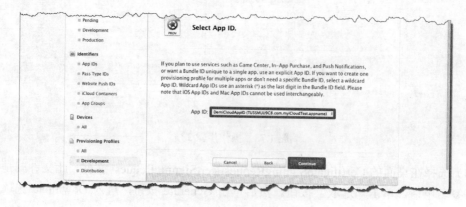

图 1.81　操作步骤 6

（8）选择 App ID（这里的 App ID 是之前创建的 App ID），然后单击 Continue 按钮，进入 Configure 选项卡的选择证书的网页中，如图 1.82 所示。

图 1.82　操作步骤 7

（9）选择 Select All 复选框或者选择某一个证书，然后单击 Continue 按钮，进入 Configure 选项卡的选择设备的网页中，如图 1.83 所示。

图 1.83　操作步骤 8

（10）选择 Select All 复选框或者选择某一个设备，然后单击 Continue 按钮，进入 Generate 选项卡的网页中，如图 1.84 所示。

图 1.84　操作步骤 9

（11）输入配置的文件名，然后单击 Generate 按钮，进入 Download 选项卡的网页中，如图 1.85 所示。

图 1.85　操作步骤 10

（12）选择 Download 按钮，对 Provisioning Profiles 进行下载，下载后的文件为 KongKong.mobileprovision。

（13）双击下载的 ios_development.cer 证书，弹出"添加证书"对话框，如图 1.86 所示。

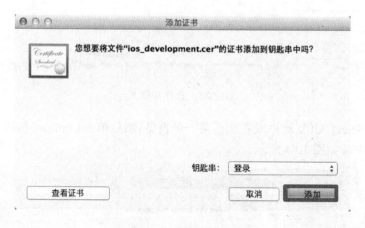

图 1.86　操作步骤 11

（14）单击"添加"按钮，将下载的 ios_development.cer 证书添加到钥匙串中。

（15）双击下载的 KongKong.mobileprovision 文件，将此文件添加到 Organizer 的 Provisioning Profiles 中。

1.4.3　实现真机测试

在进行真机测试之前，首先需要确保设备已经连在 Mac（或者 Mac 虚拟机）上了，在第 1.4.2 节开始，设备就一直连接在 Mac（或者 Mac 虚拟机）上，并且此设备就是注册过的。打开创建的工程，在运行按钮一栏中，将程序运行的设备设置为真机的名称。它会自动加载到"选择程序运行的设备"这一项中，如图 1.87 所示。单击运行按钮，就可以看到应用程序在真机上运行了。

图 1.87　设置设备

第 2 章　用户界面——视图

在 iPhone 或者 iPad 中，用户看到的都是视图。视图是用户界面的重要组成元素。例如，想要让用户实现文本输入时，需要使用输入文本的视图；想要让用户显示图像时，需要使用显示图像的视图。本章将为开发者详细讲解如何构建视图。

2.1　视　　图

在应用程序开发中，最常用的视图如表 2-1 所示。

表 2-1　常用视图

视　　图	功　　能
UIView	这是一个容器，是大部分 iOS 用户界面控件的基本对象
UIButton	按钮视图，它与.NET 中的 Button 类似
UILabel	标签视图，用于显示文本
UIImageView	图像视图，用于显示图像
UITextField	文本框视图，用于输入并显示单行文本
UITextView	文本视图，用于输入并显示多行文本
UIScrollView	滚动视图，提供了滚动视图的能力
UIPageControl	页面控件，提供了不同的页面或屏幕的导航功能
UIAlertView	警告视图，用于向用户显示一个消息框

2.2　添加和定制视图

本节将主要讲解视图的两种添加方式：一种是使用 Interface Builder，另一种是使用代码；以及定制视图等内容。

2.2.1　使用 Interface Builder 添加视图

使用 Interface Builder 添加视图是一个相当简单的工作。以下的示例将为开发者讲解该如何使用 Interface Builder 添加视图。

【示例 2-1】以下将使用 Interface Builder 添加一个视图，具体步骤如下所述。

（1）创建一个 Single View Application 类型的工程。

（2）打开 MainStoryboard.storyboard 文件，选择 Xamarin Studio 最右边的工具栏按钮，打开工具栏界面（"查看"|Pads|"工具栏"命令）。从工具栏中拖动 View 空白视图对象到主视图中，如图 2.1 所示。

☪注意：此时在视图添加了一个空白的视图。

（3）保存文件（Command+S），单击"运行"按钮，此时就会出现 iOS 模拟器运行结果，如图 2.2 所示。

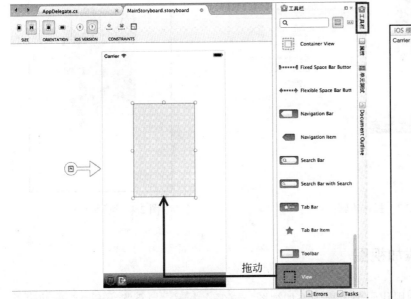

图 2.1　操作步骤

图 2.2　运行效果

由于使用 Interface Builder 添加的 View 空白视图默认的背景颜色为白色，所以在模拟器上是看不出效果的。那么该如何在模拟器上看到添加的 View 空白视图呢？开发者需要回到 MainStoryboard.storyboard 文件，选择主视图上添加的 View 空白视图对象。然后，选择 Xamarin Studio 最右边的属性按钮，打开属性界面（"查看"|Pads|"属性"命令）。将 View 的 Background 属性设置为 Scrollview Textured Background color，如图 2.3 所示。

☪注意：当改变 Background 属性后，选择的 View 视图对象的颜色也会发现相应的变化。
　　　运行效果如图 2.4 所示。

此时，就可以在 iOS 模拟器上看到添加的视图了。我们的这个示例是没有任何作用的，它只是为开发者演示了如何使用 Interface Builder 来添加一个视图。

☪注意：View 空白视图是开发者使用最频繁的，原因如下：
　　　（1）每一个可视化的视图对象都是继承自 UIView 类。
　　　（2）提供了自动调整大小的功能。
　　　（3）UIView 可以管理内容绘制。
　　　（4）由于它是一个容器，可以接受其他的视图作为其子视图。

（5）可以接受本身和其子视图的触摸事件。

（6）它的很多属性可以实现动画。

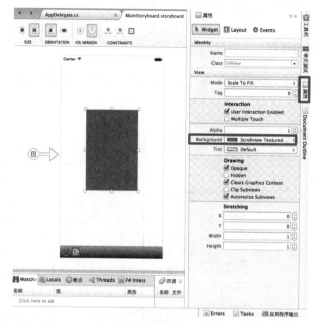

图 2.3 操作步骤

图 2.4 运行效果

2.2.2 使用代码添加视图

如果开发者想要使用代码为主视图添加视图，该怎么办呢。以下将为开发者解决这一问题。要使用代码为主视图添加视图需要 3 个步骤。

1. 实例化视图对象

每一个视图都是一个特定的类。在 C#中，经常会说，类是一个抽象的概念，而非具体的事物，所以要将类进行实例化。实例化一个视图对象的具体语法如下：

```
视图类 对象名=new 视图类();
```

以我们接触的第一个视图 View 为例，它的实例化对象如下：

```
UIView vv=new UIView();
```

其中，UIView 是空白视图的类，vv 是 UIView 类实例化出来的一个对象。

2. 设置视图的位置和大小

每一个视图都是一个区域，所以需要为此区域设置位置和大小。设置位置和大小的属性为 Frame，其语法形式如下：

```
对象名.Frame=new RectangleF (X ,Y ,Width,Height);
```

其中，X 和 Y 表示视图在主视图中的位置，Width 和 Height 表示视图的大小。以下为实例化的对象 vv 设置位置和大小：

```
vv.Frame = new RectangleF (0, 0, 320, 580);
```

其中，0 和 0 表示此视图的主视图中的位置，320 和 580 表示此视图的大小。

> **注意**：步骤 1 和步骤 2 也可以进行合并。例如，以下的代码是将 UIView 类的实例化对象和设置位置大小进行了合并：

```
UIView vv = new UIView (new RectangleF (0, 0, 200, 200));
```

3．将视图添加到当前的视图中

最后，也是最为关键的一步，就是将实例化的对象添加到主视图中。这样才可以进行显示。此时需要使用 AddSubview() 方法，其语法形式如下：

```
this.View.AddSubview (视图对象名);
```

以下将实例化的对象 vv 添加到当前的主视图中，代码如下：

```
this.View.AddSubview (vv);
```

【示例 2-2】 以下将使用代码为主视图添加一个 View 空白视图。代码如下：

```
using System;
using System.Drawing;
using MonoTouch.Foundation;
using MonoTouch.UIKit;
namespace Application
{
    public partial class __2ViewController : UIViewController
    {
    …… //这里省略了视图控制器的构造方法和析构方法（视图控制器的基本功能是负责处理
        与视图的交互，我们会在后面讲解）
        #region View lifecycle
        public override void ViewDidLoad ()
        {
            base.ViewDidLoad ();
            UIView vv = new UIView ();                    //实例化对象
            vv.Frame = new RectangleF (0, 0, 320, 580);
                                                          //设置视图对象的大小和位置
            this.View.AddSubview (vv);                    //将视图对象添加到当前视图中
        }
    …… //这里省略了视图加载和卸载前后的一些方法
        #endregion
    }
}
```

运行效果如图 2.5 所示。在此运行效果中也是看不到添加的视图的。这是因为添加的视图默认是白色的背景，如果想要看到视图，需要设置它的背景。例如以下的代码，将背景颜色设置为浅灰色：

```
vv.BackgroundColor = UIColor.LightGray;                //将背景设置为浅灰色
```

运行效果如图 2.6 所示。

图 2.5 运行效果

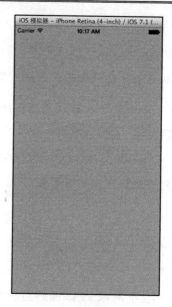

图 2.6 运行效果

2.2.3 删除视图

有视图的添加，就会有视图的删除。如果开发者不需要添加的视图，就可以使用 RemoveFromSuperview()方法删除，其语法形式如下：

```
要删除的视图对象名.RemoveFromSuperview();
```

【示例 2-3】 以下代码将在主视图中添加两个视图，然后使用 RemoveFromSuperview() 方法删除其中一个视图。代码如下：

```csharp
using System;
using System.Drawing;
using MonoTouch.Foundation;
using MonoTouch.UIKit;
namespace Application
{
    public partial class __15ViewController : UIViewController
    {
        ……      //这里省略了视图控制器的构造方法和析构方法
        #region View lifecycle
        public override void ViewDidLoad ()
        {
            base.ViewDidLoad ();

            // Perform any additional setup after loading the view, typically
                from a nib.
            //实例化并设置视图对象 vv1
            UIView vv1 = new UIView ();
            vv1.Frame = new RectangleF (0, 20, 320, 250);
            vv1.BackgroundColor = UIColor.Cyan;
            this.View.AddSubview (vv1);
            //实例化并设置视图对象 vv2
            UIView vv2 = new UIView ();
```

```
            vv2.Frame = new RectangleF (0, 300, 320, 250);
            vv2.BackgroundColor = UIColor.Orange;
            this.View.AddSubview (vv2);
    }
    ……          //这里省略了视图加载和卸载前后的一些方法
    #endregion
}
```

运行效果如图 2.7 所示。

如果想要删除视图对象 vv1 的话，需要使用 RemoveFromSuperview ()方法，代码如下：

```
vv1.RemoveFromSuperview ();                       //删除视图对象 vv1
```

运行效果如图 2.8 所示。

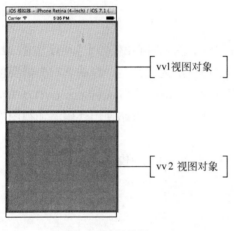

图 2.7　运行效果

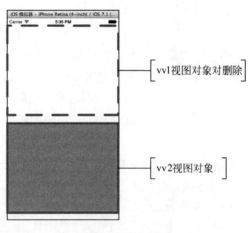

图 2.8　运行效果

2.2.4　视图的位置和大小

当一个视图使用 Interface Builder 添加到主视图后，它的位置和大小可以使用拖动的方式进行设置，也可以使用属性中的布局进行设置，如图 2.9 所示。

注意：在默认的情况下，坐标系统的原点位于左上角，并向底部和右侧延伸，如图 2.10 所示。

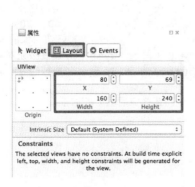

图 2.9　设置位置和大小

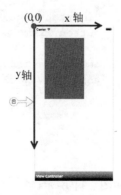

图 2.10　系统坐标

除了使用以上两种方式进行改变视图的位置和大小外，还可以通过编程的方式改变。但是需要注意，以编程的形式也不可以改变默认的坐标系统原点。

2.3 使用按钮接受用户输入

按钮是用户交互的最基础控件。即使是在 iPhone 或者 iPad 中，用户使用最多的操作就是通过触摸实现点击。而点击操作最多的控件往往是按钮控件。一般使用 UIButton 类来实现按钮。本节将主要讲解一些关于按钮的内容。

2.3.1 使用代码添加按钮

由于按钮拖放的方式比较简单，所以不再介绍。下面直接讲解在代码中如何添加按钮。使用代码为主视图添加一个按钮的方式和在 2.2.2 节中讲解的步骤是一样的。首先需要使用 UIButton 类实例化一个按钮对象，然后设置位置和大小，最后使用 AddSubview()方法将按钮对象添加到主视图中。由于视图的添加方式都一样，后面将省略使用代码添加视图这块内容。

【示例 2-4】 以下将使用代码为主视图添加一个青色的按钮。代码如下：

```
using System;
using System.Drawing;
using MonoTouch.Foundation;
using MonoTouch.UIKit;
namespace Application
{
    public partial class __16ViewController : UIViewController
    {
        ……        //这里省略了视图控制器的构造方法和析构方法
        #region View lifecycle
        public override void ViewDidLoad ()
        {
            base.ViewDidLoad ();
            // Perform any additional setup after loading the view, typically 
              from a nib.
            UIButton button = new UIButton ();        //实例化按钮对象
            button.Frame = new RectangleF (120, 261, 80, 30);
                                                //设置按钮对象的位置和大小
            button.BackgroundColor = UIColor.Cyan;//设置按钮对象的背景颜色
            this.View.AddSubview (button);        //将按钮对象添加到主视图中
        }
        ……        //这里省略了视图加载和卸载前后的一些方法
        #endregion
    }
}
```

运行效果如图 2.11 所示。

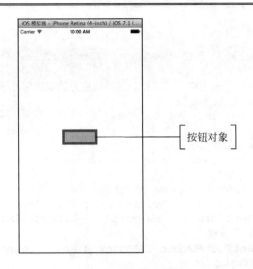

图 2.11　运行效果

注意：由于按钮视图继承了 UIView 类，所以它继承了在 UIView 类中的属性和方法。

2.3.2　按钮的格式化设置

在图 2.11 中可以看到，明明是添加了一个按钮，但是却和添加了一个空白视图一样。为了让按钮和空白视图区别开，需要对按钮进行一些设置。

1．设置按钮的外观

外观是直接区别按钮和其他视图的手段。如果是使用 Interface Builder 添加的按钮，它的外观设置方式有两种，一种是直接打开属性界面，对按钮的外观进行设置，如图 2.12 所示。

另一种就是使用代码对按钮的外观进行设置，这一种方式适用于使用代码添加的按钮中。表 2-2 列出了常用的一些外观设置属性。

表 2-2　常用属性

属　　性	功　　能
SetBackgroundImage	设置指定状态下按钮的背景图像
SetImage	设置指定状态下按钮的图像
SetTitle	设置指定状态下按钮的标题
SetTitleColor	设置指定状态下按钮的标题颜色
SetTitleShadowColor	设置指定状态下按钮标题的应用

【示例 2-5】　下面将在主视图中添加一个按钮。此按钮的标题为 I am button，标题的颜色为黑色。代码如下：

```
using System;
using System.Drawing;
using MonoTouch.Foundation;
using MonoTouch.UIKit;
```

```
namespace Application
{
    public partial class __18ViewController : UIViewController
    {
        ……          //这里省略了视图控制器的构造方法和析构方法
        #region View lifecycle
        public override void ViewDidLoad ()
        {
            base.ViewDidLoad ();
            // Perform any additional setup after loading the view, typically
               from a nib.
            UIButton button = new UIButton ();
            button.Frame = new RectangleF (107, 269, 120, 30);
            button.SetTitle ("I am button", UIControlState.Normal);
            //设置按钮的标题
            button.SetTitleColor (UIColor.Black, UIControlState.Normal);
            //设置按钮的标题颜色
            this.View.AddSubview (button);
        }
        ……          //这里省略了视图加载和卸载前后的一些方法
    endregion
    }
}
```

运行效果如图 2.13 所示。

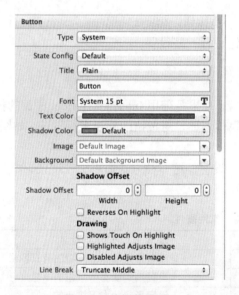

图 2.12　按钮的设置

图 2.13　运行效果

2．设置按钮的状态

在【示例 2-5】中，设置按钮的标题和颜色时，需要对按钮的状态进行设置，表示按钮在某一状态下的标题和标题颜色是什么样子。例如，UIControlState.Normal 表示按钮的一种状态。对于像按钮这类视图，即可以接受用户输入的视图也被称为控件。这些控件都有自己的状态。表 2-3 为开发者详细介绍了控件的状态。

第 2 章 用户界面——视图

表 2-3 控件的状态

按钮的状态	解释说明
Normal	常规状态
Highlighted	高亮状态
Disabled	禁用状态，不接受任何事件
Selected	选中状态
Application	应用程序标志
Reserved	为内部框架预留，可以不管他

3．设置按钮的类型

按钮的形式是多种多样的。例如，在通讯录中，添加新联系人的按钮是一个加号；查看来电的详细信息时是一个感叹号等。这些按钮的实现，可以在实例化按钮对象时使用 UIButtonType 来实现。UIButtonType 中的内容如表 2-4 所示。

表 2-4 UIButtonType的内容

按钮类型	解释说明
System	默认的风格的按钮
Custom	自定义风格的按钮
RoundedRect	具有圆角矩形风格的按钮，在 iOS 7 中使用 System 代替了 RoundedRect
DetailDisclosure	蓝色感叹号按钮，主要用于详细说明
InfoLight	亮色感叹号
InfoDark	暗色感叹号
ContactAdd	十字加号按钮，此按钮通常在添加联系人条目中显示

【示例 2-6】 以下代码将设置两个不同风格的按钮。代码如下：

```
using System;
using System.Drawing;
using MonoTouch.Foundation;
using MonoTouch.UIKit;
namespace Application
{
    public partial class __19ViewController : UIViewController
    {
        ……            //这里省略了视图控制器的构造方法和析构方法
        #region View lifecycle
        public override void ViewDidLoad ()
        {
            base.ViewDidLoad ();

            // Perform any additional setup after loading the view, typically
                from a nib.                    //实例化按钮对象并设置按钮的类型
            UIButton button1 = new UIButton (UIButtonType.DetailDisclosure);
            button1.Center = new PointF (160, 150);
                                              //设置按钮的中心位置
            this.View.AddSubview (button1);
            //实例化按钮对象并设置按钮的类型
            UIButton button2 = new UIButton (UIButtonType.ContactAdd);
            button2.Center = new PointF (160, 350);    //设置按钮的中心位置
            this.View.AddSubview (button2);
        }
```

```
        ……         //这里省略了视图加载和卸载前后的一些方法
        #endregion
    }
}
```

运行效果如图 2.14 所示。

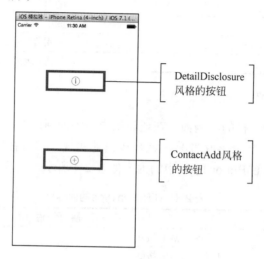

图 2.14 运行效果

4．设置按钮的发光效果

开发者可能在很多的地方见到过发光的按钮，它的实现其实很简单，就是使用了 ShowsTouchWhenHighlighted 属性来实现的。

【示例 2-7】以下代码将实现一个发光的按钮。代码如下：

```
using System;
using System.Drawing;
using MonoTouch.Foundation;
using MonoTouch.UIKit;
namespace Application
{
    public partial class __17ViewController : UIViewController
    {
        ……         //这里省略了视图加载和卸载前后的一些方法
        #region View lifecycle
        public override void ViewDidLoad ()
        {
            base.ViewDidLoad ();
            // Perform any additional setup after loading the view, typically
               from a nib.
            UIButton button = new UIButton ();
            button.Frame = new RectangleF (137, 269, 46, 30);
            button.SetTitle ("Hello", UIControlState.Normal);
            this.View.AddSubview (button);
            button.ShowsTouchWhenHighlighted = true;           //按钮发光的设置
        }
        ……         //这里省略了视图加载和卸载前后的一些方法
        #endregion
    }
}
```

运行效果如图 2.15 所示。

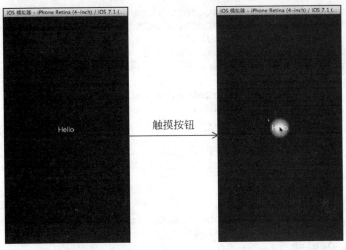

图 2.15　运行效果

2.3.3　按钮的响应

以上讲解了按钮的格式化设置，作为按钮，最重要的功能就是实现和用户的响应。它的实现事件是 TouchUpInside。其语法形式如下：

按钮对象.TouchUpInside +=触摸按钮后的方法;

或者：

按钮对象名.TouchUpInside +=(sender,e)=>{
　　……
};

其中，sender 表示事件监视的对象，e 就是事件所需要的数据。

【示例 2-8】　以下将实现按钮的响应。当用户触摸按钮后，主视图就会变色。代码如下：

```
using System;
using System.Drawing;
using MonoTouch.Foundation;
using MonoTouch.UIKit;
namespace Application
{
    public partial class __3ViewController : UIViewController
    {
        UIButton buttonChangeColor;
        bool isYellow;
        ……        //这里省略了视图控制器的构造方法和析构方法
        #region View lifecycle
        //创建按钮 buttonChangeColor
        private void CreateButton ()
        {
            RectangleF viewFrame = this.View.Frame;
            RectangleF buttonFrame = new RectangleF (10f, viewFrame.Bottom
            - 200f,viewFrame.Width - 20f, 50f);
            this.buttonChangeColor = UIButton.FromType (UIButtonType.System);
```

```
        //实例化对象
        //对按钮的格式化设置
        this.buttonChangeColor.Frame = buttonFrame;
        this.buttonChangeColor.SetTitle ("Tap to change view color",
        UIControlState.Normal);
        this.buttonChangeColor.SetTitle ("Changing color...", UIControlState.
        Highlighted);
        //实现响应
        this.buttonChangeColor.TouchUpInside += this.ButtonChange
        Color_TouchUpInside;
        this.View.AddSubview (this.buttonChangeColor);
    }
    //实现触摸按钮后改变主视图的背景颜色
    private void ButtonChangeColor_TouchUpInside (object sender,
    EventArgs e)
    {
        if (isYellow) {
            this.View.BackgroundColor = UIColor.LightGray;
            isYellow = false;
        } else {
            this.View.BackgroundColor = UIColor.Yellow;
            isYellow = true;
        }
    }
    public override void ViewDidLoad ()
    {
        base.ViewDidLoad ();

        // Perform any additional setup after loading the view, typically
            from a nib.
        this.CreateButton ();                           //调用CreateButton()方法
    }
    ……       //这里省略了视图加载和卸载前后的一些方法
    #endregion
    }
}
```

在此程序中，有一个isYellow，它是一个布尔类型的变量，当此变量为true时，将主视图的背景改为浅灰色。当此变量为 false 时，将主视图的背景改为黄色。运行效果如图2.16所示。

图 2.16　运行效果

2.4 显示图像

在主视图中显示一个图像,可以让开发者的应用程序变得更加有趣,例如,在一些应用程序开始运行时,都会通过图像来显示此应用程序的玩法或者规则等。这不仅可以使用户快速理解此应用程序的相关信息,也减少了开发者对应用软件文字的介绍。显示图像的视图被称为图像视图。以下将主要讲解图像视图的一些功能。

2.4.1 为视图显示图像

显示图像需要使用到 UIImageView 类创建的对象。

【示例 2-9】 以下就是在主视图中显示图像的具体步骤。

(1)创建一个 Single View Application 类型的工程。
(2)添加一个图像 1.jpg 到创建的工程中,添加图像的具体步骤如下:
首先,右击工程中的 Resources 文件夹,弹出一个下拉菜单,如图 2.17 所示。

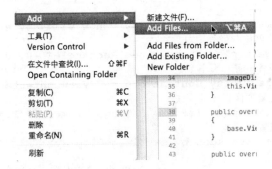

图 2.17 添加图像文件 1

其次,选择下拉菜单中的 Add|Add Files…命令,弹出"添加文件"对话框,如图 2.18 所示。

图 2.18 添加图像文件 2

接着，选择需要添加到工程中的图像文件，然后单击 Open 按钮，弹出 Add File to Folder 对话框，如图 2.19 所示。

最后，单击"确定"按钮，需要添加的图像就添加到了创建工程的 Resources 文件夹中。

（3）打开 2-4ViewController.cs 文件，编写代码，实现为主视图添加图像视图的功能。代码如下：

```
using System;
using System.Drawing;
using MonoTouch.Foundation;
using MonoTouch.UIKit;
namespace Application
{
    public partial class __4ViewController : UIViewController
    {
        ……                              //这里省略了视图控制器的构造方法和析构方法
        #region View lifecycle
        public override void ViewDidLoad ()
        {
            base.ViewDidLoad ();

            // Perform any additional setup after loading the view, typically
            //    from a nib.
            UIImageView imageDisplay = new UIImageView ();
            //实例化图像视图对象
            imageDisplay.Frame = new RectangleF (0, 0, 320, 580);
            //设置位置和大小
            imageDisplay.ContentMode = UIViewContentMode.ScaleAspectFit;
            //设置图像视图的模式
            imageDisplay.Image = UIImage.FromFile("1.jpg");
            //设置图像视图显示的图像
            this.View.AddSubview (imageDisplay);
        }
        ……                              //这里省略了视图加载和卸载前后的一些方法
        #endregion
    }
}
```

运行效果如图 2.20 所示。

图 2.19 添加图像文件 3

图 2.20 运行效果

在此示例中，需要开发者掌握两个知识点。

1．图像支持的格式

图像视图显示的图像需要使用 image 进行设置。UIImage 类是用来代表图像信息的。图像支持的格式如表 2-5 所示。

表 2-5　图像支持的格式

文 件 格 式	文件扩展名
Portable Network Graphics (PNG)	.png
Joint Photographic Experts Group (JPEG)	.jpg/.jpeg
Tagged Image File Format (TIFF)	.tiff/.tif
Graphic Interchange Format	.gif
Windows Bitmap Format	.bmp
Windows Icon Format	.ico
Windows Cursor	.cur
XWindow bitmap	.xbm

2．图像在显示时的模式

为了使图像在显示的时候可以满足各种需要，UIView 视图提供了多种显示模式。选择属性按钮后，在属性界面中设置 View 中的 Mode。或者使用代码通过 ContentMode 属性设置 Mode。Mode 选项中提供了 13 种显示模式，分别为 Scale To Fill、Aspect Fit、Aspect Fill、Redraw、Center、Top、Bottom、Left、Right、Top Left、Top Right、Bottom Left、Bottom Right。在这 13 种显示模式中，最常用的有 3 种，分别为 Scale To Fill、Aspect Fit 和 Aspect Fill，效果如图 2.21 所示。

图 2.21　效果

> **注意**：Scale To Fill 会使图像全部显示出来，但是会导致图像变形。Aspect Fit 会保证图像比例不变，而且全部显示在图像视图中，这意味着图像视图会有部分空白。AspectFill 也会保证图像比例不变，会填充整个图像视图，但是可能只有部分图像显示出来，一般不对 Mode 进行设置，默认为 Scale To Fill 模式。

2.4.2 定制特殊的图像

在主视图中显示的图像也不是一成不变的，可能会将图像变倾斜一些，或者将图像进行缩放等。以下就介绍两个常用的功能。

1．旋转图像

旋转图像其实就是将图像所在的图像视图进行旋转。需要使用到 Transform 属性，其语法形式如下：

```
图像视图对象.Transform= CGAffineTransform.MakeRotation(float angle);
```

其中，angle 表示旋转的弧度。

【**示例 2-10**】以下将实现将主视图显示的图像旋转 10 度。具体步骤如下：

（1）创建一个 Single View Application 类型的工程。

（2）添加一个图像 1.jpg 到创建工程的 Resources 文件夹中。

（3）打开 2-21ViewController.cs 文件，编写代码，实现将主视图显示的图像旋转 10 度。代码如下：

```
using System;
using System.Drawing;
using MonoTouch.CoreGraphics;           //引入 MonoTouch.CoreGraphics 命名空间
using MonoTouch.Foundation;
using MonoTouch.UIKit;
namespace Application
{
    public partial class __21ViewController : UIViewController
    {
        ……                              //这里省略了视图控制器的构造方法和析构方法
        #region View lifecycle
        public override void ViewDidLoad ()
        {
            base.ViewDidLoad ();
            // Perform any additional setup after loading the view, typically
                from a nib.
            UIImageView imageDisplay = new UIImageView ();
                                        //实例化图像视图对象
            imageDisplay.Frame = new RectangleF (0, 0, 320, 568);
                                        //设置位置和大小
            imageDisplay.Image = UIImage.FromFile("1.jpg");
                                        //设置图像视图显示的图像
            imageDisplay.Transform=CGAffineTransform.MakeRotation((float)
            (3.14 / 10));               //旋转图像
            this.View.AddSubview (imageDisplay);
        }
```

```
            ……                //这里省略了视图加载和卸载前后的一些方法
            #endregion
    }
}
```

在此程序中，我们引入了一个新的命名空间 MonoTouch.CoreGraphics，此命名空间用于图形绘制，因为 CGAffineTransform.MakeRotation 是这个命名空间的对象和方法。运行效果如图 2.22 所示。

旋转前　　　　　　　　旋转后

图 2.22　运行效果

2．缩放图像

缩放图像其实就是将图像所在的图像视图进行缩放。同样还需要使用到 Transform 属性，其语法形式如下：

```
图像视图对象.Transform= CGAffineTransform.MakeScale (float sx,float sy);
```

其中，sx 与 sy 分别表示将原来的宽度和高度缩放到多少倍。

【示例 2-11】 以下将实现将主视图显示的图像放大 2 倍。具体步骤如下：

（1）创建一个 Single View Application 类型的工程。
（2）添加一个图像 1.jpg 到创建工程的 Resources 文件夹中。
（3）打开 2-22ViewController.cs 文件，编写代码，实现将主视图显示的图像放大两倍。代码如下：

```
using System;
using System.Drawing;
using MonoTouch.CoreGraphics;
using MonoTouch.Foundation;
using MonoTouch.UIKit;
namespace Application
{
    public partial class __22ViewController : UIViewController
    {
```

```
……                              //这里省略了视图控制器的构造方法和析构方法
#region View lifecycle
public override void ViewDidLoad ()
{
    base.ViewDidLoad ();
    // Perform any additional setup after loading the view, typically
        from a nib.
    UIImageView imageDisplay = new UIImageView ();
    //实例化图像视图对象
    imageDisplay.Frame = new RectangleF (80, 158, 160, 240);
    //设置位置和大小
    imageDisplay.Image = UIImage.FromFile("1.jpg");
    //设置图像视图显示的图像
    imageDisplay.Transform = CGAffineTransform.MakeScale (2,2);
    //缩放图像
    this.View.AddSubview (imageDisplay);
}
……                              //这里省略了视图加载和卸载前后的一些方法
#endregion
}
```

运行效果如图 2.23 所示。

图 2.23　运行效果

2.5　显示和编辑文本

在一个应用程序中，文字是非常重要的。它就是这些不会说话的设备的嘴巴。通过这些文字，可以很清楚地表达这些应用程序的信息。以下将为开发者介绍 3 种关于文本的视图。

2.5.1　标签视图

标签视图（一般使用 UILabel 类实现）一般用于在应用程序中为用户显示少量的信息。

【示例2-12】 以下就是通过标签视图为开发者显示一首诗的效果。具体步骤如下：
（1）创建一个 Single View Application 类型的工程。
（2）添加图像 1.jpg 到创建工程的 Resources 文件夹中。
（3）打开 MainStoryboard.storyboard 文件，对主视图进行设置。效果如图 2.24 所示。

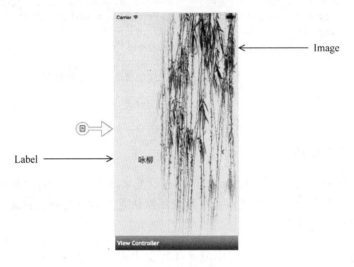

图 2.24 主视图的效果

需要添加的视图以及设置，如表 2-6 所示。

表 2-6 设置主视图

视　　图	设　　置
Image	Image：1.jpg
Label	Text：咏柳 Font：System 20 pt Alignment：居中

（4）打开 2-20ViewController.cs 文件，编写代码，实现为主视图添加标签的功能。代码如下：

```
using System;
using System.Drawing;
using MonoTouch.Foundation;
using MonoTouch.UIKit;
namespace Application
{
    public partial class __20ViewController : UIViewController
    {
        ……                          //这里省略了视图控制器的构造方法和析构方法
        #region View lifecycle
        public override void ViewDidLoad ()
        {
            base.ViewDidLoad ();
            // Perform any additional setup after loading the view, typically
               from a nib.
            UILabel label1 = new UILabel ();
            label1.Frame = new RectangleF (2, 410, 155, 28);
            label1.TextAlignment = UITextAlignment.Center;
```

```
            label1.Text = "碧玉妆成一树高,";          //设置标签文本内容的对齐方式
                                                    //设置标签的文本内容
            this.View.AddSubview (label1);
            ……
            UILabel label4 = new UILabel ();
            label4.Frame = new RectangleF (2, 500, 155, 28);
            label4.TextAlignment = UITextAlignment.Center;
            label4.Text = "二月春风似剪刀.";
            this.View.AddSubview (label4);
        }
        ……                              //这里省略了视图加载和卸载前后的一些方法
        #endregion
    }
}
```

运行效果如图 2.25 所示。

> **注意**：在此程序中，使用 TextAlignment 属性设置文本在标签中的对齐方式。使用 Text 属性设置标签中显示的文本。标签视图默认是显示一行的，但是，也可以将标签的内容显示为多行。

【**示例 2-13**】 以下将在一个标签中显示 3 行文本内容。具体步骤如下：

（1）创建一个 Single View Application 类型的工程。

（2）添加图像 1.jpg 到创建工程的 Resources 文件夹中。

（3）打开 MainStoryboard.storyboard 文件，从工具栏中拖动 Image View 图像视图到主视图中，将此视图的 Image 属性设置为 1.jpg。

（4）打开 2-23ViewController.cs 文件，编写代码，实现标签多行显示的功能。代码如下：

```
using System;
using System.Drawing;
using MonoTouch.Foundation;
using MonoTouch.UIKit;
namespace Application
{
    public partial class __23ViewController : UIViewController
    {
        ……                              //这里省略了视图控制器的构造方法和析构方法
        #region View lifecycle
        public override void ViewDidLoad ()
        {
            base.ViewDidLoad ();
            // Perform any additional setup after loading the view, typically
                from a nib.
            UILabel label = new UILabel ();
            label.Frame = new RectangleF (20, 100, 280, 64);
            label.Text = "        如何让你遇见我,在我最美丽的时刻。为这,我已在佛前
                求了五百年,求他让我们结一段尘缘.";
            label.Lines = 3;                                //设置显示文本的行数
            this.View.AddSubview (label);
        }
        ……                              //这里省略了视图加载和卸载前后的一些方法
```

```
        #endregion
    }
}
```

运行效果如图 2.26 所示。

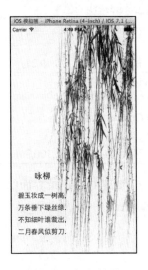 

图 2.25　运行效果　　　　　图 2.26　运行效果

2.5.2　文本框视图

与标签视图不同,文本框视图(一般使用 UITextField 类实现)可以接收用户的文本输入以及显示。

【示例 2-14】 以下将使用文本框来实现 QQ 登录界面的效果。具体步骤如下:
(1)创建一个 Single View Application 类型的工程。
(2)打开 MainStoryboard.storyboard 文件,对主视图进行设置。效果如图 2.27 所示。

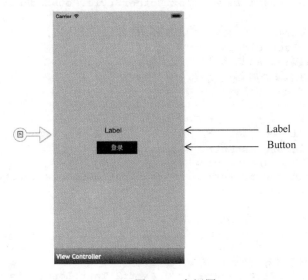

图 2.27　主视图

需要添加的视图以及设置,如表 2-7 所示。

表 2-7 设置主视图

视 图	设 置
主视图	Background:浅灰色
Label	Name:label Text:内容不全 位置和大小:(125,275,79,21)
Button	Name:button Title:登录 Text Color:白色 Background:黑色 位置和大小:(105,312,100,30)

(3)打开 2-5ViewController.cs 文件,编写代码,实现 QQ 登录界面的功能。代码如下:

```
using System;
using System.Drawing;
using MonoTouch.Foundation;
using MonoTouch.UIKit;
namespace Application
{
    public partial class __5ViewController : UIViewController
    {
        UITextField tf1;
        UITextField tf2;
        ……                              //这里省略了视图控制器的构造方法和析构方法
        #region View lifecycle
        public override void ViewDidLoad ()
        {
            base.ViewDidLoad ();
            // Perform any additional setup after loading the view, typically
                from a nib.
            //为主视图添加文本框对象 tf1
            tf1 = new UITextField ();
            tf1.BorderStyle = UITextBorderStyle.RoundedRect;
                                                        //设置文本框的边框
            tf1.Frame = new RectangleF (54, 150, 211, 30);
            tf1.Placeholder="账号";                     //设置文本框的占位符
            this.View.AddSubview (tf1);
            //为主视图添加文本框对象 tf2
            tf2 = new UITextField ();
            tf2.BorderStyle = UITextBorderStyle.RoundedRect;
            tf2.Frame = new RectangleF (54, 220, 211, 30);
            tf2.Placeholder="密码";
            tf2.SecureTextEntry = true;                 //设置文本框的文本是否隐藏
            this.View.AddSubview (tf2);
            button.TouchUpInside += this.ButtonChange_TouchUpInside;
            label.Hidden = true;
        }
        //触摸"登录"按钮后,执行的动作
        private void ButtonChange_TouchUpInside (object sender, EventArgs e)
        {
            //判断文本框对象 tf1 和 tf2 是否为空
            if (tf1.Text.Length != 0 && tf2.Text.Length != 0) {
```

第 2 章 用户界面——视图

```
            this.View.BackgroundColor = UIColor.Orange;
            label.Hidden = true;
        } else {
            label.Hidden = false;
        }
    }
    ……                          //这里省略了视图加载和卸载前后的一些方法
    #endregion
}
```

运行效果如图 2.28 所示。

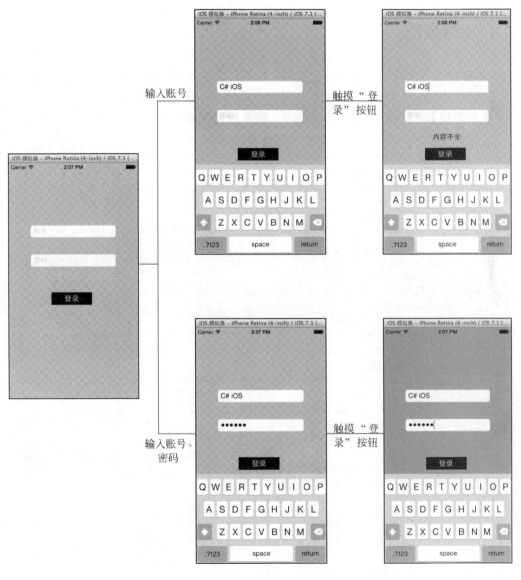

图 2.28　运行效果

在此程序中，使用了 BorderStyle 属性对文本框的边框风格进行了设置，这些边框风格如表 2-8 所示。

表 2-8 边框风格

风　　格	介　　绍
Bezel	Bezel 风格线框
Line	线框
None	无框
RoundedRect	圆角边框

SecureTextEntry 属性可以使编辑的文本隐藏，以小黑点的形式显示。此属性一般使用在输入密码时，防止被他人盗取。

2.5.3 文本视图

文本框视图只允许用户输入单行文本，如果想要输入多行文本该怎么办呢？这就需要使用文本视图解决。UITextView 类提供了一个显示文本编辑块的对象。

【示例 2-15】 以下就是使用文本视图实现多行文本的输入。代码如下：

```
using System;
using System.Drawing;
using MonoTouch.Foundation;
using MonoTouch.UIKit;
namespace Application
{
    public partial class __6ViewController : UIViewController
    {
        ……                              //这里省略了视图控制器的构造方法和析构方法
        #region View lifecycle
        public override void ViewDidLoad ()
        {
            base.ViewDidLoad ();
            // Perform any additional setup after loading the view, typically
                from a nib.
            //为主视图添加文本视图对象myTextView
            UITextView myTextView = new UITextView ();
            myTextView .Frame = new RectangleF (9, 90, 302, 180);
            this.View.AddSubview (myTextView);
            //为主视图添加文本视图对象myText
            UITextView myText = new UITextView ();
            myText .Frame = new RectangleF (9, 330, 302, 180);
            myText.Editable=false;
            this.View.AddSubview (myText);
            myText.Hidden = true;
            //为主视图添加按钮对象button
            UIButton button = new UIButton ();
            button.Frame = new RectangleF (137, 56, 46, 30);
            button.SetTitle ("完成", UIControlState.Normal);
            this.View.AddSubview (button);
            button.TouchUpInside += (sender, e) => {
                myTextView.ResignFirstResponder();           //关闭键盘
                myText.Hidden=false;
                myText.Text=myTextView.Text;
            };
```

```
            myTextView.Delegate = new MyTextViewDelegate();        //设置委托
        }
        //添加嵌套的类
        private class MyTextViewDelegate : UITextViewDelegate
        {
            //当文本视图刚开始编辑时调用
            public override void EditingStarted (UITextView textView)
            {
                Console.WriteLine ("开始编辑文本");
            }
            //当文本视图编辑结束时调用
            public override void EditingEnded (UITextView textView)
            {
                Console.WriteLine ("结束编辑文本");
            }
            //当文本视图中的内容改变时调用
            public override void Changed (UITextView textView)
            {
                Console.WriteLine ("编辑文本");
            }
        }
        ……              //这里省略了视图加载和卸载前后的一些方法
        #endregion
    }
}
```

运行效果如图 2.29 所示。

在此示例中，需要注意以下两点。

（1）键盘的消失

当用户触摸文本视图区域时，就会显示键盘；当触摸"完成"按钮后，显示的键盘就会消失。让键盘消失的方式其实很简单，就是使用 ResignFirstResponder()方法取消当前的视图的第一响应功能。

（2）文本视图的委托

当用户触摸文本框视图时，会在应用程序输出窗口输出"开始编辑文本"；当文本的内容有所改变时，会在应用程序输出窗口输出"编辑文本"；当触摸按钮后，会输出"结束文本编辑"。这些功能的实现就是通过设置文本视图的委托 delegate 实现的。我们将文本视图的委托设置为 MyTextViewDelegate 类，此类继承了 UITextViewDelegate 类。如以下的代码：

```
private class MyTextViewDelegate : UITextViewDelegate
{
    ……
}
```

在 MyTextViewDelegate 类中重写了父类 UITextViewDelegate 中的方法 EditingStarted()、EditingEnded()和 Changed()，实现了在应用程序输出窗口的字符串输出。

第 1 篇　界面构建篇

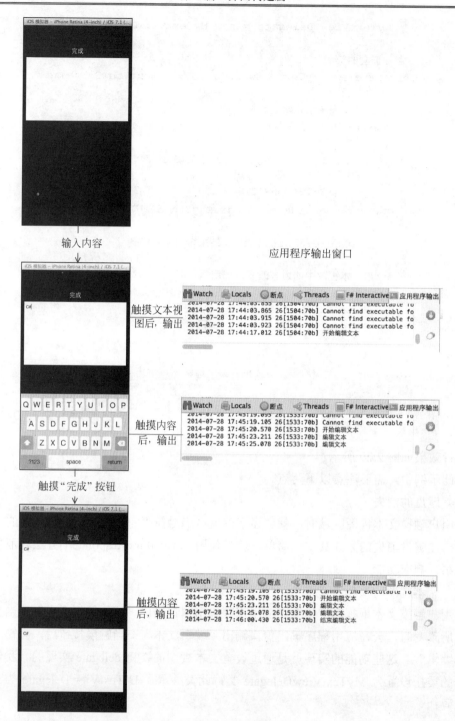

图 2.29　运行效果

2.6　使用键盘

在文本框和文本视图中可以看到，当用户在触摸这些视图后，就会弹出键盘。本节将主要讲解键盘的输入类型定义、显示键盘时改变输入视图的位置等内容。

2.6.1 定制键盘的输入类型

键盘的类型不只一种，而是有很多种。当用户要编辑一个联系人时，键盘就会随着所输入的内容不同而发生变化。例如当要输入联系人的电话号码时，键盘为数字键盘。在不同的地方使用不同类型的键盘，会使用户的操作变得简单。定制键盘的显示类型其实很简单，就是对文本框或者文本视图的第二大属性进行设置，一般称第二大属性为"输入设置"，如图 2.30 所示。

在 iOS 7 中，可以显示的键盘类型如表 2-9 所示。

表 2-9 键盘类型

设 置 类 型	设 置 项	功　　能
Capitalization	None	设置键盘输入的单词、句子，以及所有字符数据转换为大写
	Words	
	Sentences	
	All Characters	
Correction	Default	设置键盘为那些拼写错误的单词提供建议
	NO	
	YES	
Keyboard	Default	针对输入不同类型的数据选择不同类型的键盘
	ASCII Capable	
	Numbers and Punctuation	
	URL	
	Number Pad	
	Phone Pad	
	Name Phone Pad	
	E-mail Address	
	Decimal Pad	
	Twitter	
	Web Search	
Appearance	Default	设置键盘的外观
	Dark	
	Light	
Return Key	Default	键盘上显示不同类型的 Return 键
	Go	
	Google	
	Join	
	Next	
	Route	
	Search	
	Send	
	Yahoo	
	Done	
	Emergency Call	

第 1 篇　界面构建篇

续表

设　置　类　型	设　置　项	功　　能
Auto-enable Returnn Key		如果没有向文本域中输入数据，就会禁用 Return 键
Secure		将文本框的内容是为密码，并隐藏每个字符

【示例 2-16】 以下将使用代码定义一个独特的键盘。具体步骤如下：

（1）创建一个 Single View Application 类型的工程。

（2）打开 MainStoryboard.storyboard 文件，对主视图进行设置。效果如图 2.31 所示。

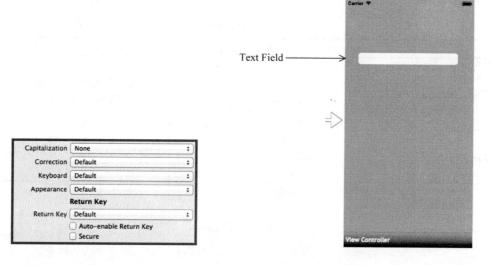

图 2.30　输入设置　　　　　　图 2.31　主视图的效果

需要添加的视图以及设置，如表 2-10 所示。

表 2-10　设置视图

视　　图	设　　置
主视图	Background：Light Gray Color
Text Field	Name：tf Text：（空）

（3）打开 2-24ViewController.cs 文件，编写代码，实现定制一个特殊的键盘。代码如下：

```
using System;
using System.Drawing;
using MonoTouch.Foundation;
using MonoTouch.UIKit;
namespace Application
{
    public partial class __24ViewController : UIViewController
    {
        ……                        //这里省略了视图控制器的构造方法和析构方法
```

```
#region View lifecycle
public override void ViewDidLoad ()
{
    base.ViewDidLoad ();

    // Perform any additional setup after loading the view, typically
       from a nib.
    tf.KeyboardType = UIKeyboardType.Url;            //设置键盘的类型
    tf.KeyboardAppearance = UIKeyboardAppearance.Dark;
                                                     //设置键盘的外观
    tf.ReturnKeyType = UIReturnKeyType.Next;//设置键盘的 Return 键
}
……                              //这里省略了视图加载和卸载前后的一些方法
#endregion
}
```

运行效果如图 2.32 所示。

图 2.32 运行效果

2.6.2 显示键盘时改变输入视图的位置

有的时候，使用应用程序的用户遇到弹出的键盘挡住了输入的文本框或者文本视图，此时该如何解决呢，这就是下面将要讲解的内容。

【示例 2-17】 以下将在弹出键盘后，将挡住的文本框改变位置，具体步骤如下：
（1）创建一个 Single View Application 类型的工程。
（2）打开 MainStoryboard.storyboard 文件，对主视图进行设置。效果如图 2.33 所示。需要添加的视图以及设置，如表 2-11 所示。

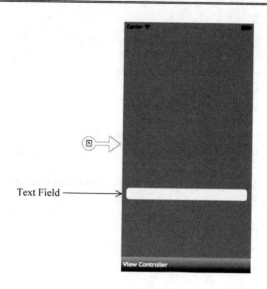

图 2.33　主视图的效果

表 2-11　视图设置

视　　图	设　　置
主视图	Background：Scrollview Textured Background color
Text Field	Text：（空） Keyboard：E-mail Address Return：Done 位置和大小：(10,400,300,30)

（3）打开 2-7ViewController.cs 文件，编写代码，实现在显示键盘时改变文本框视图的位置。代码如下：

```
using System;
using System.Drawing;
using MonoTouch.Foundation;
using MonoTouch.UIKit;
namespace Application
{
    public partial class __7ViewController : UIViewController
    {
        private NSObject kbdWillShow, kbdDidHide;
        ……                           //这里省略了视图控制器的构造方法和析构方法
        #region View lifecycle
        public override void ViewDidLoad ()
        {
            base.ViewDidLoad ();
            // Perform any additional setup after loading the view, typically
               from a nib.
            //键盘将要显示时
            kbdWillShow = UIKeyboard.Notifications.ObserveWillShow((s, e) => {
                RectangleF kbdBounds = e.FrameEnd;
                RectangleF textFrame = emailField.Frame;
                textFrame.Y -= kbdBounds.Height;
                emailField.Frame = textFrame;
            });
```

第 2 章 用户界面——视图

```
            //键盘将要隐藏时
            kbdDidHide = UIKeyboard.Notifications.ObserveDidHide((s, e) => {
                RectangleF kbdBounds = e.FrameEnd;
                RectangleF textFrame = emailField.Frame;
                textFrame.Y += kbdBounds.Height;
                emailField.Frame = textFrame;
            } );
            //触摸键盘上的 return 键
            emailField.ShouldReturn = delegate(UITextField textField) {
                return textField.ResignFirstResponder ();
            } ;
        }
        ……              //这里省略了视图加载和卸载前后的一些方法
        #endregion
    }
}
```

运行效果如图 2.34 所示。

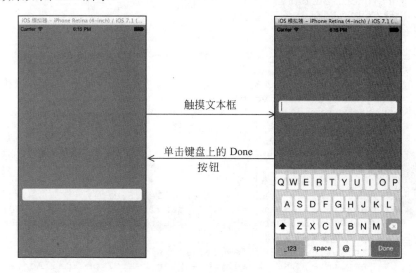

图 2.34　运行效果

需要注意的是，当用户轻拍文本框时，弹出的键盘不可以挡住需要输入文本的文本框，这是每一个开发者的责任，以确保用户可以看到自己在文本框中究竟输入了什么。在这种情况下，我们需要在默认的通知中心添加一个观察者 ObserveDid Hide（键盘将要显示时的观察者）和 ObserveDidHide（键盘将要隐藏时的观察者），代码如下：

```
this.kbdWillShow = UIKeyboard.Notifications.ObserveWillShow((s, e) => {
    ……
} );
kbdDidHide = UIKeyboard.Notifications.ObserveDidHide((s, e) => {
    ……
} );
```

通知中心是 iOS 的机制，是专门为程序中不同类间的消息通信而设置的。正常情况下，它可以通过 NSNotificationCenter.DefaultCenter 进行访问。在 Xamarin.iOS 中提供了一些 API，它们可以简化一些事情。在此示例中开发者会发现两个 API，即 ObserveWillShow 和 ObserveDidHide。通过调用 UIKeyboard.Notifications.ObserveWillShow，通知中心将会通知

• 69 •

我们键盘将要显示，同时会执行 UIKeyboard.Notifications.ObserveWillShow 中的程序。通过调用 UIKeyboard.Notifications.ObserveDidHide，通知中心将会通知我们键盘将要隐藏，同时会执行 UIKeyboard.Notifications.ObserveDidHide 中的程序。

2.6.3 为键盘添加工具栏

有的时候，为了让键盘的功能更为齐全，免不了要为它添加一个工具栏，此时需要使用到 InputAccessoryView 属性。

【示例 2-18】 以下就是为键盘添加工具栏，在工具栏中有一个"完成"的按钮，单击此按钮后，键盘就会隐藏。代码如下：

```
using System;
using System.Drawing;
using MonoTouch.Foundation;
using MonoTouch.UIKit;
namespace Application
{
    public partial class __8ViewController : UIViewController
    {
        ……                                  //这里省略了视图控制器的构造方法和析构方法
        #region View lifecycle
        public override void ViewDidLoad ()
        {
            base.ViewDidLoad ();
            // Perform any additional setup after loading the view, typically
                from a nib.
            //为主视图添加文本框对象
            UITextField emailField=new UITextField();
            emailField.Frame = new RectangleF (10, 100, 300, 30);
            emailField.BorderStyle = UITextBorderStyle.RoundedRect;
            //设置文本框的边框
            this.View.AddSubview (emailField);
            emailField.KeyboardType = UIKeyboardType.EmailAddress;
            //设置键盘的类型
            //设置工具栏
            UIToolbar toolHigh = new UIToolbar ();           //实例化工具栏对象
            toolHigh.SizeToFit();
            toolHigh.BackgroundColor = UIColor.DarkGray;     //设置背景
            //实例化栏按钮条目
            UIBarButtonItem doneHigh = new UIBarButtonItem("完成",
            UIBarButtonItemStyle.Done, (ss, ea) => {
                    emailField.ResignFirstResponder();       //关闭键盘
                }
            );
            toolHigh.SetItems(new UIBarButtonItem[] { doneHigh }, true);
                                                //为工具栏设置条目
            emailField.InputAccessoryView = toolHigh;//为键盘添加自定义视图
        }
        ……                              //这里省略了视图加载和卸载前后的一些方法
        #endregion
    }
}
```

运行效果如图 2.35 所示。

第 2 章　用户界面——视图

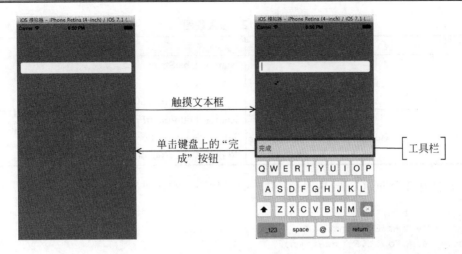

图 2.35　运行效果

2.7　进　度　条

　　进度条可以看到每一项任务现在的状态。例如，在下载的应用程序中有进度条，用户可以很方便地看到当前程序下载了多少，还剩下多少；QQ 音乐播放器中也使用到了进度条，它可以让用户看到当前音乐播放了多少，还剩多少等。在 Xamarin.iOS 中也提供用户实现进度条的类，即 UIProgressView。

　　【示例 2-19】　以下将实现进度条加载的效果。具体步骤如下：

　　（1）创建一个 Single View Application 类型的工程。

　　（2）打开 MainStoryboard.storyboard 文件，对主视图进行设置。效果如图 2.36 所示。

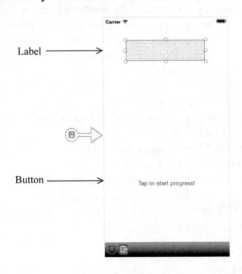

图 2.36　主视图的效果

　　需要添加的视图以及设置，如表 2-12 所示。

表 2-12 设置视图

视 图	设 置
Label	Name：labelStatus Text：（空） 位置和大小：(60, 60, 200, 50)
Button	Name：buttonStartProgress Title：Tap to start progress! 位置和大小：(60, 400, 200, 40)

(3) 打开 2-9ViewController.cs 文件，编写代码，实现进度条的加载。代码如下：

```
using System;
using System.Drawing;
using MonoTouch.Foundation;
using MonoTouch.UIKit;
using System.Threading;
using System.Threading.Tasks;
namespace Application
{
    public partial class __9ViewController : UIViewController
    {
        UIProgressView progressView;
        float incrementBy = 0f;
        ……                          //这里省略了视图控制器的构造方法和析构方法
        #region View lifecycle
        public override void ViewDidLoad ()
        {
            base.ViewDidLoad ();
            // Perform any additional setup after loading the view, typically
                from a nib.
            //触摸按钮后执行的动作
            buttonStartProgress.TouchUpInside += delegate {
                    buttonStartProgress.Enabled = false;
                    progressView.Progress = 0f;
                    Task.Factory.StartNew(this.StartProgress);
            //创建一个新的任务
            } ;
            //为主视图添加进度条对象
            progressView = new UIProgressView (new RectangleF (60f, 200f,
            200f, 50f));
            progressView.Progress = 0f;              //设置进度条的进度
            incrementBy = 1f / 10f;                  //设置进度条进度的增量值
            this.View.AddSubview(progressView);
        }
        //进度条开始加载
        public void StartProgress ()
        {
            float currentProgress = 0f;
            //判断 currentProgress 是否小于 1，如果是，执行进度条进度的加载
            while (currentProgress < 1f)
            {
                Thread.Sleep(1000);                  //1000 毫秒后暂停当前线程
```

```
            this.InvokeOnMainThread(delegate {
                progressView.Progress += this.incrementBy;
                currentProgress = this.progressView.Progress;
                labelStatus.Text=string.Format("Current value: {0}",
                Round(progressView.Progress,2));
                //判断进度条的当前进度是否为1
                if (currentProgress >= 1f)
                {
                    labelStatus.Text = "Progress completed!";
                    buttonStartProgress.Enabled = true;
                }
            } );
        }
    }
    ……                          //这里省略了视图加载和卸载前后的一些方法
    #endregion
    }
}
```

运行效果如图 2.37 所示。

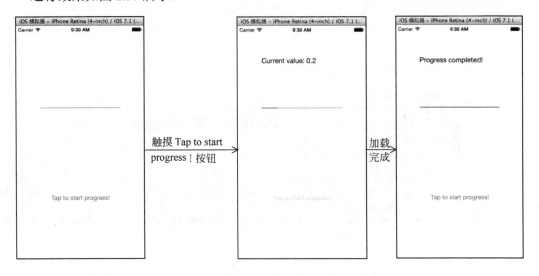

图 2.37 运行效果

在此程序中，开发者需要注意以下两个知识点。
（1）进度条进度的设置
在实例化进度条时，我们就为进度条设置了进度，使用的属性是 Progress。其语法形式如下：

进度条对象.Progress=值;

其中，值是一个浮点类型的数据，它的有效范围为 0~1。
（2）进度的增加
当触摸 Tap to start progress!按钮时，进度条就会实现自动加载进度的功能，它是通过调用 Task.Factory.StartNew()方法实现的。它的功能是创建一个 StartProgress()方法的任务，即实现加载。

2.8 滚动视图

由于iPhone或iPad屏幕边界的影响，使我们添加的控件和界面元素受到限制。但是在iPhone或iPad开发中，人们使用滚动视图解决了这一受到限制的问题。滚动视图由UIScrollView类的一个实例对象实现。

【示例2-20】 以下的代码就使用滚动视图来显示一个比屏幕还要大的图像。具体步骤如下：

（1）创建一个Single View Application类型的工程。

（2）添加图像1.jpg到创建工程的Resources文件夹中。

（3）打开2-10ViewController.cs文件，编写代码，实现通过滚动视图来观看一个比屏幕还要大的图像。代码如下：

```
using System;
using System.Drawing;
using MonoTouch.Foundation;
using MonoTouch.UIKit;
namespace Application
{
    public partial class __10ViewController : UIViewController
    {
        UIImageView imgView;
        UIScrollView scrollView;
        ……                        //这里省略了视图控制器的构造方法和析构方法
        #region View lifecycle
        public override void ViewDidLoad ()
        {
            base.ViewDidLoad ();
            // Perform any additional setup after loading the view, typically
                from a nib.
            imgView = new UIImageView (UIImage.FromFile ("1.jpg"));
            //为主视图添加滚动视图对象
            scrollView = new UIScrollView ();
            scrollView.Frame=new RectangleF(0,0,320,568) ;
            scrollView.ContentSize = imgView.Image.Size;    //滚动范围的大小
            scrollView.ContentOffset = new PointF (200f, 50f);
                                                            //目前滚动的位置
            scrollView.PagingEnabled = true;                //可以整页翻动
            scrollView.MinimumZoomScale = 0.25f;            //缩小的最小比例
            scrollView.MaximumZoomScale = 2f;               //放大的最大比例
            //获取要缩放的图像视图
            scrollView.ViewForZoomingInScrollView = delegate(UIScrollView
                scroll) {
                return this.imgView;
            } ;
            scrollView.ZoomScale = 1f;                      //设置变化比例
            scrollView.IndicatorStyle = UIScrollViewIndicatorStyle.Black;
                                                            //滚动指示器的风格设置
            scrollView.AddSubview (imgView);
            this.View.AddSubview (scrollView);
        }
        ……                        //这里省略了视图加载和卸载前后的一些方法
```

```
        #endregion
    }
}
```

运行结果如图 2.38 所示。

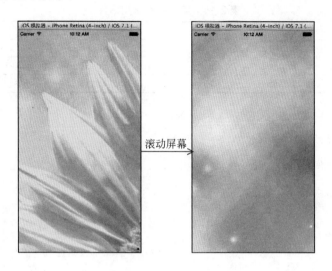

图 2.38　运行效果

在滚动视图中需要注意以下两点。
（1）常用属性
滚动视图的属性有很多，表 2-13 总结了滚动视图常用的一些属性。

表 2-13　滚动视图的属性

属　　性	功　　能
ContentOffSet	滚动视图目前滚动的位置
ContentSize	滚动范围的大小
contentInset	其他视图在滚动视图中的位置
delegate	设置委托
directionalLockEnabled	指定滚动视图是否只能在一个方向上滚动
bounces	指定滚动视图遇到边框后是否反弹
alwaysBounceVertical	指定垂直方向遇到边框是否反弹
alwaysBounceHorizontal	指定水平方向遇到边框是否反弹
pagingEnabled	是否能滚动
scrollEnabled	滚动视图是否整页翻动
showsHorizontalScrollIndicator	是否显示水平方向的滚动条
showsVerticalScrollIndicator	是否显示垂直方向的滚动条
scrollIndicatorInsets	指定滚动条在滚动视图中的位置
indicatorStyle	设定滚动条的样式
minimumZoomScale	缩小的最小比例
maximumZoomScale	放大的最大比例
zoomScale	设置变化比例
bouncesZoom	指定缩放的时候是否会反弹

（2）滚动视图常用事件

在滚动视图中一般会使用到一些事件。这里将常用到的一些事件做了总结，如表 2-14 所示。

表 2-14　滚动视图常用事件

事件	功能
Scrolled	在内容开始滚动时调用
DecelerationStarted	用户已经开始滚动内容时调用
DecelerationEnded	当用户滚动结束，并且内容也停止移动时调用

【示例 2-21】 以下将实现滚动视图的滚动，并为滚动视图添加了事件。代码如下：

```
using System;
using System.Drawing;
using MonoTouch.Foundation;
using MonoTouch.UIKit;
namespace Application
{
    public partial class __30ViewController : UIViewController
    {
        ……                              //这里省略了视图控制器的构造方法和析构方法
        #region View lifecycle
        public override void ViewDidLoad ()
        {
            base.ViewDidLoad ();
            // Perform any additional setup after loading the view, typically
            //    from a nib.
            UIScrollView scrollView = new UIScrollView ();
            scrollView.Frame = new RectangleF (0, 0, 320, 568);
            scrollView.ContentSize = new SizeF (320, 2000);
            this.View.AddSubview (scrollView);
            //滚动视图开始滚动时调用
            scrollView.Scrolled += delegate {
                Console.WriteLine ("开始滚动...");
            } ;
            //滚动视图结束滚动时调用
            scrollView.DecelerationEnded += delegate {
                Console.WriteLine ("滚动结束...");
            };
            float y = 10;
            //为滚动视图对象添加标签对象
            for (float i = 1; i < 21; i++) {
                UILabel label = new UILabel ();
                label.Frame = new RectangleF (0, y, 320, 50);
                label.BackgroundColor = UIColor.Cyan;
                label.Text = String.Format ("{0}", i);
                scrollView.AddSubview (label);
                y += 100;
            }
        }
        ……                              //这里省略了视图加载和卸载前后的一些方法
        #endregion
    }
}
```

运行效果如图 2.39 所示。

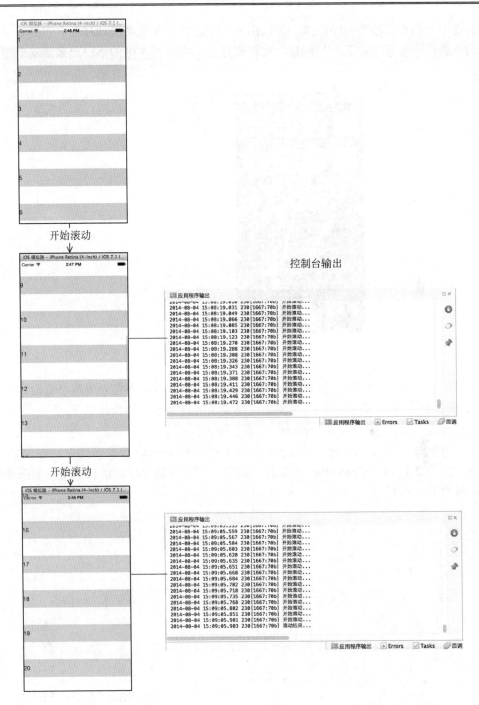

图 2.39 运行效果

2.9 页面控件

在 iPhone 手机的主界面中，经常会看到一排小白点，那就是页面控件，如图 2.40 所示。它是由小白点和滚动视图组成,可以用来控制翻页。在滚动滚动视图时可通过页面控件中的小白点来观察当前页面的位置,也可通过点击页面控件中的小白点来滚动到指定的页面。

图 2.40　页面控件

在此图中,小白点对应的当前页被高亮显示。此控件指示内容分为两个页面。

【示例 2-22】 以下将使用页面视图来控制图像的显示。具体步骤如下:

(1)创建一个 Single View Application 类型的工程。

(2)添加图像 1.jpg、2.jpg 和 3.jpg 到创建工程的 Resources 文件夹中。

(3)打开 2-11ViewController.cs 文件,编写代码,实现使用滚动视图来控制图像的显示。代码如下:

```
using System;
using System.Drawing;
using MonoTouch.Foundation;
using MonoTouch.UIKit;
namespace Application
{
    public partial class __11ViewController : UIViewController
    {
        UIImageView page1;
        UIImageView page2;
        UIImageView page3;
        UIScrollView scrollView;
        UIPageControl pageControl;
        ……                          //这里省略了视图控制器的构造方法和析构方法
        #region View lifecycle
        public override void ViewDidLoad ()
        {
            base.ViewDidLoad ();

            // Perform any additional setup after loading the view, typically
               from a nib.
            //添加滚动视图对象 scrollView
            scrollView = new UIScrollView ();
            scrollView.Frame = new RectangleF (0, 0, 320, 495);
```

```
            //滚动视图结束滚动时所调用的方法
            scrollView.DecelerationEnded += this.scrollView_
            DecelerationEnded;
            //添加页面
            pageControl = new UIPageControl ();
            pageControl.Frame = new RectangleF (0, 540, 320, 37);
            pageControl.Pages = 3;                    //设置页面控件的页数,即小白点
            //当页面控件的数值发生改变时调用
            pageControl.ValueChanged += this.pageControl_ValueChanged;
            //滚动视图的滚动事件
            scrollView.Scrolled += delegate {
                Console.WriteLine ("Scrolled!");
            } ;
            scrollView.PagingEnabled = true;
            RectangleF pageFrame = scrollView.Frame;
            scrollView.ContentSize = new SizeF (pageFrame.Width * 3,
            pageFrame.Height);
            //添加图像视图对象 page1
            page1 = new UIImageView (pageFrame);
            page1.ContentMode = UIViewContentMode.ScaleAspectFit;
            page1.Image = UIImage.FromFile ("1.jpg");
            pageFrame.X += this.scrollView.Frame.Width;
            //添加图像视图对象 page2
            page2 = new UIImageView (pageFrame);
            page2.ContentMode = UIViewContentMode.ScaleAspectFit;
            page2.Image = UIImage.FromFile ("2.jpg");
            pageFrame.X += this.scrollView.Frame.Width;
            //添加图像视图对象 page3
            page3 = new UIImageView (pageFrame);
            page3.ContentMode = UIViewContentMode.ScaleAspectFit;
            page3.Image = UIImage.FromFile ("3.jpg");
            scrollView.AddSubview (page1);
            scrollView.AddSubview (page2);
            scrollView.AddSubview (page3);
            this.View.AddSubview (scrollView);
            this.View.AddSubview (pageControl);
        }

        private void scrollView_DecelerationEnded (object sender, EventArgs e)
        {
            float x1 = this.page1.Frame.X; //获取图像视图对象 page1 的 x 位置
            float x2 = this.page2.Frame.X; //获取图像视图对象 page2 的 x 位置
            float x = this.scrollView.ContentOffset.X;
                              //获取滚动视图对象 scrollView 目前滚动的 x 位置
            //判断 x 是否和 x1 相等
            if (x == x1)
            {
                this.pageControl.CurrentPage = 0;    //设置页面控件当前的页
            } else if (x == x2)                          //判断 x 是否和 x2 相等
            {
                this.pageControl.CurrentPage = 1;
            } else
            {
                this.pageControl.CurrentPage = 2;
            }
        }

        private void pageControl_ValueChanged (object sender, EventArgs e)
        {
```

```
            PointF contentOffset = this.scrollView.ContentOffset;
            //使用 switch 语句判断当前的页数
            switch (this.pageControl.CurrentPage)
            {
            case 0:
                contentOffset.X = this.page1.Frame.X;
                this.scrollView.SetContentOffset (contentOffset, true);
                //设置滚动视图目前滚动的位置
                break;
            case 1:
                contentOffset.X = this.page2.Frame.X;
                this.scrollView.SetContentOffset (contentOffset, true);
                break;
            case 2:
                contentOffset.X = this.page3.Frame.X;
                this.scrollView.SetContentOffset (contentOffset, true);
                break;
            default:
                break;
            }
        ……              //这里省略了视图加载和卸载前后的一些方法
        #endregion
    }
}
```

运行效果如图 2.41 所示。

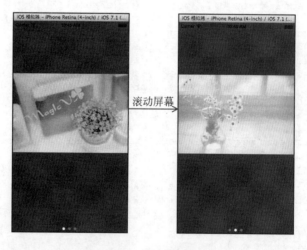

图 2.41 运行效果

在页面控件中,需要开发者注意以下两个问题。
(1) 页面控件的属性设置

页面控件属性设置并不多,一般就是设置页数以及当前页。设置页面控件的页数,需要使用 Pages 属性,其语法形式如下:

页面控件对象.Pages=页数;

其中,页数是一个整型数据。

设置页面控件的当前页,需要使用属性 CurrentPage 属性,其语法形式如下:

```
页面控件对象.CurrentPage=当前页;
```

其中，当前页是一个整型数据。

（2）页面控件的响应

页面控件的响应需要使用 ValueChanged 事件实现。【示例 2-22】中的代码如下：

```
pageControl.ValueChanged += this.pageControl_ValueChanged;
```

2.10 警告视图

如果需要向用户显示一条非常重要的消息时，警告视图（UIAlertView 类）就可以派上用场了。它的功能是把需要注意的信息显示给用户。一般显示一条信息，或者显示一条信息和几个按钮。本节将主要讲解如何为主视图添加警告视图、如何将警告视图进行显示、如何以不同的形式显示警告视图，以及响应警告视图。

2.10.1 为主视图添加警告视图

在工具栏中是没有警告视图的，开发者必须使用代码的形式在主视图中进行添加。它的添加和其他视图的添加是有区别的。具体步骤如下所述。

1. 实例化对象

UIAlertView 类提供的警告视图，在使用前必须要进行实例化。其语法形式如下：

```
UIAlertView 对象名=new UIAlertView();
```

2. 显示

在实例化对象以后，就要将此警告视图对象进行显示。此时不需要使用 AddSubview()方法，而是使用 Show()方法，其语法形式如下：

```
对象名.Show();
```

【示例 2-23】 以下将实现在主视图中显示一个警告视图的功能。代码如下：

```
using System;
using System.Drawing;
using MonoTouch.Foundation;
using MonoTouch.UIKit;
namespace Application
{
    public partial class __25ViewController : UIViewController
    {
        ……                    //这里省略了视图控制器的构造方法和析构方法
        #region View lifecycle
        public override void ViewDidLoad ()
        {
            base.ViewDidLoad ();
```

```
            // Perform any additional setup after loading the view, typically
                from a nib.
            UIAlertView alertview = new UIAlertView ();//实例化警告视图对象
            alertview.Show ();                        //显示
        }
        ……                       //这里省略了视图加载和卸载前后的一些方法
        #endregion
    }
}
```

运行效果如图 2.42 所示。

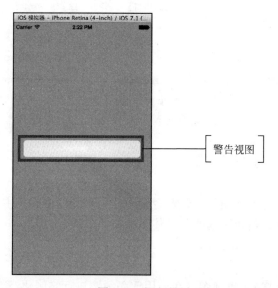

图 2.42　运行效果

2.10.2　常用的警告视图样式

在一个应用程序中，警告视图的样式是多种多样的，以下将讲解在应用程序中常见的 4 种警告视图样式。

1．无按钮的警告视图

在有的警告视图中是不需要有按钮的，当然，它不需要用户在此视图上做任何操作，这就是无按钮的警告视图。

【示例 2-24】　以下就是一个简单的无按钮的警告视图的实现。代码如下：

```
using System;
using System.Drawing;
using MonoTouch.Foundation;
using MonoTouch.UIKit;
namespace Application
{
    public partial class __26ViewController : UIViewController
    {
        ……              //这里省略了视图控制器的构造方法和析构方法
        #region View lifecycle
```

```
        public override void ViewDidLoad ()
        {
            base.ViewDidLoad ();
            // Perform any additional setup after loading the view, typically
                from a nib.
            UIAlertView alertView = new UIAlertView ();
            alertView.Title = "提示";                //设置警告视图的标题
            alertView.Message = "电量不足";          //设置警告视图向用户显示的信息
            alertView.Show ();
        }
        ……                          //这里省略了视图加载和卸载前后的一些方法
        #endregion
    }
}
```

运行效果如图 2.43 所示。

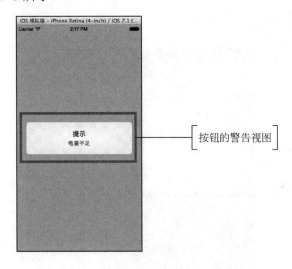

图 2.43　运行效果

2．具有一个按钮的警告视图

具有一个按钮的警告视图，一般是想要强制性的引起用户的注意，如果用户没有注意到此视图，那么它会一直存在，直到用户注意此视图，并使用此视图中提示的按钮去关闭，警告视图才会消失。如何为警告视图添加按钮呢？需要使用 AddButton()方法实现，其语法形式如下：

```
警告视图对象名.AddButton("按钮标题");
```

【示例 2-25】 以下就是一个具有一个按钮的警告视图的实现。代码如下：

```
using System;
using System.Drawing;
using MonoTouch.Foundation;
using MonoTouch.UIKit;
namespace Application
{
    public partial class __27ViewController : UIViewController
    {
        ……              //这里省略了视图控制器的构造方法和析构方法
```

```
#region View lifecycle
public override void ViewDidLoad ()
{
    base.ViewDidLoad ();
    // Perform any additional setup after loading the view, typically
       from a nib.
    UIAlertView alertView = new UIAlertView ();
    alertView.Title="提示";
    alertView.Message="内存空间不足";
    alertView.AddButton ("前去清理");              //为警告视图添加按钮
    alertView.Show ();
}
……                              //这里省略了视图加载和卸载前后的一些方法
#endregion
}
```

运行效果如图 2.44 所示。

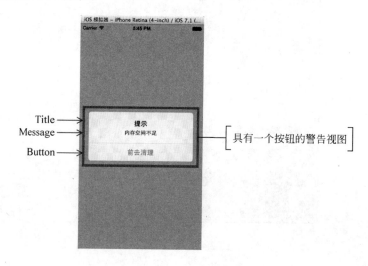

图 2.44　运行效果

3．具有多个按钮的警告视图

有时，在警告视图中只有一个按钮是无法满足需要的，需要在此视图中添加多个按钮，从而让用户实现多方面的选择。此时还是需要使用 **AddButton()** 方法。

【示例 2-26】 以下就是具有多个按钮的警告视图的效果，代码如下：

```
using System;
using System.Drawing;
using MonoTouch.Foundation;
using MonoTouch.UIKit;
namespace Application
{
    public partial class __28ViewController : UIViewController
    {
        ……                         //这里省略了视图控制器的构造方法和析构方法
        #region View lifecycle
        public override void ViewDidLoad ()
        {
```

```
            base.ViewDidLoad ();
            // Perform any additional setup after loading the view, typically
               from a nib.
            UIAlertView alertView = new UIAlertView ();
            alertView.Title="谢谢";
            alertView.Message = "亲如果你对我们的商品满意，请点亮五颗星";
            //添加多个按钮
            alertView.AddButton ("前往评论");
            alertView.AddButton ("暂不评论");
            alertView.AddButton ("残忍拒绝");
            alertView.Show ();
        }
        ……                              //这里省略了视图加载和卸载前后的一些方法
        #endregion
    }
}
```

运行效果如图 2.45 所示。

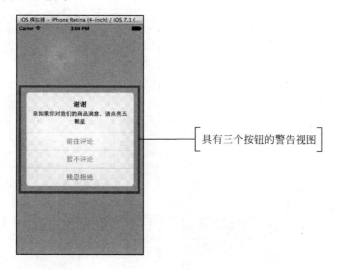

图 2.45　运行效果

4．具体文本输入框的警告视图

在警告视图中也可以添加文本框视图，此时需要使用 AlertViewStyle 属性。

【示例 2-27】　以下的代码为文本框添加两个警告视图，代码如下：

```
using System;
using System.Drawing;
using MonoTouch.Foundation;
using MonoTouch.UIKit;
namespace Application
{
    public partial class __29ViewController : UIViewController
    {
        ……                              //这里省略了视图控制器的构造方法和析构方法
        #region View lifecycle
        public override void ViewDidLoad ()
        {
            base.ViewDidLoad ();
```

```
        // Perform any additional setup after loading the view, typically
           from a nib.
        UIAlertView alertView = new UIAlertView ();
        alertView.Title="登录";
        alertView.AlertViewStyle = UIAlertViewStyle.LoginAnd
        PasswordInput;             //设置警告视图的风格
        alertView.Show ();
    }
    ......                        //这里省略了视图加载和卸载前后的一些方法
    #endregion
}
```

运行效果如图 2.46 所示。

注意：AlertViewStyle 属性一共有 4 种风格，这 4 种风格的效果如图 2.47 所示。

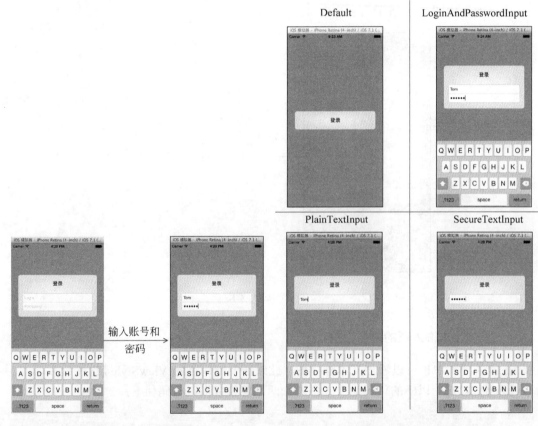

图 2.46　运行效果　　　　　　　　图 2.47　警告视图风格效果

2.10.3　响应警告视图

　　在以上的警告视图示例中，为警告视图添加了按钮，它们的功能都是一样的，那就是实现警告视图的关闭。此时，多按钮的警告视图就显得多此一举了，那么如何让警告视图中的每一个按钮都实现自己的功能呢？这就需要实现警告视图的响应，此时需要使用到

Dismissed 事件，其语法形式如下：

```
警告视图对象.Dismissed +=触摸按钮后的方法；
```

或者：

```
警告视图对象.Dismissed += (sender, e) => {
           ……
};
```

其中，sender 表示事件监视的对象，e 就是事件所需要的数据。

【示例 2-28】 以下的代码将在主视图中显示一个标题为 btnShowAlert 的按钮，触摸此按钮后将显示一个具有两个按钮的警告视图，其标题分别为 OK 和 Cancel，当用户触摸警告视图中的任意一个按钮，都会改变为 btnShowAlert 按钮的标题，代码如下：

```
using System;
using System.Drawing;
using MonoTouch.Foundation;
using MonoTouch.UIKit;
namespace Application
{
    public partial class __12ViewController : UIViewController
    {
        UIButton btnShowAler;
        ……                            //这里省略了视图控制器的构造方法和析构方法
        #region View lifecycle
        public override void ViewDidLoad ()
        {
            base.ViewDidLoad ();
            // Perform any additional setup after loading the view, typically
              from a nib.
            btnShowAler = new UIButton ();
            btnShowAler.Frame = new RectangleF (106, 269, 108, 30);
            btnShowAler.SetTitle ("Show Alert", UIControlState.Normal);
            this.View.AddSubview (btnShowAler);
            btnShowAler.TouchUpInside+=(sender,e)=>this.ShowAlert("Alert
            Message", "Tap OK or Cancel") ;
        }
        //触摸 btnShowAler 按钮后实现的功能
        private void ShowAlert(string title, string message)
        {
            // Create the alert
            UIAlertView alertView = new UIAlertView(); //实例化警告视图对象
            alertView.Title = title;                   //设置标题
            alertView.Message = message;               //设置信息
            //添加按钮
            alertView.AddButton("OK");
            alertView.AddButton("Cancel");
            //响应警告视图
            alertView.Dismissed += (sender, e) => {
                if (e.ButtonIndex == 0)
                {
                    btnShowAler.SetTitle("OK!",UIControlState.Normal);
                } else
                {
                    btnShowAler.SetTitle("Cancelled!",UIControlState.Normal);
                }
            };
```

```
            alertView.Show();
        }
        ……                    //这里省略了视图加载和卸载前后的一些方法
        #endregion
    }
}
```

在此程序中，sender 表示警告视图中的按钮，e 表示点击按钮的一些参数，如 ButtonIndex。运行效果如图 2.48 所示。

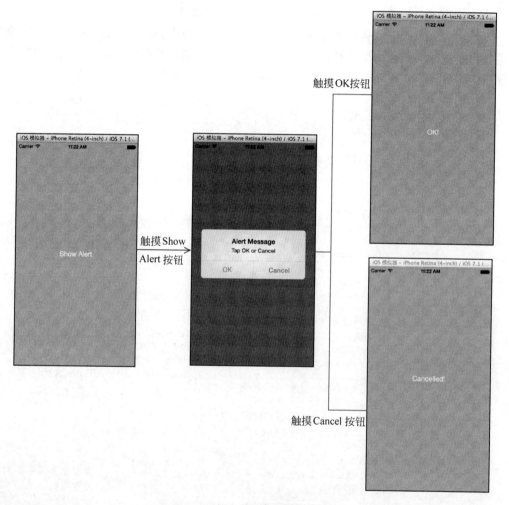

图 2.48　运行效果

2.11　自定义视图

工具栏中的视图在实际应用开发中用得很多，但是为了吸引用户的眼球，开发者可以做出一些自定义的视图。

【示例 2-29】 以下将实现一个自定义的视图，当用户触摸屏幕时，就会出现一个显示手指当前位置的标签视图，以及改变主视图的背景颜色。代码如下：

（1）创建一个 Single View Application 类型的工程。

（2）添加一个 C#的类文件，并命名为 MyView，具体步骤如下：

首先，选择菜单栏中的"文件"|New|File...命令，弹出 New File 对话框，如图 2.49 所示。

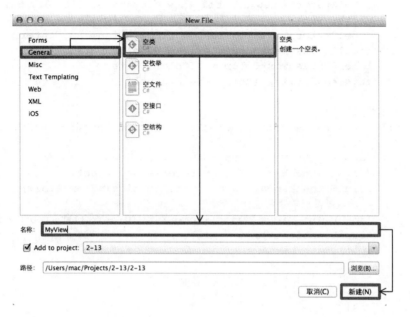

图 2.49 操作步骤 1

然后选择 General 中的"空类"选项，输入类的名称后，单击"新建"按钮，此时，一个名为 MyView 的类文件就创建好了。

（3）打开 MainStoryboard.storyboard 文件，选择主视图后，单击最右端的"属性"按钮，在属性对话框中，将 Class 设置为创建的类文件名 MyView。如图 2.50 所示。

（4）打开 MyView.cs 文件，编写代码，实现一个自定义的视图。代码如下：

图 2.50 操作步骤 2

```
using System;
using System.Drawing;
using MonoTouch.Foundation;
using MonoTouch.UIKit;
using System.CodeDom.Compiler;
namespace Application
{
    partial class MyView : UIView
    {
        private UILabel labelStatus;
        public MyView (IntPtr handle) : base(handle)
        {
            this.Initialize();
        }
        public MyView(RectangleF frame) : base(frame)
        {
            this.Initialize();
        }
```

```
//初始化方法
private void Initialize()
{
    this.BackgroundColor = UIColor.LightGray;
    //添加一个标签对象
    labelStatus = new UILabel (new RectangleF (0f, 0f, this.Frame.
    Width, 60f));
    labelStatus.TextAlignment = UITextAlignment.Center;
    labelStatus.BackgroundColor = UIColor.DarkGray;
    labelStatus.TextColor = UIColor.White;
    this.AddSubview (this.labelStatus);
}
//实现触摸事件
public override void TouchesMoved (NSSet touches, UIEvent evt)
{
    base.TouchesMoved (touches, evt);
    UITouch touch = (UITouch)touches.AnyObject;
    PointF touchLocation = touch.LocationInView (this);
    //获取触摸点的当前位置
    labelStatus.Text = String.Format ("X: {0} - Y: {1}", touchLocation.X,
    touchLocation.Y);
}
}
```

运行效果如图 2.51 所示。

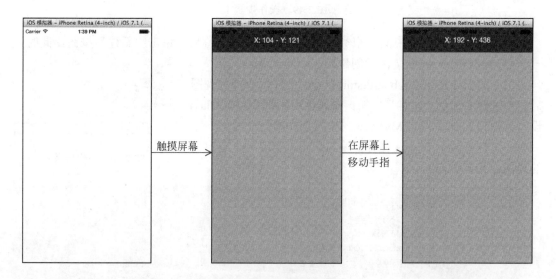

图 2.51　运行效果

需要注意的是，以下的构造器覆盖了基类的 UIView(IntPtr)构造器，此构造函数总是被当做一个通过本地化代码进行初始化的视图。

```
public MyView (RectangleF frame) : base(frame) {}
```

TouchesMoved()方法被重写，当用户的手指在主视图上进行移动时，就会执行此方法中的内容。

2.12 一次性修改相同的视图

在一个应用程序中，使用了很多相同的视图。如果想要更改这些视图的属性，并且属性都相同，该怎么办呢？可能聪明的开发者会想到，首先在一个视图对象中编写好更改的属性，然后进行复制，最后改变此属性对应的对象名就可以了。

这样的方法确实可行，但是它只适用于个数较少的视图对象。如果此应用程序中有成百上千的相同的视图对象时，这种方法还是否可行？当然是不可行的了，这样会使代码看起来冗余，并且会花费开发者相当长的时间。那么有没有方法可以一次性将相同视图的相同属性进行修改呢？答案当然是肯定的了。使用 Appearance 属性就可以实现了，它是一个类型方法，其语法形式如下：

视图类.Appearance.视图的属性=属性设置；

【示例2-30】以下的代码使用 Appearance 属性，将主视图中的所有标签改为青色背景，标题颜色为棕色的视图。代码如下：

```
using System;
using System.Drawing;
using MonoTouch.Foundation;
using MonoTouch.UIKit;
namespace Application
{
    public partial class __14ViewController : UIViewController
    {
        ……                              //这里省略了视图控制器的构造方法和析构方法
        #region View lifecycle
        public override void ViewDidLoad ()
        {
            base.ViewDidLoad ();
            // Perform any additional setup after loading the view, typically
                from a nib.
            //添加标题对象 label1
            UILabel label1 = new UILabel ();
            label1.Frame = new RectangleF (0, 90, 320, 50);
            label1.Text="红色";
            this.View.AddSubview (label1);
            //添加标题对象 label1
            UILabel label2 = new UILabel ();
            label2.Frame = new RectangleF (0, 200, 320, 50);
            label2.Text="黄色";
            this.View.AddSubview (label2);
            //添加标题对象 label1
            UILabel label3 = new UILabel ();
            label3.Frame = new RectangleF (0, 310, 320, 50);
            label3.Text="青色";
            this.View.AddSubview (label3);
            //添加标题对象 label1
            UILabel label4 = new UILabel ();
            label4.Frame = new RectangleF (0, 420, 320, 50);
            label4.Text="蓝色";
            this.View.AddSubview (label4);
```

```
        }
        ……                    //这里省略了视图加载和卸载前后的一些方法
        #endregion
    }
}
```

运行效果如图 2.52 所示。

```
UILabel.Appearance.BackgroundColor = UIColor.Cyan;  //设置所有标签的背景
UILabel.Appearance.TextColor = UIColor.Brown;       //设置所有标签的文本颜色
```

运行效果如图 2.53 所示。

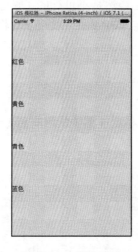
图 2.52　运行效果

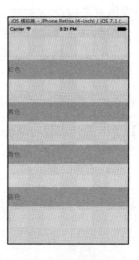
图 2.53　运行效果

第3章 用户界面——控制器

到现在为止，我们已经讲解了一些视图以及它们的使用。在真实世界的应用程序场景中，单独的视图往往是不够的。所以Apple提供了另外的基类——UIViewController，它可以用来对视图进行管理。视图控制器可以响应设备的通知。例如，当设备旋转时，可以提供不同的方法显示并关闭多个视图，甚至其他控制器。本章将主要讲解4种常用的控制器：UIViewController视图控制器、UINavigationController导航控制器、UITabBarController标签栏控制器、iPad-specific View controller分割控制器，以及如何创建一个自定义视图控制器等内容。

3.1 使用视图控制器加载视图

所有控制器的基类就是UIViewController视图控制器。因为它是整个应用程序的枢纽，每一个视图至少有一个视图控制器，所以我们在前面的程序开发中都用到了一个视图控制器。

【示例3-1】以下将在一个空类型的工程中，通过视图控制器为应用程序加载一个视图，具体操作步骤如下所述。

（1）创建一个空类型的工程，空工程的创建步骤如下：

首先，单击打开Xamarin Studio，选择New Solution…选项，弹出"新建解决方案"对话框，如图3.1所示。

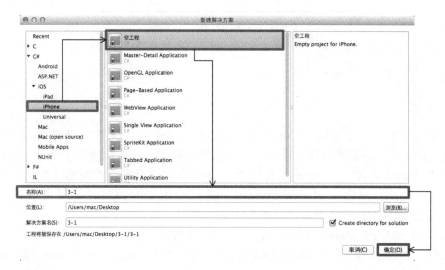

图3.1 操作步骤1

然后，选择 iOS 下的 iPhone，在 iPhone 中选择"空工程"选项。将名称输入为 3-1，然后单击"确定"按钮，此时就创建好了一个空的工程，其名称为 3-1。

💡注意：在此工程中是没有任何视图的。此时运行程序就会出现如图 3.2 所示的效果。

由于此程序没有为 UIWindow 添加根视图控制器，导致程序出现了如图 3.2 所示的效果。其中，UIWindow 是 UIView 的子类，UIWindow 的主要作用有两个，一是提供一个区域来显示 UIView，二是将事件分发给 UIView，一个应用基本上只有一个 UIWindow。根视图控制器中包含了主视图。在应用程序输出窗口会出现如下的提示：

图 3.2　运行效果

```
2014-08-04 17:57:13.945 32[2410:70b] Application windows are expected to
have a root view controller at the end of application launch
```

（2）在创建的工程中添加一个类型为 iPhone View Controller 的文件，并命名为 MainViewController。具体步骤如下：

首先，选择菜单栏中的"文件"|New|File…命令，弹出 New Files 对话框，如图 3.3 所示。

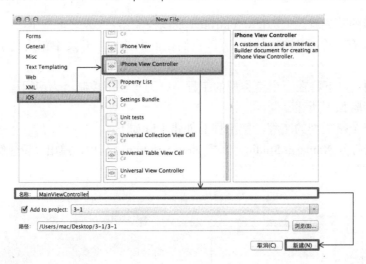

图 3.3　操作步骤 2

然后选择 iOS 下的 iPhone View Controller，并输入此文件的名称为 MainViewController。完成后，单击"新建"按钮，此时会在创建的工程中出现以下几个文件：

❑ 文件 MainViewController.cs，是一个 C#源文件，它是实现控制器的类；
❑ 文件 MainViewController.designer.cs，是自动生成的源文件，可反映控制器在 Interface Builder 中的改变，一般不用此文件；
❑ 文件 MainViewController.xib，是一个 XIB 文件，包含了一个视图控制器，可以在此文件中对此视图控制器的主视图进行设置，它是 Interface Builder 最早的一种形式。

（3）双击打开 MainViewController.xib，此时会打开 Xcode，并显示 Xcode 早期的 Interface Builer 界面，如图 3.4 所示。

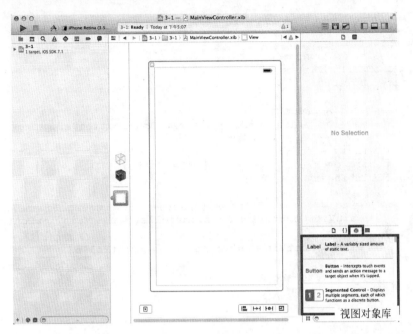

图 3.4　操作步骤 3

（4）从视图对象库中，拖动 Label 标签视图对象到主视图中，双击鼠标，将标题改为 Load a View，如图 3.5 所示。

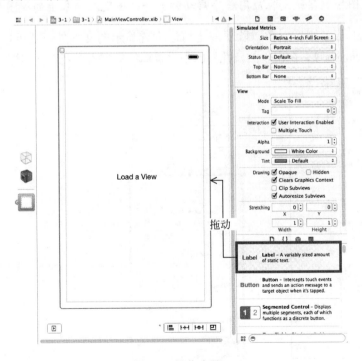

图 3.5　操作步骤 4

（5）回到 Xamarin Studio，打开 AppDelegate.cs 文件，编写代码，实现使用控制器加载视图的功能，代码如下：

```
using System;
using System.Collections.Generic;
using System.Linq;
using MonoTouch.Foundation;
using MonoTouch.UIKit;
namespace Application
{
    [Register ("AppDelegate")]
    public partial class AppDelegate : UIApplicationDelegate
    {
        UIWindow window;
        public override bool FinishedLaunching (UIApplication app,
        NSDictionary options)
        {
            window = new UIWindow (UIScreen.MainScreen.Bounds);
            MainViewController mainController = new MainViewController ();
                                                            //实例化对象
            window.RootViewController = mainController;
                                                 //为窗口设置根视图控制器
            window.MakeKeyAndVisible ();
            return true;
        }
    }
}
```

在此程序中，对根视图控制器的设置需要使用 RootViewController 属性。运行效果如图 3.6 所示。

注意：根视图控制器就是在应用程序启动时第一个出现在屏幕上的视图控制器。

对于 Label 标签视图对象，除了双击可以设置显示的文本外，还可以通过 Show the Attributes inspector 选项，在出现的属性界面中设置标签的文本以及其他属性，如图 3.7 所示。

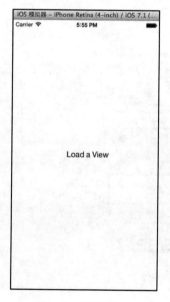

图 3.6　运行效果

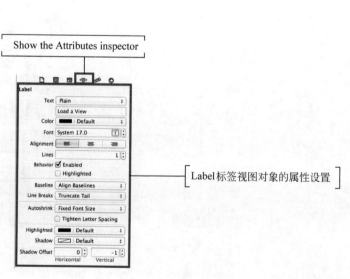

图 3.7　属性设置

此属性设置的界面和 Xamarin Studio 属性设置的界面相同，都是对视图对象的属性进行设置。除此之外，开发者还可以通过代码对 Label 标签视图对象的属性进行设置。以下将图 3.6 中标签视图对象的显示内容通过代码改为 Hello C#，具体步骤如下所述。

（1）将 Xcode 中主视图上的标题和 MainViewController.h 文件进行关联（即为主视图中的标签声明插座变量。插座变量是 Objective-C 中的说法，通过此变量，可以对主视图中关联的视图进行控制。插座变量类似于 Xamarin Studio 中为主视图中的视图对象命名，即设置 Name 属性）。关联的具体步骤如下：

首先将 Xcode 中显示的 3-1 展开，如图 3.8 所示。

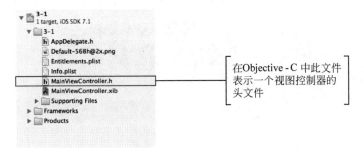

图 3.8　操作步骤 1

然后双击 MainViewController.h，将此文件在一个单独的窗口中打开，或者是使用调整窗口中的按钮进行窗口的调整，调整窗口如图 3.9 所示。

图 3.9　调整窗口

对界面调整后的样子如图 3.10 所示。

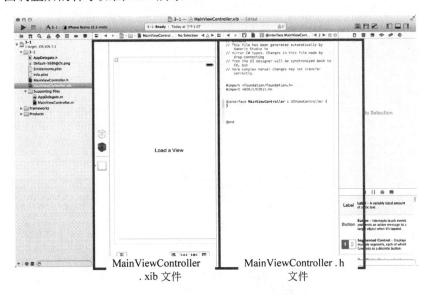

图 3.10　此时 Xcode 界面的效果

接着按住 Ctrl 键，同时拖动 MainViewController.xib 文件中主视图上的 Label 标签视图对象，如图 3.11 所示。

图 3.11　操作步骤 2

松开鼠标，此时会弹出一个对话框，如图 3.12 所示。

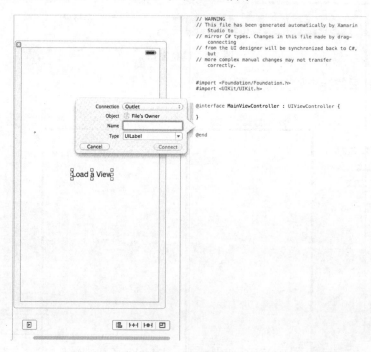

图 3.12　操作步骤 3

开发者在此对话栏中，在 Name 文本框中输入插座变量的名称，我们在这里输入的插座变量的名称为 Name。输入完毕后，会在右边，即在 MainViewController.h 文件中，显示声明的插座变量，如图 3.13 所示。

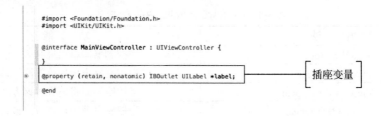

图 3.13　MainViewController.h 文件

（2）回到 Xamarin Studio 中，打开 MainViewController.cs 文件，编写代码，实现设置主视图中标签显示的文本，代码如下：

```
using System;
using System.Drawing;
using MonoTouch.Foundation;
using MonoTouch.UIKit;
namespace Application
{
    public partial class MainViewController : UIViewController
    {
        ……                              //这里省略了视图控制器的构造方法和析构方法
        public override void ViewDidLoad ()
        {
            base.ViewDidLoad ();
            label.Text="Hello C#";    //使用插座变量设置在主视图中标签的文本
        }
    }
}
```

运行结果如图 3.14 所示。

图 3.14　运行效果

UIViewController 类中包含了一些方法，可以允许我们管理视图控制器的生命周期。

这些方法可以通过系统的视图控制器去调用，并且可以实现这些方法的重写，添加我们自己的方法。这些方法如表 3-1 所示。

表 3-1 UIViewController中的方法

方　　法	功　　能
ViewWillAppear	当控制器即将出现时该方法被调用
ViewDidAppear	当控制器出现后该方法被调用
ViewWillDisappear	当控制器即将消失时该方法被调用
ViewDidDisappear	当控制器消失后该方法被调用

3.2 导航不同的视图控制器

UINavigationController 又名导航控制器，是 UIViewController 的子类。导航控制器提供了一种简单的途径，可以让用户在多个数据视图中导航，以及可以回到起点。UINavigationController 导航控制器使用栈存储结构来管理控制器，栈具有后进先出的功能。本节将主要讲解关于导航控制器的一些功能。

3.2.1 导航控制器的基本组成

通常，一个导航控制器是由 3 个部分组成的，如图 3.15 所示。

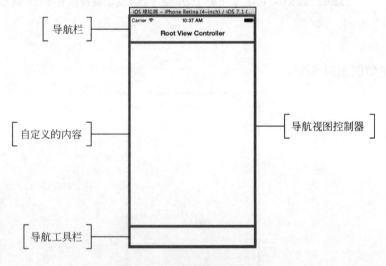

图 3.15 导航控制器的组成

- 导航栏：一个导航栏（UINavigationBar）是一个矩形视图。它出现在顶部。它显示左条目、中条目、右条目，一般情况下，中条目是一个标题。
- 自定义的内容：一个导航控制器是一个包含各个视图控制器的容器，并负责显示这些视图控制器中的一个视图。
- 导航工具栏：一个工具栏（UIToolbar）是一个显示一行条目的矩形视图，用户可以单击任何部分，基本就像一个按钮。它是一个可选的部分。

3.2.2 添加导航控制器

导航控制器该如何进行添加呢?这是本小节所要讲解的重点。要添加导航控制器需要通过以下几步实现。

1. 实例化

UINavigationController 实例化对象的语法形式如下:

```
UINavigationController 导航控制器对象=new UINavigationController();
```

2. 将视图控制器进行压栈

导航控制器使用栈存储结构来管理控制器。如果要在一个视图控制器上添加导航控制器,需要将此视图控制器以压栈的形式放入到导航控制器中。此时需要使用 PushViewController()方法。其语法形式如下:

```
导航控制器对象.PushViewController(视图控制器对象,Bool 类型值)
```

其中,Bool 类型值表示是否需要指定过渡动画效果。当此值为 true 时,表示需要指定过渡动画效果;当此值为 false 时,表示不需要指定动画效果。

注意:添加导航控制器步骤 1 和 2 可以进行合并,其语法形式如下:

```
UINavigationController 导航控制器对象=new UINavigationController(视图控制器对象);
```

3. 设置根视图控制器

对于根视图控制器的设置在上一节中已经讲解过了,其语法形式如下:

```
window.RootViewController = 导航控制器对象;
```

【示例 3-2】 以下将实现在主视图中添加导航控制器的功能,具体步骤如下所述。
(1)创建一个空类型的工程。
(2)在创建的工程中添加一个类型为 iPhone View Controller 的文件,命名为 MainViewController。
(3)打开 AppDelegate.cs 文件,编写代码,实现为 MainViewController 的主视图添加导航控制器,代码如下:

```
using System;
using System.Collections.Generic;
using System.Linq;
using MonoTouch.Foundation;
using MonoTouch.UIKit;
namespace Application
{
    [Register ("AppDelegate")]
    public partial class AppDelegate : UIApplicationDelegate
    {
        UIWindow window;
```

```
        public override bool FinishedLaunching (UIApplication app,
        NSDictionary options)
        {
            window = new UIWindow (UIScreen.MainScreen.Bounds);
            MainViewController mainController = new MainViewController ();
            mainController.Title="首页";                  //设置标题
            UINavigationController nav = new UINavigationController
            (mainController);                             //实例化
            window.RootViewController = nav;              //设置根视图
            window.MakeKeyAndVisible ();
            return true;
        }
    }
}
```

在此程序中，使用 Title 属性设置了导航栏上的标题。

（4）打开 MainViewController.xib 文件，将主视图的背景设置为浅橘黄色。

运行效果如图 3.16 所示。

3.2.3 通过导航控制器实现视图的切换

导航控制器提供了一种简单的途径，可以让用户在多个数据视图中导航，以及可以回到起点。下面就为开发者实现此功能。

【示例 3-3】 以下将通过导航控制器来实现两个视图控制器之间的视图切换。具体步骤如下所述。

（1）创建一个空类型的工程。

（2）在创建的工程中添加 3 个类型为 iPhone View Controller 的文件，分别命名为 MainViewController、ViewController1 和 ViewController2。

图 3.16 运行效果

（3）打开 AppDelegate.cs 文件，编写代码，实现通过控制器加载视图的功能，代码如下：

```
using System;
using System.Collections.Generic;
using System.Linq;
using MonoTouch.Foundation;
using MonoTouch.UIKit;
namespace Application
{
    [Register ("AppDelegate")]
    public partial class AppDelegate : UIApplicationDelegate
    {
        UIWindow window;
        public override bool FinishedLaunching (UIApplication app,
        NSDictionary options)
        {
            window = new UIWindow (UIScreen.MainScreen.Bounds);
```

```
            //实例化 MainViewController 类的对象
            MainViewController mainController = new MainViewController();
            mainController.Title = "Main View";
            //实例化导航控制器对象
            UINavigationController navController = new UINavigationController
            (mainController);
            window.RootViewController = navController;    //设置根视图控制器
            window.MakeKeyAndVisible ();
            return true;
        }
    }
}
```

(4) 打开 MainViewController.xib 文件, 对 MainViewController 的主视图进行设置, 效果如图 3.17 所示。

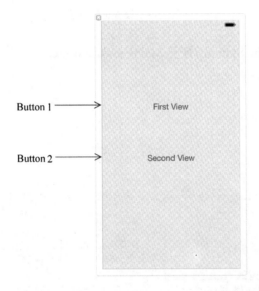

图 3.17　MainViewController 中的主视图

需要添加的视图以及设置, 如表 3-2 所示。

表 3-2　设置主视图

视　　图	设　　置
主视图	Background: 浅灰色
Button1	Title: First View Font: System 19.0 位置和大小: (102,183,117,30) 为按钮声明插座变量 buttonFirstView
Button2	Title: Second View Font: System 19.0 位置和大小: (102,183,117,30) 为按钮声明插座变量 buttonSecondView

注意: 对于位置和大小设置需要在 Show the Size inspector 选项中设置, 如图 3.18 所示。

第 1 篇　界面构建篇

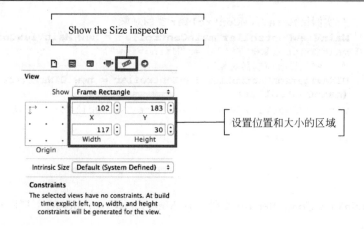

图 3.18　设置位置和大小

（5）回到 Xamarin Studio，打开 MainViewController.cs 文件，编写代码，实现在触摸按钮后，主视图的切换。代码如下：

```
using System;
using System.Drawing;
using MonoTouch.Foundation;
using MonoTouch.UIKit;
namespace Application
{
    public partial class MainViewController : UIViewController
    {
        ……
        public override void ViewDidLoad ()
        {
            base.ViewDidLoad ();
            //触摸主视图中 First View 按钮，将主视图切换到 ViewController1 的视图上
            buttonFirstView.TouchUpInside += (sender, e) => {
                ViewController1 v1 = new ViewController1();
                v1.Title = "First View";
                this.NavigationController.PushViewController(v1, true);
            } ;
            //触摸主视图中 Second View 按钮，将主视图切换到 ViewController2 的视图上
            buttonSecondView.TouchUpInside += (sender, e) => {
                ViewController2 v2 = new ViewController2();
                v2.Title = "Second View";
                this.NavigationController.PushViewController(v2, true);
            };
        }
    }
}
```

（6）打开 ViewController1.xib 文件，对 ViewController1 的主视图进行设置，效果如图 3.19 所示。

需要添加的视图以及设置，如表 3-3 所示。

第 3 章 用户界面——控制器

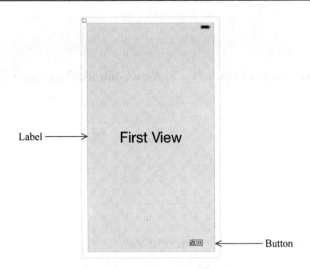

图 3.19 ViewController1 的主视图

表 3-3 设置主视图

视 图	设 置
主视图	Background：浅绿色
Label	Text：First View Font：System 37.0 Alignment：居中 位置和大小：(66,251,189,66)
Button	Title：返回 Font：System 18.0 Text Color：蓝色 位置和大小：(235,532,77,30) 为按钮声明插座变量 backbutton1

（7）回到 Xamarin Studio，打开 ViewController1.cs 文件，编写代码，实现在触摸"返回"按钮后，将当前 ViewController1 的视图切换到 MainViewController 的视图。代码如下：

```
using System;
using System.Drawing;
using MonoTouch.Foundation;
using MonoTouch.UIKit;
namespace Application
{
    public partial class ViewController1 : UIViewController
    {
        ……
        public override void ViewDidLoad ()
        {
            base.ViewDidLoad ();
            //触摸"返回"按钮后，将当前ViewController1的视图切换到
              MainViewController的视图
            backbutton1.TouchUpInside += (sender, e) => {
                this.NavigationController.PopToRootViewController(true);
            };
```

            }
        }
}
```

（8）打开 ViewController2.xib 文件，对 ViewController2 的主视图进行设置，效果如图 3.20 所示。

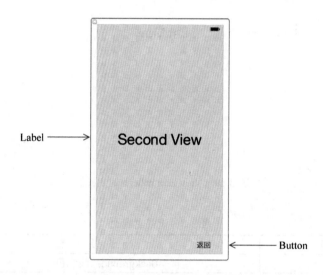

图 3.20  ViewController2 的主视图

需要添加的视图以及设置，如表 3-4 所示。

表 3-4  设置主视图

| 视 图 | 设 置 |
| --- | --- |
| 主视图 | Background：浅紫色 |
| Label | Text：Second View<br>Font：System 37.0<br>Alignment：居中<br>位置和大小：(54,251,213,66) |
| Button | Title：返回<br>Font：System 18.0<br>Text Color：蓝色<br>位置和大小：(235,532,77,30)<br>为按钮声明插座变量 backbutton2 |

（9）回到 Xamarin Studio，打开 ViewController2.cs 文件，编写代码，实现在触摸"返回"按钮后，将当前 ViewController2 的视图切换到 MainViewController 的视图。代码如下：

```
using System;
using System.Drawing;
using MonoTouch.Foundation;
using MonoTouch.UIKit;
namespace Application
{
    public partial class ViewController2 : UIViewController
```
```

```
{
    ……                          //这里省略了视图控制器的构造方法和析构方法
    public override void ViewDidLoad ()
    {
        base.ViewDidLoad ();
        backbutton2.TouchUpInside += (sender, e) => {
            this.NavigationController.PopToRootViewController(true);
            //弹出视图控制器
        };
    }
}
```

运行效果如图 3.21 所示。

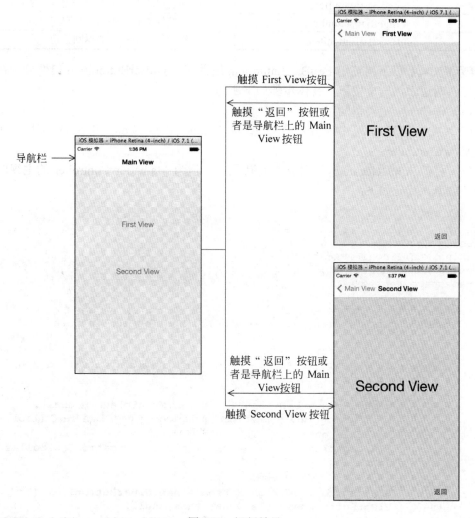

图 3.21　运行效果

需要注意的是，在触摸 First View 按钮或者 Second View 按钮后，切换到对应的视图控制器，使用了压栈的方法 PushViewController()。触摸"返回"按钮后，回到应用程序刚开始运行时的视图中，此时需要使用到 PopToRootViewController()方法，将压入栈中的视图控制器弹出。其语法形式如下：

```
PopToRootViewController(Bool 类型值);
```

其中，Bool 类型值表示是否需要指定过渡动画效果，当此值为 true 时，表示需要指定过渡动画效果；当此值为 false 时，表示不需要指定动画效果。

3.2.4 管理导航栏上的按钮

在导航栏上可以添加按钮。添加按钮的方法如表 3-5 所示。

表 3-5 添加按钮的属性

属 性	功 能
LeftBarButtonItem	在导航栏的左边添加一个自定义的按钮
RightBarButtonItem	在导航栏的右边添加一个自定义的按钮
BackBarButtonItem	改变导航栏中默认的返回按钮

【示例 3-4】以下将使用 LeftBarButtonItem 属性和 RightBarButtonItem 属性分别在导航栏的左右两边添加一个按钮，具体步骤如下所述。

（1）创建一个空类型的工程。

（2）在创建的工程中添加一个类型为 iPhone View Controller 的文件，命名为 MainViewController。

（3）打开 AppDelegate.cs 文件，编写代码，实现为 MainViewController 的主视图添加导航控制器，代码如下：

```
using System;
using System.Collections.Generic;
using System.Linq;
using MonoTouch.Foundation;
using MonoTouch.UIKit;
namespace Application
{
    [Register ("AppDelegate")]
    public partial class AppDelegate : UIApplicationDelegate
    {
        UIWindow window;
        public override bool FinishedLaunching (UIApplication app,
        NSDictionary options)
        {
            window = new UIWindow (UIScreen.MainScreen.Bounds);
            MainViewController mainViewController = new MainViewController ();
            mainViewController.Title="首页";
            UINavigationController nav = new UINavigationController
            (mainViewController);
            //为导航栏添加右边的按钮
            UIBarButtonItem rightButton = new UIBarButtonItem ("下一页",
            UIBarButtonItemStyle.Done, this, null);
            mainViewController.NavigationItem.RightBarButtonItem = rightButton;
            //为导航栏添加左边的按钮
            UIBarButtonItem leftButton = new UIBarButtonItem ("上一页",
            UIBarButtonItemStyle.Done, this, null);
            mainViewController.NavigationItem.LeftBarButtonItem = leftButton;
            window.RootViewController = nav;
            window.MakeKeyAndVisible ();
            return true;
```

```
            }
        }
}
```

运行效果如图 3.22 所示。

图 3.22　运行效果

需要注意的是，对于添加按钮的设置，需要使用 **UIBarButtonItem** 类。它实例化后，就可以对按钮进行简单的设置，其语法形式如下：

```
UIBarButtonItem 对象名=new UIBarButtonItem(按钮的标题,按钮的风格,目标对象,触摸
此按钮后的动作);
```

其中，按钮的风格 **UIBarButtonItemStyle** 有 3 种，如表 3-6 所示。

表 3-6　按钮的风格

风　　格	功　　能
Bordered	显示边框
Done	完成按钮
Plain	在触摸时会发光

注意：在 iOS 7 中，这 3 种风格的按钮几乎是不明显的。

在【示例 3-3】的运行结果中，可以看到，在触摸 First View 按钮或者 Second View 按钮后，切换到对应的视图控制器。在此视图控制器的主视图上有一个导航栏，在导航栏中有一个带有箭头的按钮，被称为返回按钮或者后退按钮，如图 3.23 所示。

图 3.23　返回按钮

此按钮可以将视图控制器导航到栈中的前一个视图控制器中。如果栈中前一个控制器设置了 Title 属性，返回按钮将显示设置的标题；如果没有设置 Title 属性，此按钮将显示视图控制器的名称。如果视图控制器没有设置名称，此按钮将会显示 Back。

【示例 3-5】以下将【示例 3-3】做了一些修改，将返回按钮的标题设置为"返回首页"。代码如下：

```
using System;
using System.Collections.Generic;
using System.Linq;
using MonoTouch.Foundation;
using MonoTouch.UIKit;
namespace Application
{
    [Register ("AppDelegate")]
    public partial class AppDelegate : UIApplicationDelegate
    {
        UIWindow window;
        public override bool FinishedLaunching (UIApplication app,
        NSDictionary options)
        {
            window = new UIWindow (UIScreen.MainScreen.Bounds);
            MainViewController mainController = new MainViewController();
            mainController.Title = "Main View";
            UINavigationController navController =
            new UINavigationController(mainController);
            UIBarButtonItem backButton = new UIBarButtonItem ("返回首页",
            UIBarButtonItemStyle.Bordered, this, null);
            mainController.NavigationItem.BackBarButtonItem = backButton;
            //设置返回按钮
            window.RootViewController = navController;
            window.MakeKeyAndVisible ();
            return true;
        }
    }
}
```

运行效果如图 3.24 所示。

图 3.24 运行效果

3.3 在标签栏中提供控制器

在标签栏中提供的控制器是 UITabBarController 标签栏控制器。标签栏控制器也是 UIViewController 视图控制器的子类。该控制器管理多个具有并列关系的控制器。多个标签位于屏幕下方。每一个标签关联一个视图控制器。

3.3.1 添加标签栏控制器

添加标签栏控制器的方式有 3 种：第一种是在 Single View Application 类型的工程中使用拖动的方式添加标签栏控制器；第二种是使用代码实现；第三种是在 Tabbed Application 类型的工程中直接实现。本节将为开发者讲解后两种。

1．使用代码添加标签栏控制器

使用代码添加标签栏控制器需要实现以下 3 个步骤。
（1）实例化。UITabBarController 实例化对象的语法形式如下：

```
UITabBarController tabController = new UITabBarController();
```

（2）设置标签栏控制器中需要添加的视图控制器。设置标签栏控制器中需要添加的视图控制器可使用 SetViewControllers()方法，其语法形式如下：

```
标签栏控制器对象.SetViewControllers(new UIViewController[] {
    视图控制器对象1,
    视图控制器对象2,
    ……
}, Bool 类型值);
```

其中，视图控制器对象 1、视图控制器对象 2 等是添加到标签栏控制器的视图控制器对象。Bool 类型值表示是否需要动画效果，如果值为 true，表示有动画效果；如果值为 false，表示没有动画效果。

（3）设置根视图控制器。设置根视图控制器的语法形式如下：

```
window.RootViewController = 标签栏对象;
```

【示例 3-6】 以下将使用代码的形式添加一个标签栏控制器。具体步骤如下所述。
（1）创建一个空类型的工程。
（2）在创建的工程中添加两个类型为 iPhone View Controller 的文件，分别命名为 FirstViewController 和 SecondViewController。
（3）打开 AppDelegate.cs 文件，编写代码，实现的 FirstViewController 和 SecondViewController 主视图添加标签栏控制器，代码如下：

```
using System;
using System.Collections.Generic;
using System.Linq;
using MonoTouch.Foundation;
using MonoTouch.UIKit;
```

```
namespace Application
{
    [Register ("AppDelegate")]
    public partial class AppDelegate : UIApplicationDelegate
    {
        UIWindow window;
        public override bool FinishedLaunching (UIApplication app,
        NSDictionary options)
        {
            window = new UIWindow (UIScreen.MainScreen.Bounds);
            //实例化对象
            FirstViewController firstViewController = new FirstView
            Controller();
            SecondViewController secondViewController = new SecondView
            Controller();
            UITabBarController tabController = new UITabBarController();
            //实例化对象tabController
            //设置在标签栏控制器中的视图控制器
            tabController.SetViewControllers(new UIViewController[] {
                firstViewController,
                secondViewController
            }, true);
            //设置标题
            tabController.TabBar.Items[0].Title = "First";
            tabController.TabBar.Items[1].Title = "Second";
            window.RootViewController = tabController; //设置根视图控制器
            window.MakeKeyAndVisible();
            return true;
        }
    }
}
```

（4）打开 FirstViewController.xib 文件，在主视图中放入一个标题，将 Text 设置为 First View，将 Font 设置为 System 30.0，将 Alignment 设置为居中对象。位置及大小设置为（89,256,142,55）。

（5）打开 SecondViewController.xib 文件，在主视图中放入一个标题，将 Text 设置为 Second View，将 Font 设置为 System 30.0，将 Alignment 设置为居中对象。位置及大小设置为（74,256,173,55）。运行效果如图 3.25 所示。

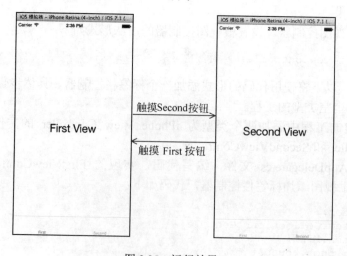

图 3.25　运行效果

2. 使用Tabbed Application类型的工程添加标签控制器

创建一个 Tabbed Application 类型的工程，同样可以实现添加标签栏控制器的功能。创建好之后，默认是一个具有标签栏控制器的工程。以下是具体的操作步骤。

（1）单击 Xamarin Studio，弹出"新建解决方案"对话框，如图 3.26 所示。

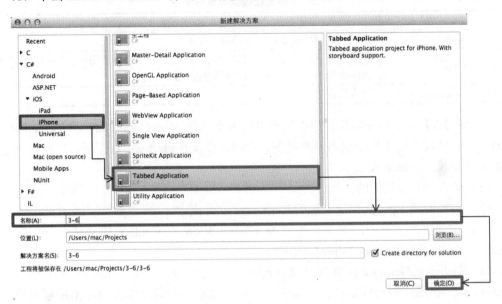

图 3.26　操作步骤

（2）选择 iOS 下的 iPhone 选项，在 iPhone 选项中选择 Tabbed Application 选项。将名称输入为 3-7，然后单击"确定"按钮，此时就创建好了一个 Tabbed Application 类型的工程。创建好工程会自动带有两个实现视图控制器类的文件，分别为 FirstViewController.cs 和 SecondViewController.cs。单击运行按钮后的运行效果如图 3.27 所示。

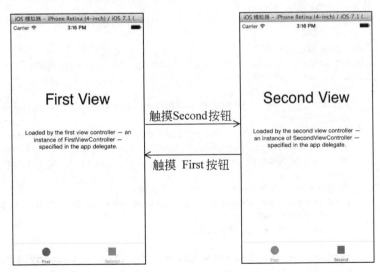

图 3.27　运行效果

3.3.2 标签栏控制器的常用属性

表 3-7 是标签栏控制器常用的属性。

表 3-7 标签栏控制器常用属性

属 性	功 能
ViewControllers	设置标签栏控制器中包含的视图控制器
SelectedIndex	通过该属性可以获得当前选中的视图控制器，设置该属性，可以显示多个视图控制器中对应的索引的视图控制器
SelectedViewController	设置当前选择的视图控制器

【示例 3-7】以下将使用标签栏控制器中的两个属性 ViewControllers 和 SelectedViewController，实现为标签栏控制器添加视图控制器，以及在启动应用程序后，改变当前选择的视图控制器。具体步骤如下：

（1）创建一个空类型的工程。

（2）添加图像 1.jpg 到创建工程的 Resources 文件夹中。

（3）在创建的工程中添加两个类型为 iPhone View Controller 的文件，分别命名为 FirstViewController 和 SecondViewController。

（4）打开 AppDelegate.cs 文件，编写代码，实现将 FirstViewController 和 SecondViewController 主视图添加到标签栏控制器中，并将应用程序启动后选择的视图控制器设置为 SecondViewController 的对象。代码如下：

```
using System;
using System.Collections.Generic;
using System.Linq;
using MonoTouch.Foundation;
using MonoTouch.UIKit;
namespace Application
{
    [Register ("AppDelegate")]
    public partial class AppDelegate : UIApplicationDelegate
    {
        UIWindow window;
        public override bool FinishedLaunching (UIApplication app,
        NSDictionary options)
        {
            window = new UIWindow (UIScreen.MainScreen.Bounds);
            //实例化对象
            FirstViewController firstViewController=new FirstViewController();
            SecondViewController secondViewController=new SecondViewController();
            UITabBarController tabController = new UITabBarController ();

            //设置标签栏控制器中包含的视图控制器
            tabController.ViewControllers = new UIViewController[] {
                firstViewController,
                secondViewController
            };
            tabController.SelectedViewController = secondViewController;
            window.RootViewController = tabController;
            //设置标题
            tabController.TabBar.Items[0].Title = "First";
```

```
            tabController.TabBar.Items[1].Title = "Second";
            window.MakeKeyAndVisible ();
            return true;
        }
    }
}
```

（5）打开 SecondViewController.cs 文件，编写代码，为 SecondViewController 的主视图添加图片。代码如下：

```
using System;
using System.Drawing;

using MonoTouch.Foundation;
using MonoTouch.UIKit;
namespace Application
{
    public partial class SecondViewController : UIViewController
    {
        ……                    //这里省略了视图控制器的构造方法和析构方法
        public override void ViewDidLoad ()
        {
            base.ViewDidLoad ();
            //实例化并设置 imageView 对象
            UIImageView imageView = new UIImageView ();
            imageView.Frame = new RectangleF (0, 0, 320, 568);
            imageView.Image = UIImage.FromFile ("1.jpg");
            this.View.AddSubview (imageView);
        }
    }
}
```

运行效果如图 3.28 所示。

3.3.3 标签栏控制器的响应

ViewControllerSelected 事件可以实现标签栏控制器的响应，其语法形式如下：

```
标签栏控制器对象名.ViewControllerSelected +=触摸按钮后的方法；
```

或者：

```
标签栏控制器对象名.ViewControllerSelected +=(sender,e)
=>{
    ……
};
```

图 3.28 运行效果

【示例 3-8】 以下将使用 ViewControllerSelected 事件实现标签栏控制器的响应。在每一次切换视图控制器时，都会弹出警告视图。具体的步骤如下。
（1）创建一个空类型的工程。
（2）添加图像 1.jpg 和 2.jpg 到创建工程的 Resources 文件夹中。
（3）在创建的工程中添加两个类型为 iPhone View Controller 的文件，分别命名为 FirstViewController 和 SecondViewController。

（4）打开 AppDelegate.cs 文件，编写代码，实现的 FirstViewController 和 SecondViewController 主视图添加到标签栏控制器中，并实现标签栏控制器的响应。为代码如下：

```
using System;
using System.Collections.Generic;
using System.Linq;
using MonoTouch.Foundation;
using MonoTouch.UIKit;
namespace Application
{
    [Register ("AppDelegate")]
    public partial class AppDelegate : UIApplicationDelegate
    {
        UIWindow window;
        public override bool FinishedLaunching (UIApplication app,
        NSDictionary options)
        {
            window = new UIWindow (UIScreen.MainScreen.Bounds);
            //实例化对象
            FirstViewController firstViewController=new FirstViewController();
            SecondViewController secondViewController=new SecondViewController();
            UITabBarController tabController = new UITabBarController ();
            //设置标签栏控制器中包含的视图控制器
            tabController.ViewControllers = new UIViewController[] {
                firstViewController,
                secondViewController
            };
            window.RootViewController = tabController;
            //设置标题
            tabController.TabBar.Items[0].Title = "First";
            tabController.TabBar.Items[1].Title = "Second";
            //响应标签栏控制器
            tabController.ViewControllerSelected+=(sender, e) => {
                //判断视图控制器对象是否为 firstViewController
                if (e.ViewController==firstViewController) {
                    UIAlertView alertView=new UIAlertView();
                    alertView.Title="打开 firstViewController 对象的视图";
                    alertView.AddButton("Cancel");
                    alertView.Show();
                }else{
                    UIAlertView alertView=new UIAlertView();
                    alertView.Title="打开 secondViewController 对象的视图";
                    alertView.AddButton("Cancel");
                    alertView.Show();
                }
            };
            window.MakeKeyAndVisible ();
            return true;
        }
    }
}
```

（5）打开 FirstViewController.cs 文件，编写代码，实现为主视图添加图像 1.jpg。代码如下：

```
using System;
using System.Drawing;
using MonoTouch.Foundation;
using MonoTouch.UIKit;
```

```
namespace Application
{
    public partial class FirstViewController : UIViewController
    {
        ……                              //这里省略了视图控制器的构造方法和析构方法
        public override void ViewDidLoad ()
        {
            base.ViewDidLoad ();
            UIImageView imageView = new UIImageView ();
            imageView.Frame = new RectangleF (0, 0, 320, 568);
            imageView.Image = UIImage.FromFile ("1.jpg");
            this.View.AddSubview (imageView);
        }
    }
}
```

（6）打开 SecondViewController.cs 文件，编写代码，实现为主视图添加图像 2.jpg。由于此代码实现的功能和 FirstViewController.cs 文件中实现代码的功能相似，所以这里就省略了。开发者可以参考源代码。运行效果如图 3.29 所示。

图 3.29　运行效果

3.4 模型视图控制器

ModalViewController 模型视图控制器并不像 UINavigationController 是一个专门的类。它是优先于其他视图或者控制器出现的，它类似于在显示一个窗口对话框时，对界面进行控制并且不允许访问应用程序的其他窗口。使用 UIViewController 的 PresentViewControllerAsync()方法指定的视图控制器就是 ModalViewController 了。其此方法的语法形式如下：

视图控制器对象1.PresentViewControllerAsync（视图控制器对象2,Bool 类型值）；

其中，视图控制器对象 2 就是一个 ModalViewController 模型视图控制器。Bool 类型值表示是否需要指定过渡动画效果，当此值为 true 时，表示需要指定过渡动画效果；当此值为 false 时，表示不需要指定动画效果。

当然，一个视图控制器设置为 ModalViewController 模型视图控制器，也可以使用 DismissViewControllerAsync()方法取消，其语法形式如下：

视图控制器对象2.DismissViewControllerAsync(Bool 类型值)

其中，Bool 类型值表示是否需要指定过渡动画效果。

【示例 3-9】 以下将使用 ModalViewController 模型视图控制器实现视图控制器的视图切换。具体步骤如下所述。

（1）创建一个空类型的工程。

（2）添加图像 1.jpg 和 2.jpg 到创建的工程的 Resource 文件夹中。

（3）在创建的工程中添加两个类型为 iPhone View Controller 的文件，分别命名为 MainController 和 ModalController。

（4）打开 AppDelegate.cs 文件，编写代码，实现使用 MainController 控制器加载视图的功能。代码如下：

```
using System;
using System.Collections.Generic;
using System.Linq;
using MonoTouch.Foundation;
using MonoTouch.UIKit;
namespace Application
{
    [Register ("AppDelegate")]
    public partial class AppDelegate : UIApplicationDelegate
    {
        UIWindow window;
        public override bool FinishedLaunching (UIApplication app,
        NSDictionary options)
        {
            window = new UIWindow (UIScreen.MainScreen.Bounds);
            MainController mainController = new MainController();
                                            //实例化对象 mainController
            window.RootViewController = mainController;
                                            //设置根视图控制器
            window.MakeKeyAndVisible ();
```

```
            return true;
        }
    }
}
```

（5）打开 MainController.xib 文件，对 MainController 的主视图进行设置，效果如图 3.30 所示。

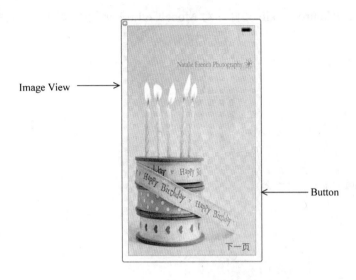

图 3.30 主视图

需要添加的视图以及设置，如表 3-8 所示。

表 3-8 设置主视图

视 图	设 置
Image View	Image：1.jpg 位置和大小：(0,0,320,568)
Button	Title：下一页 Font：System 21.0 位置和大小：(241,533,68,30) 为按钮声明插座变量 button

（6）打开 MainController.cs 文件，编写代码，实现触摸"下一页"按钮后，视图转为 ModalController 控制器的主视图。代码如下：

```
using System;
using System.Drawing;
using MonoTouch.Foundation;
using MonoTouch.UIKit;
namespace Application
{
    public partial class MainController : UIViewController
    {
        ……                         //这里省略了视图控制器的构造方法和析构方法
        public override void ViewDidLoad ()
        {
            base.ViewDidLoad ();
            button.TouchUpInside += async (sender, e) => {
```

```
                ModalController modalController = new ModalController ();
                await this.PresentViewControllerAsync (modalController,
                true);    //设置模型视图控制器
            };
        }
    }
}
```

（7）打开 ModalController.xib 文件，对 ModalController 的主视图进行设置，效果如图 3.31 所示。

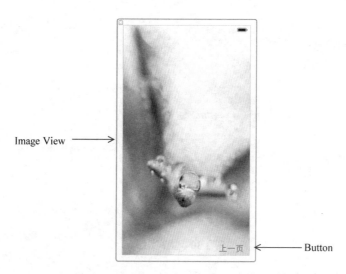

图 3.31 主视图

需要添加的视图以及设置，如表 3-9 所示。

表 3-9 设置主视图

视 图	设 置
Image View	Image：2.jpg 位置和大小：(0,0,320,568)
Button	Title：上一页 Font：System 21.0 位置和大小：(241,533,68,30) 为按钮声明插座变量 button

（8）打开 ModalController.xib 文件，编写代码，实现触摸"下一页"按钮后，视图转为 ModalController 控制器的主视图。代码如下：

```
using System;
using System.Drawing;
using MonoTouch.Foundation;
using MonoTouch.UIKit;
namespace Application
{
    public partial class ModalController : UIViewController
    {
        ……            //这里省略了视图控制器的构造方法和析构方法
        public override void ViewDidLoad ()
```

```
        {
            base.ViewDidLoad ();
            button.TouchUpInside += async (sender, e) => {
                await this.DismissViewControllerAsync(true);
                                        //将设置的模型视图控制器取消
            };
        }
}
```

运行结果如图 3.32 所示。

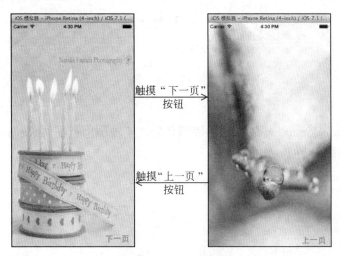

图 3.32　运行效果

3.5　创建自定义视图控制器

以上内容讲解了 UIViewController 视图控制器以及其子类的使用，这些控制器都是 Apple 给指定好的。如果想要使自己的应用程序突出，在使用这些控制器时，需要加入有自身特色的视图控制器。本节将主要讲解如何创建一个自定义的视图控制器。

【示例 3-10】以下将创建一个自定义的视图控制器，当用户触摸屏幕时，就会在标签中显示手指在屏幕上的当前位置。具体的操作步骤如下所述。

（1）创建一个空类型的工程。
（2）添加图像 1.jpg 到创建的工程的 Resource 文件夹中。
（3）在创建的工程中添加一个 C#的类文件，命名为 BaseController。
（4）在创建的工程中添加一个类型为 iPhone View Controller 的文件，命名为 DerivedController。
（5）打开 AppDelegate.cs 文件，编写代码，实现使用 DerivedController 控制器加载视图的功能。代码如下：

```
using System;
using System.Collections.Generic;
using System.Linq;
using MonoTouch.Foundation;
using MonoTouch.UIKit;
```

```
namespace Application
{
    [Register ("AppDelegate")]
    public partial class AppDelegate : UIApplicationDelegate
    {
        UIWindow window;
        public override bool FinishedLaunching (UIApplication app, NSDictionary options)
        {
            window = new UIWindow (UIScreen.MainScreen.Bounds);
            DerivedController derivedController = new DerivedController();
                //实例化对象
            window.RootViewController = derivedController;
                //设置根视图
            window.MakeKeyAndVisible ();
            return true;
        }
    }
}
```

（6）打开 BaseController.cs 文件，编写代码，实现图像视图对象和标签视图对象的创建，以及移动触摸的功能。代码如下：

```
using System;
using MonoTouch.UIKit;
using MonoTouch.Foundation;
using System.Drawing;
namespace Application
{
    public class BaseController:UIViewController
    {
        UILabel label;
        public BaseController ()
        {
        }
        //构造器
        public BaseController (string nibName, NSBundle bundle) : base(nibName, bundle) {
            //添加图像视图对象 imageView
            UIImageView imageView = new UIImageView ();
            imageView.Frame = new RectangleF (0, 0, 320, 568);
            imageView.Image = UIImage.FromFile ("1.jpg");
            this.View.AddSubview (imageView);
            //添加标签视图对象 label
            label = new UILabel ();
            label.TextAlignment = UITextAlignment.Center;
            label.Frame = new RectangleF (0, 500,320 , 50);
            this.View.AddSubview (label);
        }
        //实现当手指在屏幕上移动时，显示手指当前位置的功能
        public override void TouchesMoved (NSSet touches, UIEvent evt)
        {
            base.TouchesMoved (touches, evt);
            UITouch touch = touches.AnyObject as UITouch;
            //判断 touch 是否为空
            if (null != touch) {
                PointF locationInView = touch.LocationInView (this.View);
                label.Text = string.Format ("Touch coordinates: {0}", locationInView);
            }
```

（7）打开 DerivedController.cs 文件，将 DerivedController 基于的类改为 BaseController，代码如下：

```
using System;
using System.Drawing;
using MonoTouch.Foundation;
using MonoTouch.UIKit;
namespace Application
{
    public partial class DerivedController : BaseController
    {
        ……          //这里省去了构造器方法、析构方法，以及加载视图的方法
    }
}
```

运行效果如图 3.33 所示。

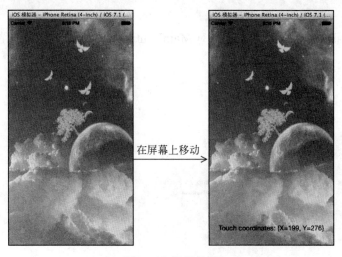

图 3.33　运行效果

3.6　利用视图控制器的有效性

iOS 具有非常严格的内存使用。如果一个应用程序使用了太多的内存，iOS 将发出内存警告。如果我们不做出相应的响应（释放不需要的资源），很有可能会终止程序的执行。本节将主要讲解如何避免这些情况。具体步骤如下所述。

（1）创建一个空类型的工程。

（2）在创建的工程中添加一个类型为 iPad View Controller 的文件，命名为 MainController。

（3）打开 AppDelegate.cs 文件，编写代码，实现根视图控制器的设置。代码如下：

```
using System;
using System.Collections.Generic;
```

```
using System.Linq;
using MonoTouch.Foundation;
using MonoTouch.UIKit;
namespace Application
{
    [Register ("AppDelegate")]
    public partial class AppDelegate : UIApplicationDelegate
    {
        UIWindow window;
        public override bool FinishedLaunching (UIApplication app, NSDictionary options)
        {
            window = new UIWindow (UIScreen.MainScreen.Bounds);
            MainController mainController = new MainController();
            //实例化
            window.RootViewController = mainController;
            //设置根视图控制器
            window.MakeKeyAndVisible ();
            return true;
        }
    }
}
```

（4）打开 MainController.xib 文件，对 MainController 的主视图进行设置，效果如图 3.34 所示。

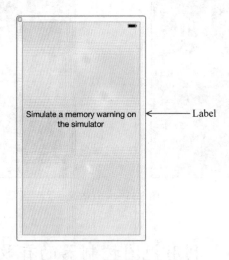

图 3.34　运行效果

需要添加的视图以及设置，如表 3-10 所示。

表 3-10　设置主视图

视　　图	设　　置
主视图	Background：浅灰色
Label	Text：Simulate a memory warning on the simulator Font：System 22.0 Alignment：居中 Lines：2 位置和大小：(0,221,320,77)

(5)打开 MainController.cs 文件，编写代码，实现在应用程序输出窗口输出字符串"Main controller received memory warning!"。代码如下：

```
using System;
using System.Drawing;
using MonoTouch.Foundation;
using MonoTouch.UIKit;
namespace Application
{
    public partial class MainController : UIViewController
    {
    ……        //这里省略了视图控制器的构造方法
        public override void DidReceiveMemoryWarning ()
        {
            base.DidReceiveMemoryWarning ();
            Console.WriteLine("Main controller received memory warning!");
        }
    ……        //这里省略了视图控制器在加载视图时的方法
    }
}
```

运行效果如图 3.35 所示。选择菜单中的"硬件"命令，如图 3.36 所示。

图 3.35　运行效果

图 3.36　菜单

在出现的下拉菜单中选择"模拟内存警告"命令后，就会在应用程序输出窗口输出 "Main controller received memory warning!"字符串，效果如图 3.37 所示。

图 3.37　应用程序输出窗口输出

3.7　iPad 视图控制器

以上讲解的所有控制器可以在 iPhone 和 iPad 应用程序中使用，但是有两种控制器只可以在 iPad 应用程序中使用，一种是 UISplitViewController 控制器；另一种是 UIPopoverController 控制器。以下将主要讲解 UISplitViewController 控制器。UIPopoverController 控制器一般不常用，这里就不做介绍了。UISplitViewController 控制器又名分割控制器，它充分发挥了 iPad 大屏幕的优势。它提供了在相同的屏幕区域中同时显示两种不同的视图的解决方案。

【示例 3-11】 以下将使用 UISplitViewController 控制器来实现一个应用程序。具体的步骤如下所述。

（1）创建一个空类型的工程。在创建此工程时需要注意，在弹出的"新建解决方案"对话框中，需要选择 iOS 下的 iPad 选项，如图 3.38 所示。

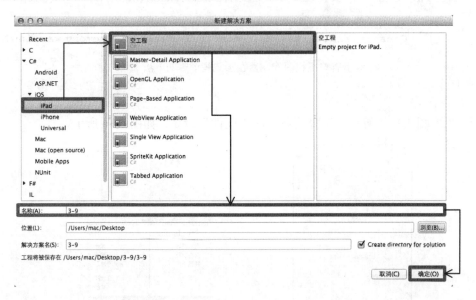

图 3.38　操作步骤 1

（2）在创建的工程中添加两个类型为 iPad View Controller 的文件，分别命名为 FirstController 和 SecondController。

（3）打开 FirstController.xib 文件，将 FirstController 的主视图的背影颜色设置为青色。

（4）打开 SecondController.xib 文件，对 SecondController 的主视图进行设置，效果如图 3.39 所示。

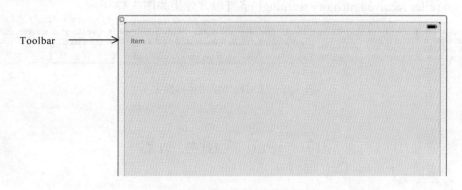

图 3.39　主视图

需要添加的视图以及设置，如表 3-11 所示。

表 3-11　设置主视图

视　　图	设　　置
主视图	Background：橘黄色
Toolbar	Bar Tint：浅灰色 位置和大小：(0,21,768,44) 为按钮声明插座变量 myToolbar

（5）打开 SecondController.cs 文件，编写代码，实现分割控制器的委托功能，即在一个屏幕中显示两个视图。代码如下：

```
using System;
using System.Drawing;
using MonoTouch.Foundation;
using MonoTouch.UIKit;
namespace Application
{
    public partial class SecondController : UIViewController
    {
        ……              //这里省略了视图控制器的构造方法、析构方法和视图加载时的方法
        public UIToolbar MyToolbar {
            get { return myToolbar;}
        }
    }
    public class SplitControllerDelegate:UISplitViewControllerDelegate{
        public SplitControllerDelegate (SecondController controller)
        {
            this.secondController = controller;
        }
        private SecondController secondController;
        //视图控制器将要隐藏时调用
        public override void WillHideViewController
        (UISplitViewController svc,
            UIViewController aViewController,
            UIBarButtonItem barButtonItem, UIPopoverController pc)
        {
            barButtonItem.Title = "First";           //设置工具栏的标题
            this.secondController.MyToolbar.SetItems (new
                UIBarButtonItem[] { barButtonItem }, true);
        }
        //视图控制器将要显示时调用
        public override void WillShowViewController
        (UISplitViewController svc,
            UIViewController aViewController,
            UIBarButtonItem button)
        {
            this.secondController.MyToolbar.SetItems (new
                UIBarButtonItem[0], true);
        }
    }
}
```

打开 AppDelegate.cs 文件，编写代码，实现根视图控制器的设置。代码如下：

```
using System;
using System.Collections.Generic;
```

```
using System.Linq;
using MonoTouch.Foundation;
using MonoTouch.UIKit;
namespace Application
{
    [Register ("AppDelegate")]
    public partial class AppDelegate : UIApplicationDelegate
    {
        UIWindow window;
        public override bool FinishedLaunching (UIApplication app,
        NSDictionary options)
        {
            window = new UIWindow (UIScreen.MainScreen.Bounds);
            //实例化视图控制器对象
            FirstController firstController = new FirstController();
            SecondController secondController = new SecondController();
            UISplitViewController splitController = new UISplitView
            Controller();
            //为 UISplitViewController 视图控制器的对象添加另外两个视图控制器对象
            splitController.ViewControllers = new UIViewController[] {
                firstController,
                secondController
            };
            splitController.Delegate = new SplitControllerDelegate
            (secondController);                              //设置委托
            window.RootViewController = splitController;
                                                    //设置根视图控制器
            window.MakeKeyAndVisible ();
            return true;
        }
    }
}
```

运行效果如图 3.40 所示。

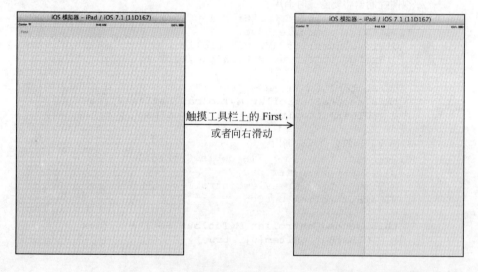

图 3.40　运行效果

⚠️**注意**：如果模拟器在垂直模式下，只显示了一个视图控制器，按住 Command+任意方向键将模拟器变为水平模式，这时会自动在模拟器上显示两个视图控制器的视图。运行效果如图 3.41 所示。

图 3.41　运行效果

3.8　使用故事面板设计 UI

在 iOS 5 之后，Apple 推出了故事面板，在开发 App 界面时可以极大地节省时间。在创建一个 Single View Application 类型的工程后就会自动带有一个故事面板，也就是 MainStoryboard.storyboard 文件中的内容。使用故事面板设计 UI 就是使用拖动的形式将需要的控制器拖动到故事面板的画布中去。我们在第 2 章中讲解过使用 Interface builder 添加视图，即使用故事面板将各种需要的视图添加到视图控制器中。

【示例 3-12】以下将使用故事面板设置一个应用程序的界面，实现两个视图的切换。具体步骤如下所述。

（1）创建一个 Single View Application 类型的工程。

（2）打开 MainStoryboard.storyboard 文件，删除画布中的 View Controller 控制器。

（3）在工具栏中拖动 Navigation Controller 导航控制器到画布中，此时画布的效果如图 3.42 所示。

在此图中需要注意，从工具栏拖动到画布中的 Navigation Controller 导航控制器并不是只有一个导航控制器，而是由一个导航控制器和一个视图控制器组成，即在导航控制器中压入一个视图控制器。

此时，就为主视图添加了一个导航控制器，单击运行按钮后的运行结果如图 3.43 所示。

（4）在与导航控制器关联的视图控制器的主视图中添加一个 Button 按钮对象。

（5）从工具栏中拖动一个 View Controller 视图控制器到画布中，按住 Ctrl 键，同时拖动按钮到新的视图中，如图 3.44 所示。

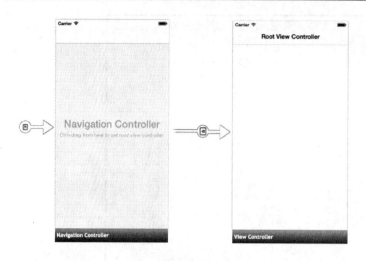

图 3.42 画布的效果　　　　　图 3.43 运行效果

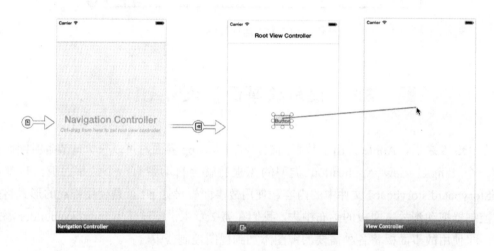

图 3.44 操作步骤 1

（6）松开鼠标后，就会弹出一个小的对话框，如图 3.45 所示。
（7）选择其中的 Push 选项，此时画布中的效果如图 3.46 所示。

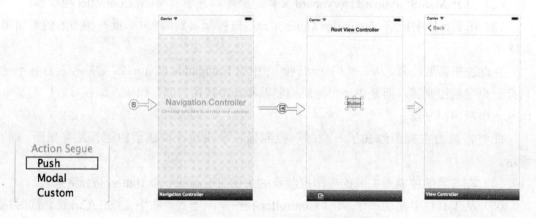

图 3.45 操作步骤　　　　　　图 3.46 画布效果

第 3 章　用户界面——控制器

> 🔔 **注意**：从步骤 5 到步骤 7 是为视图控制器的主视图建立 Segue（Segue 描述了视图之间的动画）。这些动画就是图 3.46 所看到的带箭头的连线，它们各自的功能如表 3-12 所示。

表 3-12　Segue 动画

Segue 动画	功　　能
Push	一般是需要头一个界面是 Navigation Controller 导航控制器
Modal	模态转换，一般用在视图的切换中
Custom	自定义跳转方式

（8）单击运行按钮，运行结果如图 3.47 所示。

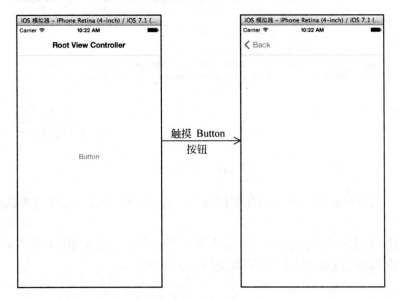

图 3.47　运行效果

如果在一个工程中没有故事面板该如何实现添加呢？

【示例 3-13】以下将会在一个空类型的工程中添加故事面板，并实现两个视图的切换。具体步骤如下所述。

（1）创建一个空类型的工程。

（2）在创建的工程中添加两个 C# 的类文件，分别命名为 FirstController 和 SecondController。

（3）在创建的工程中添加一个类型为 Empty iPhone Storyboard 的文件，命名为 MainStoryboard。此文件的添加步骤如下：

首先，选择菜单栏中的"文件"|New|File…命令，弹出 New Files 对话框，如图 3.48 所示。

然后选择 iOS 下的 Empty iPhone Storyboard，并输入此文件的名称 MainStoryboard。完成后，单击"新建"按钮，此时会在创建的工程中出现一个名为 MainStoryboard.storyboard 的文件。

（4）打开 MainStoryboard.storyboard 文件，可以看到在画布中是没有任何控制器的。

从工具栏中拖动 Navigation Controller 导航控制器到画布中。在与导航控制器关联的视图控制器的主视图中添加 Button 按钮对象，将 Title 设置为"下一页"。

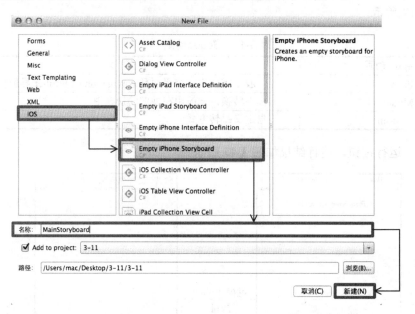

图 3.48 操作步骤 1

（5）从工具栏中拖动一个新的视图控制器。将此控制器的主视图的背景颜色设置为绿色。

（6）按住 Ctrl 键，同时拖动"下一页"按钮到新的视图控制器的主视图中。在弹出的对话框中选择 Push 选项。此时画布的效果如图 3.49 所示。

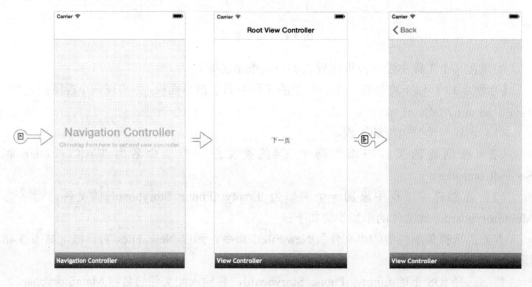

图 3.49 画布的效果

（7）设置背景为白色的视图控制器关联的类。具体步骤如下：
选择与导航控制器关联的视图控制器，单击下方的 View Controller 图标，如图 3.50 所

· 132 ·

示。选择"属性"按钮,将 Class 改为 FirstController,如图 3.51 所示。

图 3.50　操作步骤 1　　　　　　　图 3.51　操作步骤 2

此时在 FristController.cs 文件中的代码如下:

```
using System;
using MonoTouch.Foundation;
using MonoTouch.UIKit;
using System.CodeDom.Compiler;
namespace Application
{
    partial class FirstController : UIViewController
    {
        public FirstController (IntPtr handle) : base (handle)
        {
        }
    }
}
```

(8) 设置背景为绿色的视图控制器关联的类,即将 Class 设置为 SecondController。
(9) 打开 AppDelegate.cs 文件,编写代码,实现对根视图控制器的设置。代码如下:

```
using System;
using System.Collections.Generic;
using System.Linq;
using MonoTouch.Foundation;
using MonoTouch.UIKit;
namespace Application
{
    [Register ("AppDelegate")]
    public partial class AppDelegate : UIApplicationDelegate
    {
        UIWindow window;
        public override bool FinishedLaunching (UIApplication app,
        NSDictionary options)
        {
            window = new UIWindow (UIScreen.MainScreen.Bounds);
            UIStoryboard storyboard = UIStoryboard.FromName("MainStoryboard",
            NSBundle.MainBundle);                          //实例化故事面板
            UINavigationController navController =
                (UINavigationController)storyboard.
                InstantiateInitialViewController();        //实例化导航控制器
            window.RootViewController = navController;    //设置根视图控制器
            window.MakeKeyAndVisible ();
            return true;
```

```
            }
        }
}
```

（10）打开 FirstController.cs 文件，编写代码，实现数据的传递。代码如下：

```
using System;
using MonoTouch.Foundation;
using MonoTouch.UIKit;
using System.CodeDom.Compiler;
namespace Application
{
    partial class FirstController : UIViewController
    {
        ……        //这里省略了视图控制器的构造方法
                  //参数传递
        public override void PrepareForSegue (UIStoryboardSegue segue,
        NSObject sender)
        {
            base.PrepareForSegue (segue, sender);
            Console.WriteLine ("Preparing for segue {0}.", segue.
            Identifier);
            if (segue.Identifier == "MyPushSegue") {
                SecondController secondController = (SecondController)
                segue.DestinationViewController;
            }
        }
    }
}
```

运行效果如图 3.52 所示。

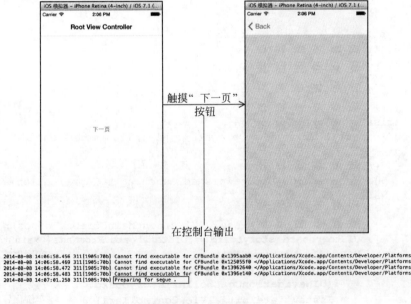

图 3.52 运行效果

在此应用程序中需要注意以下两个问题。

（1）设置根视图控制器

由于在此示例中使用了故事面板，所以在设置根视图控制器时，不可以使用以前所讲解的方法，即实例化一个控制器对象，然后使用 RootViewController 属性进行根视图控制器的设置。那么它该如何实现设置根视图控制器呢？需要 3 个步骤：

② 实例化故事面板

```
UIStoryboard 对象名=UIStoryboard.FromName(故事面板名称,包含故事板文件及其相关资源的捆绑);
```

其中，第一个参数故事面板名称不需要添加扩展名。第二个参数可以设置为 null。

② 实例化控制器

对故事面板实例化后，才可以对控制器进行实例化。此时需要使用 InstantiateInitialViewController()方法。此方法实现的功能就是返回第一个界面，即主视图。以此程序中的代码为例，代码如下：

```
UINavigationController navController = (UINavigationController)storyboard.InstantiateInitialViewController();
```

③ 设置根视图

完成以上两步后，才可以对根视图控制器进行设置。

（2）数据传递

在程序中，当触摸"下一页"按钮时，会在应用程序输出窗口输出"Preparing for segue"字符串，此功能的实现就使用了数据传递，需要重写 PrepareForSegue()方法。

3.9 故事面板中的 Unwind Segue

在故事面板中有一个非常重要的功能就是 Unwind Segue。使用它，开发者可以很容易实现 segue 到一个指定的视图上。

【示例 3-14】 以下将使用 Unwind Segue 功能来实现视图的切换。具体步骤如下所述。

（1）创建一个空类型的工程。

（2）在创建的工程中添加 3 个 C#的类文件，分别命名为 FirstController、SecondController 和 ModalController。

（3）在创建的工程中添加一个类型为 Empty iPhone Storyboard 的文件，命名为 MainStory。

（4）打开 MainStory.storyboard 文件，从工具栏中拖动 Navigation Controller 导航控制器到画布中。在与导航控制器关联的视图控制器的主视图中添加 Button 按钮对象，将 Title 设置为"第一页"，将背景颜色设置为粉色。

（5）从工具栏中拖动一个新的视图控制器。在此控制器的主视图中添加 Button 按钮对象，将 Title 设置为"第二页"，将背景颜色设置为绿色。

（6）从工具栏中拖动一个新的视图控制器。在此控制器的主视图中添加 Button 按钮对象，将 Title 设置为"第三页"，将背景颜色设置为紫色。

（7）按住 Ctrl 键，同时拖动"第一页"按钮到背影为绿色的视图控制器的主视图中。在弹出的对话框中选择 Push 选项。

（8）按住 Ctrl 键，同时拖动"第二页"按钮到背景为紫色的视图控制器的主视图中。在弹出的对话框中选择 Modal 选项。此时画布的效果如图 3.53 所示。

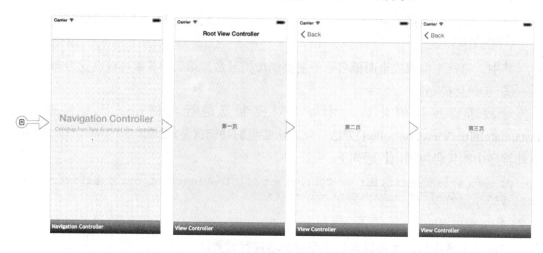

图 3.53　画布效果

（9）设置背景为粉色的视图控制器关联的类，即将 Class 设置为 FirstController；设置背景为绿色的视图控制器关联的类，即将 Class 设置为 SecondController；设置背景为紫色的视图控制器关联的类，即将 Class 设置为 ModalController。

（10）打开 AppDelegate.cs 文件，编写代码，实现对根视图控制器的设置。代码如下：

```
using System;
using System.Collections.Generic;
using System.Linq;
using MonoTouch.Foundation;
using MonoTouch.UIKit;
namespace Application
{
    [Register ("AppDelegate")]
    public partial class AppDelegate : UIApplicationDelegate
    {
        UIWindow window;
        public override bool FinishedLaunching (UIApplication app,
        NSDictionary options)
        {
            window = new UIWindow (UIScreen.MainScreen.Bounds);
            //实例化对象
            UIStoryboard storyboard = UIStoryboard.FromName("MainStory",
            NSBundle.MainBundle);
            UINavigationController navController =
                (UINavigationController)storyboard.InstantiateInitialView
                Controller();
            window.RootViewController = navController;  //设置根视图控制器
            window.MakeKeyAndVisible ();
            return true;
```

        }
    }
}

（11）打开 FirstController.cs 文件，编写代码，实现 UnwindFromModalController()方法的功能，即在触摸"第三页"按钮后，使用 Unwind Segue 功能将主视图切换到粉色的视图上。代码如下：

```
using System;
using MonoTouch.Foundation;
using MonoTouch.UIKit;
using System.CodeDom.Compiler;
namespace Application
{
    partial class FirstController : UIViewController
    {
    ……                                          //这里省略视图控制器的构造方法
        public override void ViewDidLoad ()
        {
            base.ViewDidLoad ();
            this.Title = "First Controller";          //设置标题
        }
        [Action("unwindFromModalController:")]        //标记方法
        public void UnwindFromModalController(UIStoryboardSegue segue)
        {
            Console.WriteLine("Unwind from modal controller!");
            ModalController modalController =
                (ModalController)segue.SourceViewController;
            //实例化对象
        }
    }
}
```

（12）回到 MainStory.storyboard 文件，按住 Ctrl 键，同时拖动"第三页"按钮到此视图控制器下方的 Scene Exit 图标上，如图 3.54 所示。

（13）松开鼠标，在弹出的对话框中选择 UnwindFromModalController，如图 3.55 所示。

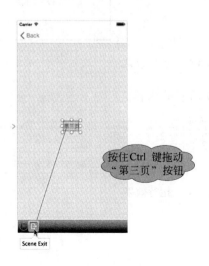

图 3.54　操作步骤 1　　　　　　　　图 3.55　操作步骤 2

运行效果如图3.56所示。

图3.56　运行效果

在应用程序输出窗口输出如图3.57所示的内容。

```
2014-08-12 13:54:02.048 313[1295:70b] Cannot find executable for CFBundle 0xe2d7690 </Applications/Xcode.app/Contents/Developer/
2014-08-12 13:54:02.055 313[1295:70b] Cannot find executable for CFBundle 0xe2e17c0 </Applications/Xcode.app/Contents/Developer/
2014-08-12 13:54:02.057 313[1295:70b] Cannot find executable for CFBundle 0x123f37d0 </Applications/Xcode.app/Contents/Developer
2014-08-12 13:54:02.070 313[1295:70b] Cannot find executable for CFBundle 0x13b0cb00 </Applications/Xcode.app/Contents/Developer
2014-08-12 13:54:06.674 313[1295:70b] Unwind from modal controller!
```

图3.57　运行效果

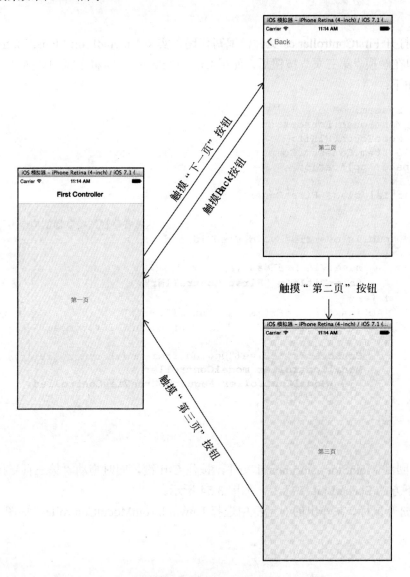

第2篇 资源使用篇

- 第4章 数据管理
- 第5章 显示数据
- 第6章 网络服务
- 第7章 多媒体资源
- 第8章 内置应用程序
- 第9章 与外部设备交互
- 第10章 位置服务和地图

第 4 章 数 据 管 理

基本上每一个应用程序都需要在文件系统上有永久性的数据存储。那么这些数据该如何进行存储呢？本章将为开发者解决这一问题，实现使用 3 种不同的方式对数据进行存储。

4.1 文 件 管 理

文件是存储数据最常用的方式。所以在 iOS 应用程序的开发中，避免不了对文件的操作，如文件的创建和删除等。本节将讲解对文件的创建、写入/读取内容，以及删除文件等。

4.1.1 创建文件

创建一个文件，需要使用 File 类中的 Create()方法来实现，其语法形式如下：

`File.Create (文件路径);`

其中，文件路径是一个字符串。

【示例 4-1】 以下将创建一个文件名为 TXTFile.txt 的文件。具体的操作步骤如下所述。
（1）创建一个 Single View Application 类型的工程。
（2）打开 MainStoryboard.storyboard 文件，对主视图进行设置，效果如图 4.1 所示。

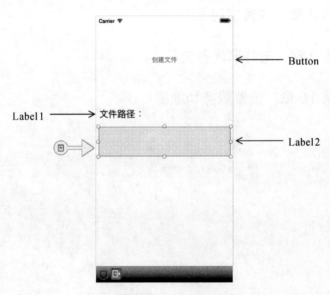

图 4.1　主视图的效果

需要添加的视图以及设置,如表 4-1 所示。

表 4-1 设置主视图

视图	设置
Button	Name:createButton Title:创建文件 位置和大小:(110,84,100,30)
Label1	Text:文件路径: Font:System 21 pt 位置和大小:(8,212,109,21)
Label2	Name:showLabel Text:(空) Alignment:居中 Lines:3 位置和大小:(8,253,306,65)

(3)打开 4-1ViewController.cs 文件,编写代码,实现文件的创建。代码如下:

```
using System;
using System.Drawing;
using MonoTouch.Foundation;
using MonoTouch.UIKit;
using System.IO;
namespace Application
{
    public partial class __1ViewController : UIViewController
    {
        ……
        #region View lifecycle
        public override void ViewDidLoad ()
        {
            base.ViewDidLoad ();
            createButton.TouchUpInside += (sender, e) => {
                string filePath = Path.Combine (Environment.GetFolderPath
                    (Environment.SpecialFolder.Personal),
                                "TXTFile.txt");       //设置文件保存的路径
                showButton.Text = filePath;
                File.Create (filePath);               //创建文件
            };
        }
        ……
        #endregion
    }
}
```

运行效果如图 4.2 所示。

在此程序中,引入了 System.IO 命名空间,此命名空间包含了允许读写文件和数据流的类型,以及提供基本文件和目录支持的类型。在创建文件时,最关键的一步就是为文件设置保存的路径,此时使用的 Path 类中的 Combine 方法,其语法形式如下:

```
string 对象名=Path.Combine(路径1,路径2);
```

第 2 篇 资源使用篇

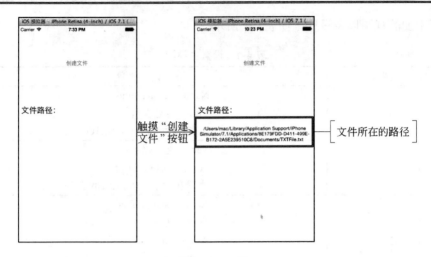

图 4.2 运行效果

其中，路径 1 和路径 2 都为 string 类型。在此程序中，设置文件保存路径的代码如下：

```
string filePath = Path.Combine (Environment.GetFolderPath
                (Environment.SpecialFolder.Personal),
                "TXTFile.txt");
```

在此代码中，Environment.GetFolderPath 方法可以通过 Environment.SpecialFolder 枚举参数获取系统特定的目录，如 Personal、Desktop 和 System 等。

在图 4.2 所示的运行结果中可以看到文件所在的路径，开发者可以通过此路径找到创建的空文件，查找文件具体的做法如下：

（1）选择"前往"|"前往文件夹"命令，弹出"前往文件夹"对话框，如图 4.3 所示。

（2）在文本框中输入文件所在的路径后，单击"前往"按钮，打开 Documents 文件夹，此时可以看到在【示例 4-1】中创建的文件，如图 4.4 所示。

图 4.3 操作步骤 1　　　　　　　　图 4.4 操作步骤 2

（3）双击此文件，便可以打开了。在此文件中是没有任何内容的，如图 4.5 所示。

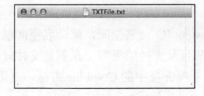

图 4.5 文件中的内容

> 注意：使用 Create()方法创建的文件是一个空文件。

4.1.2 写入/读取内容

本小节将要讲解如何在空的文件中写入内容以及读取内容。

1. 写入内容

如果想要在文件中写入内容需要使用到 **StreamWriter** 类。要使用此类，首先需要对此类进行实例化，其语法形式如下：

```
StreamWriter 对象名 = new StreamWriter (文件路径);
```

写入内容需要使用 **StreamWriter** 类中的 **Write()**方法，其语法形式如下：

```
StreamWriter 类的对象.Write(内容);
```

其中，内容一般为字符串。

2. 读取内容

如果想要读取文件中的内容需要使用到 **StreamReader** 类。要使用此类，首先需要对此类进行实例化，其语法形式如下：

```
StreamReader 对象名 = new StreamReader (文件路径)
```

读取内容需要使用 **StreamReader** 类中的 **ReadToEnd()**方法，此方法实现的功能是读取文件中所有的字符串，其语法形式如下：

```
StreamReader 类的对象.ReadToEnd ();
```

【**示例 4-2**】 以下将创建一个文件名为 **MyFile.txt** 的文件，并在此文件中写入并读取内容。具体的操作步骤如下所述。

（1）创建一个 Single View Application 类型的工程。

（2）打开 MainStoryboard.storyboard 文件，对主视图进行设置，效果如图 4.6 所示。

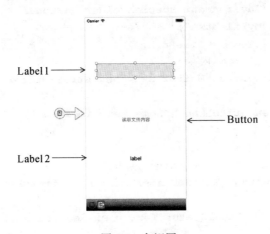

图 4.6 主视图

需要添加的视图以及设置,如表 4-2 所示。

表 4-2 设置主视图

视 图	设 置
Label1	Name:labelStatus Text:(空) Alignment:居中 位置和大小:(38,145,244,41)
Label2	Name:label Alignment:居中 Lines:3 位置和大小:(9,373,300,139)
Button	Name:btnShow Title:读取文件内容 位置和大小:(105,305,111,30)

(3)打开 4-1ViewController.cs 文件,编写代码,实现文件的创建,以及内容的写入和读取。代码如下:

```
using System;
using System.Drawing;
using MonoTouch.Foundation;
using MonoTouch.UIKit;
using System.IO;                    //引入 System.IO 命名空间
namespace Application
{
    public partial class __1ViewController : UIViewController
    {
        ……                          //这里省略了视图控制器的构造方法和析构方法
        #region View lifecycle
        public override void ViewDidLoad ()
        {
            base.ViewDidLoad ();
            string filePath = Path.Combine (Environment.GetFolderPath
                (Environment.SpecialFolder.Personal),
                "MyFile.txt");
            label.Text = filePath;
            //在文件中写入内容
            using (StreamWriter sw = new StreamWriter (filePath))
            {
                sw.WriteLine ("Some text in file!");
            }
            //读取内容
            btnShow.TouchUpInside += (s, e) => {
                using (StreamReader sr = new StreamReader (filePath))
                {
                    labelStatus.Text = sr.ReadToEnd ();
                }
```

```
                };
        }
......                              //这里省略了视图加载和卸载前后的一些方法
        #endregion
    }
}
```

运行效果如图 4.7 所示。

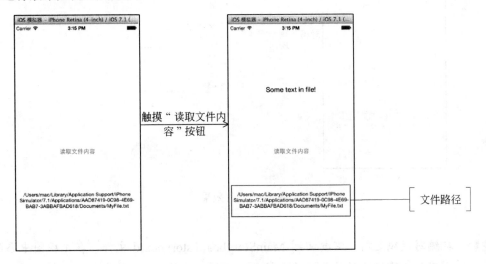

图 4.7　运行效果

> 注意：在此程序中，没有使用创建文件的方法 Create()，因为在写入内容到指定的文件时，就实现了文件的创建，也可以将写入内容到文件称为创建一个具有内容的文件。写入的内容信息如图 4.8 所示。

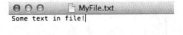

图 4.8　MyFile.txt 文件

4.1.3　删除文件

由于存储空间有限，如果不再使用此文件，需要及时删除它，以免造成空间的浪费。要实现文件删除，需要使用 File 文件中的 Delete() 方法。

【示例 4-3】 以下将以【示例 4-2】为基础，对文件进行删除。代码如下：

```
btndele.TouchUpInside+=(s,e)=>{
    File.Delete(filePath);             //删除指定文件
};
```

运行效果如图 4.9 所示。

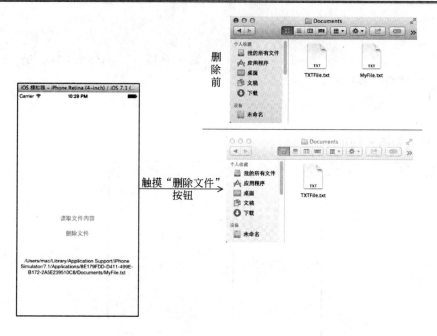

图 4.9 运行效果

> **注意**：在编写代码之前，需要回到 MainStoryboard.storyboard 文件，在主视图中添加一个用于删除操作的按钮，其中按钮的 Title 设置为"删除文件"，按钮的 Name 设置为 btndele，按钮的位置和大小为(110,343,100,30)。要检查指定的文件是否删除，除了可以使用查看的方法外（到文件对应的路径中去查看文件是否被删除），还可以使用 File 类中的 Exists()方法来判断指定的文件是否存在。

【示例 4-4】 以下将以【示例 4-2】为基础，对文件进行删除，并且判断删除的文件是否存在。代码如下：

```
btnCheck.TouchUpInside+=(s,e)=>{
    //判断文件是否存在
    if(File.Exists(filePath)){
        UIAlertView alert=new UIAlertView();
        alert.Title="指定文件存在";
        alert.AddButton("Cancel");
        alert.Show();
    }
    else{
        UIAlertView alert=new UIAlertView();
        alert.Title="指定文件不存在";
        alert.AddButton("Cancel");
        alert.Show();
    }
};
```

运行效果如图 4.10 所示。

第 4 章　数据管理

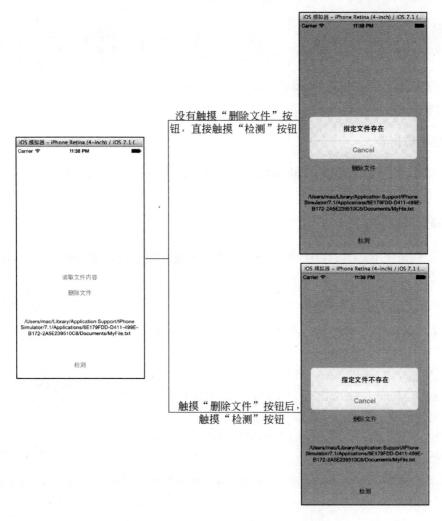

图 4.10　运行效果

> 注意：在编写代码之前，需要回到 MainStoryboard.storyboard 文件，在主视图中添加一个用于检测文件是否存在的按钮，其中按钮的 Title 设置为"检测"，按钮的 Name 设置为 btnCheck，按钮的位置和大小为(137,520,46,30)。

4.2　使用 SQLite 数据库

SQLite 是一款开源的嵌入式关系数据库，其特点是易于管理，易于使用，易于嵌入其他大型程序，易于维护和配置。本节将讲解如何使用 SQLite 数据库。

4.2.1　创建数据库

一般对数据库的操作，都需要使用到 SqliteConnection 类。要创建数据库需要实现以下

3 个步骤。

1. 打开数据库

数据库的打开需要使用 Open ()方法，其语法形式如下：

`SqliteConnection 类的对象.Open();`

2. 创建数据表

对于表的创建需要使用到 SqliteCommand 类，此类不仅可以创建数据表，还可以实现数据的插入和索引等功能。

3. 释放资源

最后，就是使用 Close ()方法将数据库中的资源进行释放或者关闭数据库，其语法形式如下：

`SqliteConnection 类的对象.Close ();`

【示例 4-5】以下将创建一个数据库。具体的操作步骤如下所述。

（1）创建一个 Single View Application 类型的工程。

（2）添加 Mono.Data.Sqlite 和 System.Data 引用到创建功能的"引用"文件夹中，具体的操作步骤如下：

首先，选择"引用"文件夹，右击鼠标，弹出快捷菜单，如图 4.11 所示。

然后选择 Edit References…命令，弹出 Edit References 对话框，如图 4.12 所示。

图 4.11 操作步骤 1

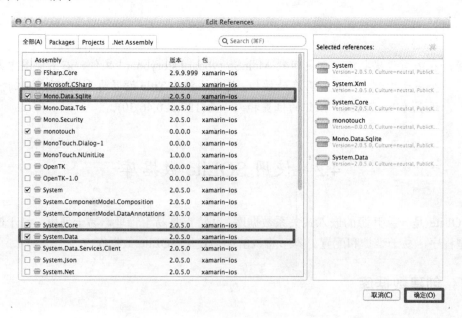

图 4.12 操作步骤 2

最后选择 Mono.Data.Sqlite 和 System.Data 引用，单击"确定"按钮后，Mono.Data.Sqlite 和 System.Data 引用就添加到了"引用"文件夹中。

（3）打开 MainStoryboard.storyboard 文件，对主视图进行设置，效果如图 4.13 所示。

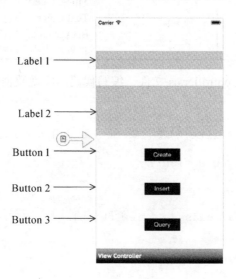

图 4.13　主视图的效果

需要添加的视图以及设置，如表 4-3 所示。

表 4-3　设置主视图

视　　图	设　　置
Label1	Name：lblStatus Text：（空） Alignment：居中 Lines：2 Background：浅灰色 位置和大小：(0,55,320,80)
Label2	Name：label Text：（空） Alignment：居中 Lines：3 Background：浅灰色 位置和大小：(0,166,320,124)
Button1	Name：btnCreate Title：Create Text Color：白色 Background：黑色 位置和大小：(122,322,90,30)
Button2	Name：btnInsert Title：Insert Text Color：白色 Background：黑色 位置和大小：(122,406,90,30)

续表

视 图	设 置
Button3	Name：btnQuery Title：Query Text Color：白色 Background：黑色 位置和大小：(122,494,90,30)

（4）打开 4-2ViewController.cs 文件，编写代码，实现数据库的创建。代码如下：

```
using System;
using System.Drawing;
using Mono.Data.Sqlite;                      //引入 Mono.Data.Sqlite 命名空间
using System.IO;
using MonoTouch.Foundation;
using MonoTouch.UIKit;
namespace Application
{
    public partial class __2ViewController : UIViewController
    {
        ……
        #region View lifecycle
        public override void ViewDidLoad ()
        {
            base.ViewDidLoad ();
            string sqlitePath = Path.Combine (Environment.GetFolderPath
            (Environment.SpecialFolder.Personal), "MyDBBdcc.db3");
            label.Text = sqlitePath;
            this.btnCreate.TouchUpInside += (s, e) => this.CreateSQLite-
            Database (sqlitePath);
            this.btnInsert.TouchUpInside += (s, e) => this.InsertData
            (sqlitePath);
            this.btnQuery.TouchUpInside += (s, e) => this.QueryData
            (sqlitePath);
        }
        //创建数据库
        private void CreateSQLiteDatabase (string databaseFile)
        {
            try
            {
                //判断数据库文件是否存在，如果不存在，则创建数据库文件
                if (!File.Exists (databaseFile))
                {
                    SqliteConnection.CreateFile (databaseFile);
                    //创建数据库文件
                    //创建数据库
                    using (SqliteConnection sqlCon = new SqliteConnection
                    (String.Format ("Data Source = {0};", databaseFile)))
                    {
                        sqlCon.Open ();                  //打开数据库
                        //创建表
                        using (SqliteCommand sqlCom = new SqliteCommand
                        (sqlCon))
                        {
                            sqlCom.CommandText = "CREATE TABLE Customers
                            (ID INTEGER PRIMARY KEY, FirstName VARCHAR(20),
                            LastName VARCHAR(20))";
                            sqlCom.ExecuteNonQuery ();
```

```
                        }
                        sqlCon.Close ();                    //关闭数据库
                    }
                    this.lblStatus.Text = "Database created!";
                } else {
                    this.lblStatus.Text = "Database already exists!";
                }
            } catch (Exception ex) {
                this.lblStatus.Text = String.Format ("Sqlite error: { 0}",
                ex.Message);
            }
        }
        ……
        #endregion
    }
}
```

运行效果如图 4.14 所示。

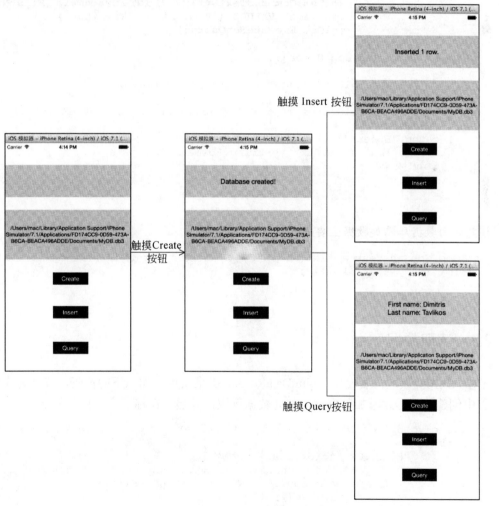

图 4.14　SQLite 的创建、插入和查询

注意：在此代码中引入了新的命名空间 Mono.Data.Sqlite，此命名空间提供了对数据库的支持。

4.2.2 插入数据

创建好数据库后，此时的数据库是一个空的数据库，那么该如何为数据库添加数据呢？这就是本小节将要讲解的内容，如以下的代码，就是为【示例 4-5】中创建的数据库添加数据。代码如下：

```
private void InsertData(string databaseFile) {
    try {
        if (File.Exists(databaseFile)) {
            using (SqliteConnection sqlCon = new SqliteConnection(String.
            Format("Data Source = {0};", databaseFile))) {
                sqlCon.Open();
                //创建表
                using (SqliteCommand sqlCom = new SqliteCommand(sqlCon)) {
                    sqlCom.CommandText = "INSERT INTO Customers (FirstName,
                    LastName) VALUES ('Dimitris', 'Tavlikos')";
                    sqlCom.ExecuteNonQuery();
                }
                sqlCon.Close();
            }
            this.lblStatus.Text = "Inserted 1 row.";
        } else {
            this.lblStatus.Text = "Database file does not exist!";
        }
    } catch (Exception ex) {
        this.lblStatus.Text = String.Format("Sqlite error: {0}",
        ex.Message);
    }
}
```

> 注意：为数据库添加数据主要是以下这一行代码：

```
sqlCom.CommandText = "INSERT INTO Customers (FirstName, LastName) VALUES
('Dimitris', 'Tavlikos')";
```

运行效果如图 4.14 所示。

4.2.3 读取数据

对于数据的读取需要使用 SqliteDataReader 类来实现。如以下的代码，就是为【示例 4-5】中创建的数据库添加数据后，将此数据读取，并显示在标签中。代码如下：

```
private void QueryData(string databaseFile) {
    try {
        if (File.Exists(databaseFile)) {
            using (SqliteConnection sqlCon = new SqliteConnection(String.
            Format("Data Source = {0};", databaseFile))) {
                sqlCon.Open();
                using (SqliteCommand sqlCom = new SqliteCommand(sqlCon)) {
                    //读取数据
                    sqlCom.CommandText = "SELECT * FROM Customers WHERE
                    FirstName='Dimitris'";
                    using (SqliteDataReader dbReader = sqlCom.
                    ExecuteReader()) {
```

```
                    while (dbReader.Read()) {
                        this.lblStatus.Text = String.Format("First
                        name: {0}\nLast name: {1}", dbReader
                        ["FirstName"], dbReader["LastName"]);
                    }
                }
            }
        } else {
            this.lblStatus.Text = "Database file does not exist!";
        }
    } catch (Exception ex) {
        this.lblStatus.Text = String.Format("Sqlite error: {0}",
        ex.Message);
    }
}
```

运行效果如图 4.14 所示。

4.2.4 查看数据库

数据库不像在第 4.1 节中提到的文件,通过显示的路径找到后,双击就可以打开对应的内容进行查看。它需要专门的查看工具。在本小节中我们将使用 SQLite Database Browser 工具对【示例 4-5】中所创建的数据库进行查看,具体的操作步骤如下所述。

(1) 找到文件,如图 4.15 所示。
(2) 双击,此时会弹出一个提醒对话框,如图 4.16 所示。
(3) 单击 "选取应用程序..." 按钮,弹出 "选取应用程序" 对话框,如图 4.17 所示。
(4) 选择应用程序中的 SQLite Database Browser 2.0 b1,单击 "打开" 按钮,打开 MyDB.db3,如图 4.18 所示。

需要注意的是,打开 MyDB.db3 文件后,首先看到的是此文件的结构。如果想要查看数据,需要选择 Browser Data 选项,如图 4.19 所示。

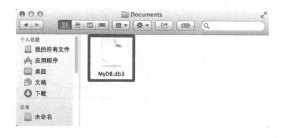

图 4.15　操作步骤 1

图 4.16　操作步骤 2

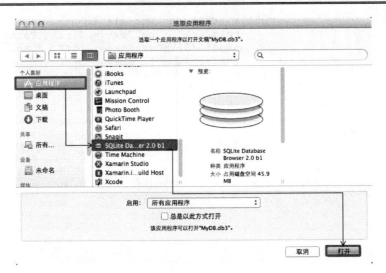

图 4.17　操作步骤 3

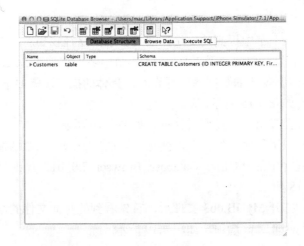

图 4.18　操作步骤 4

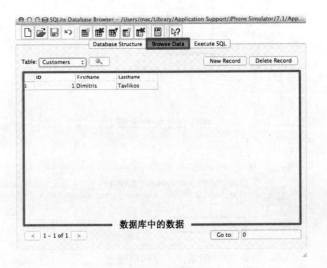

图 4.19　操作步骤 5

4.3 使用 iCloud

在 iOS 5 中提供了一个新的功能,就是 iCloud。iCloud 是苹果公司所提供的云端服务,让用户可以免费储存 5GB 的资料。免费的云服务 iCloud 将以全新的方式存储并访问用户的音乐、照片、应用程序、日历、文档及更多内容,并以无线方式推送到用户的所有设置,一切都能自动完成。对于应用程序开发,可以使用 iCloud 来存储需要保存的数据,以便应用程序在不同的设备上运行时,可以共享信息。本节将主要讲解如何使用 iCloud。

4.3.1 启动 iCloud 服务

在使用 iCloud 编写应用程序时,首先需要做的事情就是启动 iCloud 服务。启动 iCloud 服务的具体操作步骤如下所述。

(1)在创建的工程中,选择 Entitlements.plist,双击将其打开。

(2)选择 Enable iCloud 复选框和 Use key-value store 复选框。在 Ubiquity Containers 输入框中输入 com.myiCloudTest.appname。如图 4.20 所示。

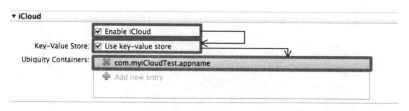

图 4.20 操作步骤 1

> 注意:这里的 com.myiCloudTest.appname 是在注册一个 App ID 时,在 Bundle ID 文本框中输入的内容。如图 4.21 所示。

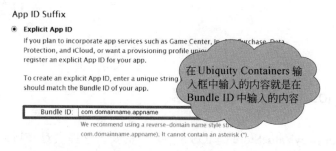

图 4.21 注册一个 App ID

(3)双击"引用"文件夹上方的 4-3,打开"工程选项"对话框,如图 4.22 所示。

(4)选择 iOS Bundle Signing 选项,将 Custom entitlements 中的内容设置为 Entitlements.plist(此项一般是默认的)。选择 iOS Application 选项,如图 4.23 所示。

(5)在 Bundle Identifier 文本框中输入在 Ubiquity Containers 输入框中输入的内容,然后单击"确定"按钮,退出"工程选项"对话框。

第 2 篇　资源使用篇

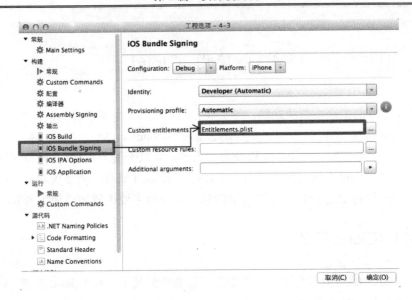

图 4.22　操作步骤 2

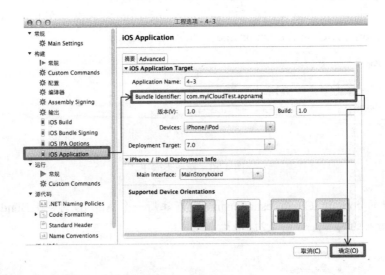

图 4.23　操作步骤 3

4.3.2　在 iCloud 中存储键/值数据

　　NSUbiquitousKeyValueStore 是 iOS 5 中新增的类，它提供对 iCloud 进行 Key/Value 方式的存储。一般用于保存跨设备的用户配置，如阅读软件保存当前用户阅读的位置。如果用户更换其他设备，可以使用用户无缝地从上次的位置继续阅读。

　　【示例 4-6】　以下将在 iCloud 中存储键/值数据。具体的操作步骤如下所述。

　　（1）创建一个 Single View Application 类型的工程。

　　（2）启动 iCloud 服务。

　　（3）打开 MainStoryboard.storyboard 文件，对主视图进行设置，效果如图 4.24 所示。

第 4 章 数据管理

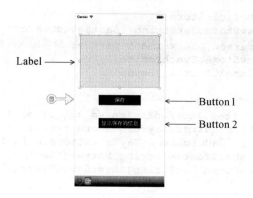

图 4.24 主视图的效果

需要添加的视图以及设置，如表 4-4 所示。

表 4-4 设置主视图

视 图	设 置
Label	Name：label Text：（空） Font：System 20 pt Alignment：居中 Lines：3 位置和大小：(20,76,280,181)
Button1	Name：saveButton Title：保存 Font：System 19 pt Text Color：白色 Background：黑色 位置和大小：(85,284,150,35)
Button2	Name：showButton Title：显示保存的信息 Font：System 19 pt Text Color：白色 Background：黑色 位置和大小：(85,365,150,35)

（4）打开 4-3ViewController.cs 文件，编写代码，实现数据的存储。代码如下：

```
using System;
using System.Drawing;
using MonoTouch.Foundation;
using MonoTouch.UIKit;
namespace Application
{
    public partial class __3ViewController : UIViewController
    {
    ……                              //这里省略了视图控制器的构造方法和析构方法
        #region View lifecycle
        public override void ViewDidLoad ()
        {
            base.ViewDidLoad ();
            //保存键/值数据
            saveButton.TouchUpInside += (s, e) => {
                NSUbiquitousKeyValueStore kvStore = NSUbiquitousKeyValueStore.
```

· 157 ·

```
                DefaultStore;
            kvStore.SetString("LastSavedSearch", "The most known
            person,the warmest partner 最懂的人，最暖的伴");
            kvStore.Synchronize();
            label.Text = "Saved!";
        };
        //显示键/值数据
        showButton.TouchUpInside += (s, e) => {
            NSUbiquitousKeyValueStore kvStore =
                NSUbiquitousKeyValueStore.DefaultStore;
            label.Text = string.Format("Last saved search is: {0}",
                kvStore.GetString("LastSavedSearch"));
        };
    }
    ……                          //这里省略了视图加载和卸载前后的一些方法
    #endregion
}
```

运行效果如图 4.25 所示。

图 4.25　运行效果

> **注意**：如果开发者有两个设备，在 iCloud 账号相同的情况下，在第一个设备上触摸"保存"按钮，在第二个设备上触摸"显示保存的信息"按钮，可以看到保存的信息显示在了标签中，效果和图 4.25 类似。

第 5 章 显 示 数 据

在上一章中讲解了数据的管理。那么这些被管理的数据该如何显示给用户呢？为了解决这一问题，Xamarin.iOS 提供了几种专门用来显示数据的视图，如表视图、网页视图和集合视图等。本章将讲解这些用来显示数据的视图。

5.1 选 择 列 表

在 iOS 应用程序中，选择操作是用户常用的，例如对日期的选择等。本节将讲解有关选择的两个视图，即日期选择器和自定义选择器。

5.1.1 日期选择器

日期选择器是为用户在输入时间和日期时能快速操作而提供的选择器。用户只要滚动日期选择器，就可以在其中找到对应的时间。一般使用 UIDatePicker 类来实现日期选择器。

【示例 5-1】 以下将使用日期选择器实现对日期的选择，并且将选择的日期显示在标签中。具体的操作步骤如下所述。

（1）创建一个 Single View Application 类型的工程。

（2）打开 MainStoryboard.storyboard 文件，对主视图进行设置，效果如图 5.1 所示。

图 5.1 主视图的效果

需要添加的视图以及设置，如表 5-1 所示。

表 5-1 设置视图

视图	设置
Label	Name：label Text：请设置当前时间 Font：System 26 pt Alignment：居中 位置和大小：(60, 104, 200, 55)
Date Picker	Name：datePicker 位置和大小：(0, 234, 320, 216)

（3）打开 5-1ViewController.cs 文件，编写代码，实现日期选择器的选择功能。代码如下：

```
using System;
using System.Drawing;
using MonoTouch.Foundation;
using MonoTouch.UIKit;
namespace Application
{
    public partial class __1ViewController : UIViewController
    {
        ……                          //这里省略了视图控制器的构造方法和析构方法
        #region View lifecycle
        public override void ViewDidLoad ()
        {
            base.ViewDidLoad ();
            //实现日期的选择
            datePicker.ValueChanged += (sender, e) => {
                NSDateFormatter formatter=new NSDateFormatter();
                formatter.DateFormat="YYYY/MM/dd HH:mm";    //设置日期的格式
                string a=formatter.ToString(datePicker.Date);
                                                            //将日期转换为字符串
                label.Text=a;
            };
        }
        ……                          //这里省略了视图加载和卸载前后的一些方法
        #endregion
    }
}
```

在此代码中，NSDateFormatter 类用来实现 NSDate 日期类的转换。运行效果如图 5.2 所示。

> 📌 **注意**：日期选择器在选择日期后，将选择的日期显示在标签中。这是通过 ValueChanged 事件实现的，此事件在值发生变化时被调用。

日期选择器的格式这可以发生改变，具体使用的属性如下所述。

1. 设置显示模式

日期选择器的显示模式可以通过 Mode 属性进行设置，其语法形式如下：

日期选择器对象.Mode=显示模式;

第 5 章 显示数据

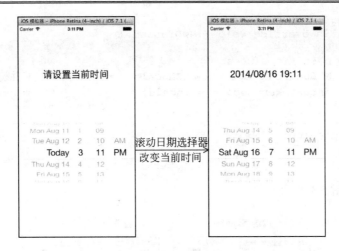

图 5.2 运行效果

其中,显示模式有 4 种,如表 5-2 所示。

表 5-2 显示模式

显 示 模 式	效 果
CountDownTimer	显示类似于时钟的界面,用于选择持续延长时间
Date	显示日期
DateAndTime	显示日期和时间
Time	显示时间

注意:如果不设置此项,默认模式为 DateAndTime。

2. 设置语言环境

每一个用户当然是对自己的母语比较熟悉了,那么如何将日期选择器的语言改为用户熟悉的语言呢?这时就需要使用 Locale 属性。

【示例 5-2】 以下使用代码将日期选择器的语言环境变为对于中国用户来说比较熟悉的汉语。具体的操作步骤如下。

(1)创建一个 Single View Application 类型的工程。

(2)打开 MainStoryboard.storyboard 文件,拖动 Date Picker 日期选择器对象到主视图中,将它的 Name 设置为 datePicker。

(3)打开 5-18ViewController.cs 文件,编写代码,实现日期选择器语言的改变。代码如下:

```
using System;
using System.Drawing;
using MonoTouch.Foundation;
using MonoTouch.UIKit;
namespace Application
{
    public partial class __18ViewController : UIViewController
    {
        ……
```

```
#region View lifecycle
public override void ViewDidLoad ()
{
    base.ViewDidLoad ();
    NSLocale locale = new NSLocale ("zh_CN");//创建并初始化对象locale
    datePicker.Locale = locale;              //设置语言环境
}
……
#endregion
}
```

运行效果如图 5.3 所示。

设置语言环境前　　　　设置语言环境后

图 5.3　运行效果

3. 设置时间间隔

在日期选择器中，秒与秒之间间隔的秒数也是可以进行设置的，此时需要使用 MinuteInterval 属性，其语法形式如下：

日期选择器对象.MinuteInterval=时间间隔的秒数;

5.1.2　自定义选择器

日期选择器主要的功能是针对日期进行选择。但是如果要想实现其他的选择该如何做呢？这时就需要使用到自定义选择器，UIPickerView 类提供了自定义选择器的实现。它是 UIDatePicker 类的父类，它非常灵活，选取器中行的个数可以设定，每一行的内容也可以设定。

【示例 5-3】以下将使用自定义选择器实现选择项目，并将选择的内容显示在标签中。具体的操作步骤如下：

（1）创建一个 Single View Application 类型的工程。

（2）打开 MainStoryboard.storyboard 文件，对主视图进行设置，效果如图 5.4 所示。

第 5 章 显示数据

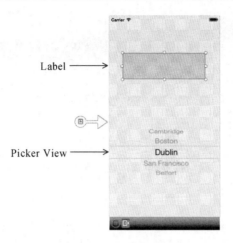

图 5.4 主视图的效果

需要添加的视图以及设置,如表 5-3 所示。

表 5-3 设置视图

视 图	设 置
主视图	Background:浅灰色
Label	Name:lblStatus Text:(空) Alignment:居中 位置和大小:(42, 107, 236, 71)
Picker View	Name:pickerView 位置和大小:(0, 277, 320, 216)

(3)打开 5-2ViewController.cs 文件,编写代码,实现自定义选择器的内容填充以及选择。代码如下:

```
using System;
using System.Drawing;
using MonoTouch.Foundation;
using MonoTouch.UIKit;
using System.Collections.Generic;//引入 System.Collections.Generic 命名空间
namespace Application
{
    public partial class __2ViewController : UIViewController
    {
        ……                    //这里省略了视图控制器的构造方法和析构方法
        #region View lifecycle
        public override void ViewDidLoad ()
        {
            base.ViewDidLoad ();
            pickerView.Model = new PickerModel(this);
        }
        private class PickerModel : UIPickerViewModel
        {
            private __2ViewController parentController;
            private List<string> transportList;
            private List<string> distanceList;
            private List<string> unitList;
            string transportSelected;
```

```
string distanceSelected;
string unitSelected;
public PickerModel (__2ViewController controller)
{
    parentController = controller;
    transportList = new List<string>() { "On foot", "Bicycle",
"Motorcycle", "Car", "Bus" };
    distanceList = new List<string>() { "0.5",     "1", "5",
"10", "100" };
    unitList = new List<string>() { "mi", "km" };
    transportSelected = this.transportList[0];
    distanceSelected = this.distanceList[0];
    unitSelected = this.unitList[0];
}
//设置列数
public override int GetComponentCount (UIPickerView picker)
{
    return 3;
}
//设置行数
public override int GetRowsInComponent (UIPickerView picker,
int component)
{
    switch (component)
    {
        case 0:
            return transportList.Count;
        case 1:
            return distanceList.Count;
        default:
            return unitList.Count;
    }
}
//为自定义选择器中设置内容
public override string GetTitle (UIPickerView picker, int row,
int component)
{
    switch (component)
    {
        case 0:
            return transportList[row];
        case 1:
            return distanceList[row];
        default:
            return this.unitList[row];
    }
}
//对自定义选择器实现响应
public override void Selected (UIPickerView picker, int row, int
component)
{
    switch (component)
    {
        case 0:
            transportSelected = this.transportList[row];
            break;
        case 1:
            distanceSelected = this.distanceList[row];
            break;
        default:
            unitSelected = this.unitList[row];
```

```
                    break;
            }
            parentController.lblStatus.Text =
                String.Format ("Transport: {0}\nDistance: {1}{2}",
                transportSelected, distanceSelected, unitSelected);
        }
    }
    ......                          //这里省略了视图加载和卸载前后的一些方法
    #endregion
    }
}
```

运行效果如图 5.5 所示。

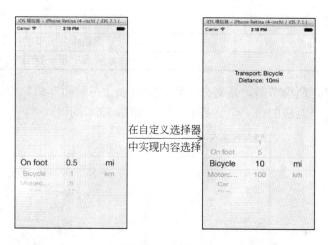

图 5.5　运行效果

在此示例中，需要注意两个方面：自定义选择器的内容显示和选择功能。要实现这两个功能需要使用 UIPickerViewDataSource 和 UIPickerViewDelegate 协议。Xamarin.iOS 提供了 UIPickerModel 类来实现这两个协议。

（1）自定义选择器内容的显示

内容的显示需要重写如表 5-4 中的方法。

表 5-4　方法

方　　法	功　　能
GetComponentCount	设置自定义选择器中的列数
GetRowsInComponent	设置自定义选择器中的行数
GetTitle	设置自定义选择器中每一行的文本内容

（2）自定义选择器的选择功能

对于自定义选择器的选择功能，需要重写 Selected()方法。

5.2　在表中显示数据

表可以用于显示大量的数据，如通讯录。Xamarin.iOS 提供了一个 UITableView 类实现表的功能。本节将讲解在表中如何显示内容，以及表单元格的设置等内容。

5.2.1 表中内容的显示

表可以用来显示大量的数据，这些数据该如何存储在表上，并且向用户显示呢？此时需要使用到 UITableViewSource 协议。在此协议中重写如表 5-5 所示的方法就可以实现表中数据的存储了。

表 5-5 方法

方　　法	功　　能
NumberOfSections	设置表中的结束
RowsInSection	设置表中每一节的行数
GetCell	设置表中每一行的内容

需要注意的是，所谓节数，是对分组表（分组表会在后面介绍）所说的，意思是在分组表中要分为几组，对应的每一组就是一个节，如图 5.6 所示。

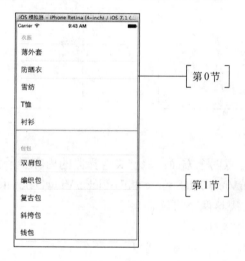

图 5.6　分组表

> 注意：在一个分组表中，最开始的节被称为第 0 节。一般不对此方法进行设置，默认为是只有 1 节的，也就是不分组的表视图。

【示例 5-4】 以下将使用表实现数据的存储与显示。具体步骤如下所述。

（1）创建一个 Single View Application 类型的工程。

（2）打开 MainStoryboard.storyboard 文件，从工具栏中拖动 Table View 表视图对象到主视图中，将它的位置和大小设置为（0,0,320,568）；将它的 Name 设置为 tableView。

（3）打开 5-3ViewController.cs 文件，编写代码，实现为表视图对象添加内容。代码如下：

```
using System;
using System.Drawing;
using MonoTouch.Foundation;
using MonoTouch.UIKit;
```

第 5 章 显示数据

```
using System.Collections.Generic;
namespace Application
{
    public partial class __3ViewController : UIViewController
    {
        private List<string> premTeams;
        ……                        //这里省略了视图控制器的构造方法和析构方法
        #region View lifecycle
        public override void ViewDidLoad ()
        {
            base.ViewDidLoad ();
           premTeams = new List<string>(){"LIVERPOOL","EVERTON","ARSENAL",
            "SWANSEA","CARDIFF","NEWCASTLE","ABRAHAM","ALEXANDER",
            "ANTOINE","ARMSTRONG","BARTHOLOMEW"};
            tableView.Source = new myViewSource(premTeams);
        }
        private class myViewSource : UITableViewSource {
            private List<string>dupPremTeams;
            public myViewSource(List<string>prems) {
                dupPremTeams = prems;
            }
            //设置行数
            public override int RowsInSection(UITableView table,
                int section) {
                return dupPremTeams.Count;
            }
            //设置每一行的内容
            public override UITableViewCell GetCell(UITableView tableView,
NSIndexPath index) {
                UITableViewCell theCell = new UITableViewCell();
                theCell.TextLabel.Text = dupPremTeams[index.Row];
    //设置内容
                return theCell;
            }
        }
        ……                //这里省略了视图加载和卸载前后的一些方法
        #endregion
    }
}
```

运行效果如图 5.7 所示。

图 5.7　运行效果

5.2.2 设置表

为了让表中的内容更加丰富,可以在表中添加一些图像、页眉和页脚等内容。

1. 添加图像

如果要在表中添加图像,可以对表单元格的 imageView 属性进行设置。

【示例 5-5】 以下以【示例 5-4】为基础,在表中添加图像。代码如下:

```
public override UITableViewCell GetCell(UITableView tableView, NSIndexPath index) {
    UITableViewCell theCell = new UITableViewCell(UITableViewCellStyle.Value1,null);
    theCell.TextLabel.Text = dupPremTeams[index.Row];
    theCell.ImageView.Image = UIImage.FromFile ("1.jpg");
    return theCell;
}
```

运行效果如图 5.8 所示。

> 注意:在编写代码之前,需要在创建工程的 Resources 文件夹中添加一个名为 1.jpg 的图像。

2. 添加页眉和页脚

页眉和页眉通常用来显示文档的附加信息,如时间、日期、页码和单位名称等。页眉还可以实现添加文档注释等内容。其中,页眉在页面的顶部,页脚在页面的底部。在表中如果想要插入页眉需要重写 TitleForHeader()方法;如果想要插入页脚需要重写 TitleForFooter()方法。

【示例 5-6】 以下以【示例 5-4】为基础,在表中添加页眉和页脚。代码如下:

```
//添加页眉
public override string TitleForHeader (UITableView tableView, int section)
{
    return "表头";
}
//添加页脚
public override string TitleForFooter (UITableView tableView, int section)
{
    return "表尾";
}
```

运行效果如图 5.9 所示。

3. 设置行高

当表中每一行内容的字体的高度远远大于表中每一行的高度时,所显示的内容看起来相当的乱,而且显示的只有部分内容,如图 5.10 所示。

为了解决这一难题,Xamarin.iOS 提供了设置表中行高的方法,开发者只需要重写 GetHeightForRow()方法,就可以对行高进行设置了。

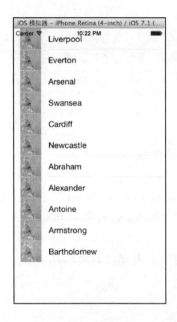

图 5.8 运行效果　　　　　　图 5.9 运行效果

【示例 5-7】 以下以【示例 5-4】为基础,实现表中行高的改变。代码如下:

```
public override float GetHeightForRow (UITableView tableView, NSIndexPath indexPath)
{
    return 60;
}
```

运行效果如图 5.11 所示。

设置高度前　　　　　　　　设置高度后

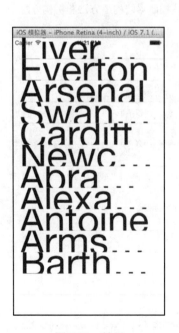

图 5.10 效果　　　　　　　图 5.11 运行效果

4. 设置表的风格

以上这些示例显示的都是表的默认风格 Plain，即普通表。除此之外，表还有另外一种风格 Grouped，即分组表。它的形式在 5.2.1 节中看到过了。要对表的风格进行设置需要使用到 Style 属性，其语法形式如下：

表对象.Style=表的风格;

【示例 5-8】以下将实现一个分组表的效果。具体操作步骤如下所述。

（1）创建一个 Single View Application 类型的工程。

（2）打开 MainStoryboard.storyboard 文件，从工具栏中拖动 Table View 表视图对象到主视图中，将它的位置和大小设置为（0,0,320,568）；将它的 Name 设置为 tableView；将 Style 属性设置为 Grouped。

（3）打开 5-4ViewController.cs 文件，编写代码，实现为表视图对象添加内容。代码如下：

```
using System;
using System.Drawing;
using MonoTouch.Foundation;
using MonoTouch.UIKit;
using System.Collections.Generic;
namespace Application
{
    public partial class __4ViewController : UIViewController
    {
        private List<string> titleTeams;
        private List<string> premTeams1;
        private List<string> premTeams2;
        private List<string> premTeams3;
        ……                          //这里省略了视图控制器的构造方法和析构方法
        #region View lifecycle

        public override void ViewDidLoad ()
        {
            base.ViewDidLoad ();
            titleTeams = new List<string>(){"衣服","包包","配饰"};
            premTeams1 = new List<string>(){"薄外套","防晒衣","雪纺","T恤",
                "衬衫"};
            premTeams2 = new List<string>(){"双肩包","编织包","复古包","斜挎
                包","钱包","手提包"};
            premTeams3 = new List<string>(){"项链","脚链","戒指","手镯","佛
                珠手串","毛衣链","耳饰","手链"};
            tableView.Source = new myViewSource(premTeams1,premTeams2,
                premTeams3,titleTeams);
        }
        private class myViewSource : UITableViewSource {
            private List<string>dupPremTeams1;
            private List<string>dupPremTeams2;
            private List<string>dupPremTeams3;
            private List<string>dupTitleTeams;
            public myViewSource(List<string>prems1,List<string>prems2,
                List<string>prems3,List<string>title)
```

```
{
            dupPremTeams1 = prems1;
            dupPremTeams2=prems2;
            dupPremTeams3=prems3;
            dupTitleTeams=title;
        }
        //设置节数
        public override int NumberOfSections (UITableView tableView)
        {
            return 3;
        }
        //设置每节中的行数
        public override int RowsInSection(UITableView table,int section) {
            switch (section) {
                case 0:
                    return dupPremTeams1.Count;
                case 1:
                    return dupPremTeams2.Count;
                default:
                    return dupPremTeams3.Count;
            }
        }
        //设置每行中的内容
        public override UITableViewCell GetCell(UITableView tableView,
        NSIndexPath index) {
            UITableViewCell theCell = new UITableViewCell();
            if (index.Section == 0)
            {
                theCell.TextLabel.Text = dupPremTeams1[index.Row];
            }
            if (index.Section == 1)
            {
                theCell.TextLabel.Text = dupPremTeams2[index.Row];
            }
            if (index.Section == 2)
            {
                theCell.TextLabel.Text = dupPremTeams3[index.Row];
            }
            return theCell;
        }
        //为每一节添加标题
        public override string TitleForHeader (UITableView tableView,
        int section)
        {
            return dupTitleTeams[section];
        }
    }
    ……
    #endregion
}
```

运行效果如图5.12所示。

> **注意**：此程序中的TitleForHeader()方法，原本是添加页眉的，但是在分组表中出现就是为表的每一节设置内容。

图 5.12 运行效果

5.2.3 设置表单元格

表可以进行设置,表单元格也可以进行设置。以下讲解对表单元格的一些设置。

1. 设置表单元格的风格

表单元格的风格不是一成不变的。在 Xamarin.iOS 中为开发者提供了 4 种表单元格的风格,如表 5-6 所示。这些风格需要靠表单元的 UILabel 实现。每一个表格元格都是由两个 UILable 组成,这两个 UILable 分别为 TextLable 和 DetaiTextLable。在上一小节中所讲的内容使用的只要是一个 TextLable,使用的风格是 Default。

表 5-6 表单元格的风格

单元格的风格	风格描述
Default	该风格提供了一个简单的左对齐的文本标签 TextLabel 和一个可选的图像 imageView。如果显示图像,那么图像将在最左边
Subtitle	该风格增加了对 DetailTextLabel 的支持,该标签将会显示在 TextLabel 标签的下面,字体相对较小
Value1	该风格居左显示 TextLabel,居右显示 DetailTextLabel
Value2	该风格居左现实一个小型蓝色 TextLabel,在其右边显示一个小型黑色副标题 DetailTextLabel

【示例 5-9】以下以【示例 5-4】为基础,对表单元格的风格进行设置。代码如下:

```
public override UITableViewCell GetCell(UITableView tableView, NSIndexPath
    index) {
    UITableViewCell theCell = new UITableViewCell(UITableViewCellStyle.
```

第 5 章 显示数据

```
        Value1,null);
   theCell.TextLabel.Text = dupPremTeams[index.Row];
   theCell.DetailTextLabel.Text = dupPremTeams [index.Row];
   return theCell;
}
```

运行效果如图 5.13 所示。

2. 设置表单元格的标记

使用 iPhone 手机的开发者一定知道，在打开设置，或者设置中的通用后，会看到在表的右侧，会有如图 5.14 所示的图标，这个图标被称为标记。

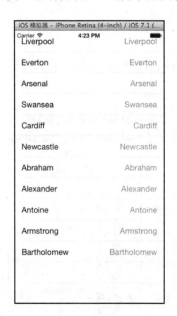

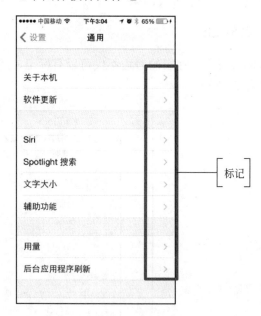

图 5.13　运行效果　　　　　　　　图 5.14　标记

这些标记可以通过 Accessory 属性进行修改，如语法形式如下：

`表单元格对象.Accessory=标记;`

其中，标记有 4 种，如表 5-7 所示。

表 5-7　标记

标　　记	格　　式
Checkmark	✓
DetailButton	ⓘ
DetailDisclosureButton	ⓘ >
DisclosureIndicator	>
None	无

• 173 •

【示例 5-10】以下以【示例 5-4】为基础，为表单元格添加风格为 DetailButton 的标记。代码如下：

```
public override UITableViewCell GetCell(UITableView tableView, NSIndexPath index) {
    UITableViewCell theCell = new UITableViewCell(UITableViewCellStyle.Value1,null);
    theCell.TextLabel.Text = dupPremTeams[index.Row];
    theCell.Accessory = UITableViewCellAccessory.DetailButton;
    return theCell;
}
```

运行效果如图 5.15 所示。

3. 添加其他视图

在实际的开发中，在表的右边不仅可以有标记，还可以有按钮、标签、开关按钮等其他视图，如图 5.16 所示。

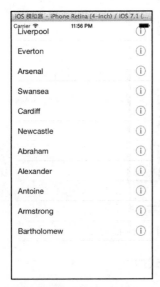
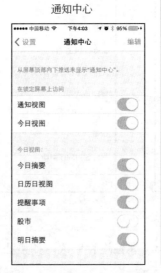

图 5.15　运行效果　　　　　　　　　　　　　图 5.16　表

这时视图的添加可以使用 AccessoryView 属性实现，其语法形式如下：

表单元格对象.AccessoryView=其它的视图对象;

【示例 5-11】以下以【示例 5-4】为基础，在表单元格的右边添加一个开关视图对象。代码如下：

```
public override UITableViewCell GetCell(UITableView tableView, NSIndexPath index) {
    UITableViewCell theCell = new UITableViewCell(UITableViewCellStyle.Value1,null);
    theCell.TextLabel.Text = dupPremTeams[index.Row];
    UISwitch myswitch=new UISwitch();
    myswitch.Frame = new RectangleF (0, 0, 50, 20);
    theCell.AccessoryView = myswitch;
    return theCell;
}
```

运行效果如图 5.17 所示。

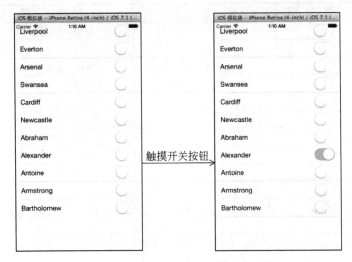

图 5.17　运行效果

4．自定义表单元格

如果表单元格的风格不是你想要的，没关系，我们还可以自定义表单元格。

【示例 5-12】　以下将实现一个自定义的表单元格。具体步骤如下所述。

（1）创建一个 Single View Application 类型的工程。

（2）添加图像 1.jpg 到创建工程的 Resources 文件夹中。

（3）在创建的工程中添加一个 C#的类文件，命名为 CustomCell。

（4）打开 MainStoryboard.storyboard 文件，对主视图进行设置，效果如图 5.18 所示。

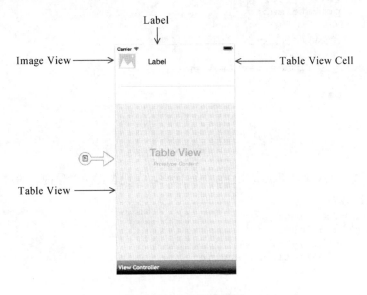

图 5.18　主视图的效果

需要添加的视图以及设置，如表 5-8 所示。

表 5-8 设置视图

视　　图	设　　置
Table View	Name：tableView 位置和大小：(0, 0, 320, 568)
Table View Cell	Class：CustomCell Identifier：CustomCell
Image View	Name：imgView 位置和大小：(10, 0, 45, 40)
Label	Name：lblTitle Font：System 20 pt 位置和大小：(88, 7, 206, 29)

注意：选择表视图对象中的任意一个表单元，就可以实现对 Table View Cell 的设置。

（5）打开 CustomCell.cs 文件，编写代码，对表单元格的属性进行设置。代码如下：

```
using System;
using MonoTouch.Foundation;
using MonoTouch.UIKit;
using System.CodeDom.Compiler;
namespace Application
{
    partial class CustomCell : UITableViewCell
    {
        ……//这里省略了构造器方法
        public const string CELLID = "CustomCell";
        //设置表单元格的lblTitle标签属性
        [Outlet("lblTitle")]
        public UILabel LabelTitle
        {
            get;
            private set;
        }
        //设置表单元的imgView图像视图属性
        [Outlet("imgView")]
        public UIImageView ImgView
        {
            get;
            private set;
        }
    }
}
```

（6）打开 5-5ViewController.cs 文件，编写代码，实现表视图对象内容的填充。代码如下：

```
using System;
using System.Drawing;
using MonoTouch.Foundation;
using MonoTouch.UIKit;
using System.Collections.Generic;
namespace Application
{
    public partial class __5ViewController : UIViewController
    {
        private Dictionary<int, string> tableData;
        ……                    //这里省略了视图控制器的构造方法和析构方法
```

```
#region View lifecycle
public override void ViewDidLoad ()
{
    base.ViewDidLoad ();
    tableData = new Dictionary<int, string> () {
        {0, "Music"} ,
        {1, "Videos"} ,
        {2, "Images"} ,
        {3, "books"} ,
        {4, "Food"} ,
    } ;
    tableView.Source = new myViewSource(tableData);
}
private class myViewSource : UITableViewSource {
    private Dictionary<int, string> duptableData;
    public myViewSource(Dictionary<int, string> duptable) {
        duptableData=duptable;
    }
    //设置行数
    public override int RowsInSection (UITableView tableview, int section)
    {
        return duptableData.Count;
    }
    //设置每一行的内容
    public override UITableViewCell GetCell (UITableView tableView, NSIndexPath indexPath)
    {
        int rowIndex = indexPath.Row;
        CustomCell cell = (CustomCell)tableView.DequeueReusableCell(CustomCell.CELLID);
        cell.LabelTitle.Text = duptableData[rowIndex];
        cell.ImgView.Image = UIImage.FromBundle("1.jpg");
        return cell;
    }
}
......                      //这里省略了视图加载和卸载前后的一些方法
#endregion
}
```

运行效果如图 5.19 所示。

图 5.19 运行效果

5.3 编 辑 表

在表中，用户可以进行行（在表中一行就代表一个表单元格）的选择、删除、添加和移动等操作，这些功能被称为编辑表。本节将主要讲解这些内容。

5.3.1 选取行

如果想要实现行的选择，需要重写 RowSelected()方法。

【示例 5-13】 以下将重写 RowSelected()方法，实现对行的选取。具体操作步骤如下所述。

（1）创建一个 Single View Application 类型的工程。

（2）打开 MainStoryboard.storyboard 文件，从工具栏中拖动 Table View 表视图对象到主视图中，将它的位置和大小设置为（0,0,320,568）；将它的 Name 设置为 tableView。

（3）打开 5-6ViewController.cs 文件，编写代码，实现为表视图对象添加内容，以及选取行的功能。代码如下：

```
using System;
using System.Drawing;
using MonoTouch.Foundation;
using MonoTouch.UIKit;
using System.Collections.Generic;
namespace Application
{
    public partial class __6ViewController : UIViewController
    {
        private List<string> premTeams;
        ……                    //这里省略了视图控制器的构造方法和析构方法
        #region View lifecycle
        public override void ViewDidLoad ()
        {
            base.ViewDidLoad ();
            premTeams = new List<string>(){"1","2","3","4","5","6","7","8"};
            tableView.Source = new myViewSource(premTeams);
        }
        private class myViewSource : UITableViewSource {
            private List<string>dupPremTeams;
            public myViewSource(List<string>prems) {
                dupPremTeams = prems;
            }
            //设置表的行数
            public override int RowsInSection(UITableView table,
                int section) {
                return dupPremTeams.Count;
            }
            //设置每行的内容
            public override UITableViewCell GetCell(UITableView tableView,
                NSIndexPath index) {
                UITableViewCell theCell = new UITableViewCell();
                theCell.TextLabel.Text = dupPremTeams[index.Row];
```

```
            return theCell;
        }
        //选取行
        public override void RowSelected (UITableView tableView,
NSIndexPath indexPath)
        {
            int rowIndex = indexPath.Row;
            UITableViewCell cellView = tableView.CellAt (indexPath);
            //判断当前行是否有标记
            if (cellView.Accessory == UITableViewCellAccessory.None) {
                cellView.Accessory = UITableViewCellAccessory.
                Checkmark;
            } else {
                cellView.Accessory = UITableViewCellAccessory.None;
                tableView.DeselectRow (indexPath, true);
            }
        }
    }
    ……                    //这里省略了视图加载和卸载前后的一些方法
    #endregion
    }
}
```

运行效果如图 5.20 所示。

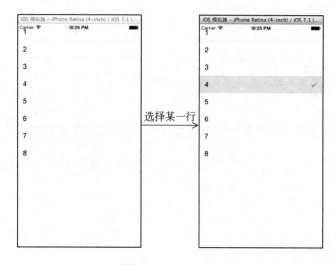

图 5.20　运行效果

5.3.2　删除行

要实现行的删除，需要重写 CommitEditingStyle()方法。此方法实现的功能是使用手指滑动删除的效果删除行。具体要删除哪一行，需要使用 deleteRows()方法实现。

【示例 5-14】　以下将实现删除选中行的功能，具体的操作步骤如下所述。

（1）创建一个 Single View Application 类型的工程。

（2）打开 MainStoryboard.storyboard 文件，从工具栏中拖动 Table View 表视图对象到主视图中，将它的位置和大小设置为（0,0,320,568）；将它的 Name 设置为 tableView。

（3）打开 5-7ViewController.cs 文件，编写代码，实现为表视图对象添加内容，以及删

除功能。代码如下:

```
using System;
using System.Drawing;
using MonoTouch.Foundation;
using MonoTouch.UIKit;
using System.Collections.Generic;
namespace Application
{
    public partial class __7ViewController : UIViewController
    {
        private List<string> premTeams;
        ……
        #region View lifecycle
        public override void ViewDidLoad ()
        {
            base.ViewDidLoad ();
            premTeams = new List<string>(){"Liverpool","Everton","Arsenal",
            "Swansea","Cardiff","Newcastle","Abraham","Alexander","Antoine",
            "Armstrong","Bartholomew     "};
            tableView.Source = new myViewSource(premTeams);
        }
        private class myViewSource : UITableViewSource {
            private List<string>dupPremTeams;
            public myViewSource(List<string>prems) {
                dupPremTeams = prems;
            }
            //
            public override int RowsInSection(UITableView table,
                int section) {
                return dupPremTeams.Count;
            }
            //
            public override UITableViewCell GetCell(UITableView tableView,
            NSIndexPath index) {
                UITableViewCell theCell = new UITableViewCell();
                theCell.TextLabel.Text = dupPremTeams[index.Row];
                return theCell;
            }
            //删除某一行
            public override void CommitEditingStyle (UITableView tableView,
            UITableViewCellEditingStyle editingStyle, NSIndexPath
            IndexPath)
            {
                dupPremTeams.RemoveAt(indexPath.Row);    //删除选择行的元素
                tableView.DeleteRows(new NSIndexPath[] { indexPath },
                    UITableViewRowAnimation.Automatic);      //删除行
            }
        }
        ……
        #endregion
    }
}
```

运行效果如图 5.21 所示。

第 5 章 显示数据

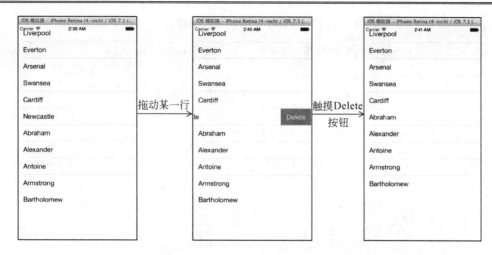

图 5.21 运行效果

5.3.3 插入行

以下将讲解两种插入行的方法。

1. 重写CommitEditingStyle()方法

要实现插入行需要重写 CommitEditingStyle()方法。具体要插入哪一行，需要使用 InsertRows()方法实现。

【示例 5-15】 以下将重写 CommitEditingStyle()方法，实现在表中插入行的功能。具体操作步骤如下所述。

（1）创建一个 Single View Application 类型的工程。

（2）打开 MainStoryboard.storyboard 文件，从工具栏中拖动 Table View 表视图对象到主视图中，将它的位置和大小设置为（0,0,320,568）；将它的 Name 设置为 tableView。

（3）打开 5-8ViewController.cs 文件，编写代码，实现为表视图对象添加内容，以及删除功能。代码如下：

```
using System;
using System.Drawing;
using MonoTouch.Foundation;
using MonoTouch.UIKit;
using System.Collections.Generic;
namespace Application
{
    public partial class __8ViewController : UIViewController
    {
        private List<string> myData;
        ……
        #region View lifecycle
        public override void ViewDidLoad ()
        {
            base.ViewDidLoad ();
            myData = new List<string> () {"A","B","C","D"} ;
            tableView.Source = new MyTableSource (myData);
```

```csharp
            tableView.SetEditing (true, true);        //设置编辑状态
        }
        private class MyTableSource : UITableViewSource
        {
            int i=0;
            public MyTableSource(List<string>prems) {
                TableData = prems;
            }
            public List<string> TableData;
            //设置表的行数
            public override int RowsInSection (UITableView tableview, int section)
            {
                return this.TableData.Count;
            }
            //设置每一行的内容
            public override UITableViewCell GetCell (UITableView tableView, NSIndexPath indexPath)
            {
                int rowIndex = indexPath.Row;
                UITableViewCell cell = new UITableViewCell();
                cell.TextLabel.Text = this.TableData[rowIndex];
                return cell;
            }
            //设置编辑风格
            public override UITableViewCellEditingStyle EditingStyleForRow (UITableView tableView, NSIndexPath indexPath)
            {
                return UITableViewCellEditingStyle.Insert;
            }
            //重写CommitEditingStyle()方法实现插入行的功能
            public override void CommitEditingStyle (UITableView tableView, UITableViewCellEditingStyle editingStyle, NSIndexPath indexPath)
            {
                i = i + 1;
                string str=string.Format("Inserted item: {0}", i);
                TableData.Insert (indexPath.Row, str);
                //插入行
                tableView.InsertRows(new NSIndexPath[] {
                    NSIndexPath.FromRowSection(indexPath.Row, 0)
                } , UITableViewRowAnimation.Automatic);
            }
        }
        ……
        #endregion
    }
}
```

运行效果如图 5.22 所示。

在此程序中需要注意以下两点：

（1）表的编辑状态的设置。在此程序中，使用了 SetEditing()方法，实现对表的编辑状态。

（2）设置表的编辑风格。

编辑风格的设置，需要重写 EditingStyleForRow()方法，其中编辑风格有 3 种，如表 5-9 所示。

第 5 章 显示数据

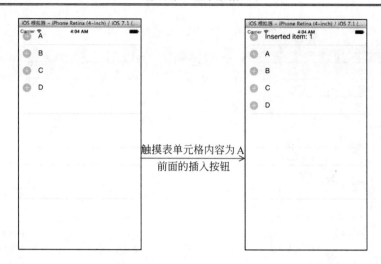

图 5.22 运行效果

表 5-9 编辑风格

风　　格	用　　途
Delete	进行删除
Insert	进行插入
None	无

注意：一般不对这个方法进行重写，默认的风格是 Delete。

2. 直接使用InsertRows()方法

如果不重写 CommitEditingStyle()方法，是否可以实现行的插入呢？答案是肯定的，开发者只需要使用 InsertRows()方法就可以实现。

【示例 5-16】 以下将只使用 InsertRows()方法实现表中行的插入。具体操作步骤如下所述。

（1）创建一个 Single View Application 类型的工程。

（2）打开 MainStoryboard.storyboard 文件，从工具栏中拖动 Navigation Controller 导航控制器对象到画布中。选择 Is Initial View Controller 复选框，删除与导航控制器关联的视图，将其改为原先在画布中的视图控制器的视图，即设置导航控制器的根视图控制器，具体步骤如下：

按住 Ctrl 键，同时拖动导航控制器对象到视图控制器的主视图上，如图 5.23 所示。松开鼠标后，弹出一个小的对话框，如图 5.24 所示。

选择其中的 Root 选项，此时就将导航控制器关联的主视图设置为原先视图控制器的主视图。此时的画布效果如图 5.25 所示。

（3）从工具栏中拖动 TableView 表视图对象到主视图中，将它的位置和大小设置为（0,0,320,568）；将它的 Name 设置为 tableView。

（4）打开 5-21ViewController.cs 文件，编写代码，实现为表视图对象添加内容，以及插入行的功能。代码如下：

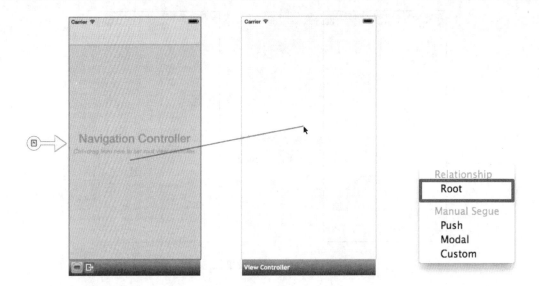

图 5.23　操作步骤 1　　　　　　　　图 5.24　操作步骤 2

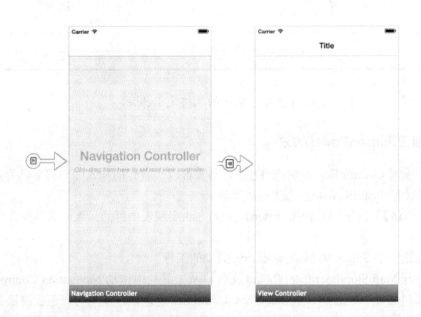

图 5.25　画布效果

```
using System;
using System.Drawing;
using MonoTouch.Foundation;
using MonoTouch.UIKit;
using System.Collections.Generic;
namespace Application
{
    public partial class __21ViewController : UIViewController
    {
        private List<string> myData;
        UIBarButtonItem btnAdd;
        ……
```

```csharp
#region View lifecycle
public override void ViewDidLoad ()
{
    base.ViewDidLoad ();
    myData = new List<string> () {"0","1","2","3"};
    tableView.Source = new MyTableSource (myData);
    //触摸"添加"按钮,实现行的插入
    this.btnAdd = new UIBarButtonItem(UIBarButtonSystemItem.Add,
         (s, e) => {
            MyTableSource tableSource = (MyTableSource)this.
            tableView.Source;
            int itemCount = tableSource.TableData.Count;
            tableSource.TableData.Add(string.Format("Inserted
            item: {0}", itemCount));
            this.tableView.InsertRows(new NSIndexPath[] {
                NSIndexPath.FromRowSection(itemCount, 0)
            } , UITableViewRowAnimation.Automatic);
        } );
    this.NavigationItem.SetRightBarButtonItem(this.btnAdd, false);
}
private class MyTableSource : UITableViewSource
{
    public MyTableSource(List<string>prems) {
        TableData = prems;
    }
    //设置 TableData
    public List<string> TableData {
        get;
        private set;
    }
    //设置表中的行
    public override int RowsInSection (UITableView tableview, int
    section)
    {
        return this.TableData.Count;
    }
    //设置每一行的内容
    public override UITableViewCell GetCell (UITableView tableView,
    NSIndexPath indexPath)
    {
        int rowIndex = indexPath.Row;
        UITableViewCell cell = new UITableViewCell();
        cell.TextLabel.Text = this.TableData[rowIndex];
        return cell;
    }
    ......
    #endregion
}
```

运行效果如图 5.26 所示。

5.3.4 移动行

行的移动需要重写 MoveRow()方法。即使在重写的方法中没有任何内容也可以实现行的移动。

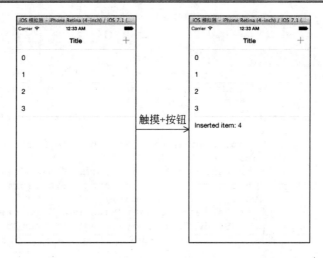

图 5.26 运行效果

【示例 5-17】 以下将重写 MoveRow()方法，从而实现行的移动，并在应用程序输出窗口输出需要移动的行的当前位置，以及移动后的当前位置。具体操作步骤如下所述。

（1）创建一个 Single View Application 类型的工程。

（2）打开 MainStoryboard.storyboard 文件，从工具栏中拖动 Table View 表视图对象到主视图中，将它的位置和大小设置为（0,0,320,568）；将它的 Name 设置为 tableView。

（3）打开 5-9ViewController.cs 文件，编写代码，实现为表视图对象添加内容，以及移动功能。代码如下：

```
using System;
using System.Drawing;
using MonoTouch.Foundation;
using MonoTouch.UIKit;
using System.Collections.Generic;
namespace Application
{
    public partial class __9ViewController : UIViewController
    {
        private List<string> myData;
        ……
        #region View lifecycle
        public override void ViewDidLoad ()
        {
            base.ViewDidLoad ();
            myData=new List<string>(){"1","2","3","4","5",};
            tableView.Source = new myViewSource (myData);
            tableView.SetEditing (true, true);
        }
        private class myViewSource : UITableViewSource {
            private List<string>dupPremTeams;
            public myViewSource(List<string>prems) {
                dupPremTeams = prems;
            }
            //设置表中的行数
            public override int RowsInSection(UITableView table,
                int section) {
                return dupPremTeams.Count;
            }
```

```
            //设置每一行的内容
            public override UITableViewCell GetCell(UITableView tableView,
        NSIndexPath index) {
                UITableViewCell theCell = new UITableViewCell();
                theCell.TextLabel.Text = dupPremTeams[index.Row];
                return theCell;
            }
            //设置编辑风格
            public override UITableViewCellEditingStyle EditingStyleForRow
        (UITableView tableView, NSIndexPath indexPath) {

                return UITableViewCellEditingStyle.None;
            }
            //行的移动
            public override void MoveRow (UITableView tableView,
        NSIndexPath sourceIndexPath, NSIndexPath destinationIndexPath)
        {
                Console.WriteLine ("拖动第{0}行", sourceIndexPath.Row);
                Console.WriteLine ("到第{0}行", destinationIndexPath.Row);
            }
        }
        ......
        #endregion
    }
}
```

运行效果如图 5.27 所示。

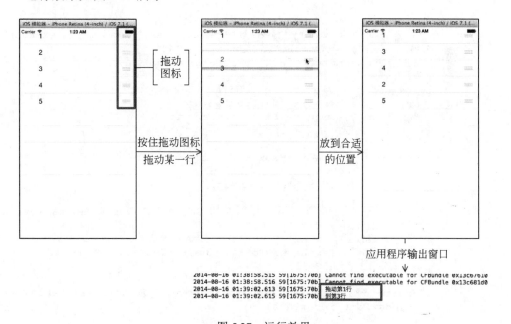

图 5.27　运行效果

5.3.5　缩进

为了让表看起来美观，还可以让表单元格进行缩进，要实现缩进的效果需要重写 IndentationLevel()方法。

【示例 5-18】 以下将实现表单元格中内容的缩进效果。具体的操作步骤如下所述。

(1)创建一个 Single View Application 类型的工程。

(2)打开 MainStoryboard.storyboard 文件,从工具栏中拖动 Table View 表视图对象到主视图中,将它的位置和大小设置为(0,0,320,568);将它的 Name 设置为 tableView。

(3)打开 5-10ViewController.cs 文件,编写代码,实现为表视图对象添加内容,以及缩进功能。代码如下:

```
using System;
using System.Drawing;
using MonoTouch.Foundation;
using MonoTouch.UIKit;
using System.Collections.Generic;
namespace Application
{
    public partial class __10ViewController : UIViewController
    {
        private List<string> myData;
        ……                          //这里省略了视图控制器的构造方法和析构方法
        #region View lifecycle
        public override void ViewDidLoad ()
        {
            base.ViewDidLoad ();


            // Perform any additional setup after loading the view, typically
            from a nib.
            myData = new List<string>(){"Liverpool","Everton",
            "Arsenal","Swansea","Cardiff","Newcastle","Abraham","Alexander",
            "Antoine","Armstrong","Bartholomew"};
            tableView.Source = new myViewSource(myData);
        }
        private class myViewSource : UITableViewSource {
            private List<string>dupPremTeams;
            public myViewSource(List<string>prems) {
                dupPremTeams = prems;
            }
            public override int RowsInSection(UITableView table,
                int section) {
                return dupPremTeams.Count;
            }
            public override UITableViewCell GetCell(UITableView tableView,
            NSIndexPath index) {
                UITableViewCell theCell = new UITableViewCell();
                theCell.TextLabel.Text = dupPremTeams[index.Row];
                return theCell;
            }
            public override int IndentationLevel (UITableView tableView,
            NSIndexPath indexPath)
            {
                return indexPath.Row%3;
            }
        }
        ……                          //这里省略了视图加载和卸载前后的一些方法
        #endregion
    }
}
```

运行效果如图 5.28 所示。

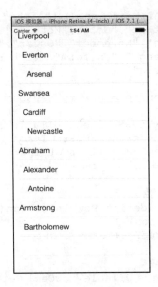

图 5.28　运行效果

⚠ 注意：在此程序中表单元格的缩进是在 0 和 2 之间变换的。

5.4　索　引　表

使用索引可快速浏览 UITableView 中的行。索引是对表中一列或多列的值进行排序的一种结构。要实现索引的添加，需要重写 SectionIndexTitles()方法。

【示例 5-19】　以下将通过对方法 SectionIndexTitles()的重写，为表添加索引。具体的操作步骤如下所述。

（1）创建一个 Single View Application 类型的工程。

（2）打开 MainStoryboard.storyboard 文件，从工具栏中拖动 Table View 表视图对象到主视图中，将它的位置和大小设置为（0,0,320,568）；将它的 Name 设置为 tableView。

（3）打开 5-11ViewController.cs 文件，编写代码，实现为表视图对象添加内容，以及索引。代码如下：

```
using System;
using System.Drawing;
using MonoTouch.Foundation;
using MonoTouch.UIKit;
using System.Collections.Generic;
using System.Linq;
namespace Application
{
    public partial class __11ViewController : UIViewController
    {
        private List<string>myData;
        ……                         //这里省略了视图控制器的构造方法和析构方法
        #region View lifecycle
```

```
public override void ViewDidLoad ()
{
    myData = new List<string>() {"Alpha","Bravo","Charlie","Delta",
    "Echo","Foxtrot","Golf","Hotel","India","Juliet","Kilo",
    "Lima","Mike","November","Oscar","Papa","Quebec","Romeo",
    "Sierra","Tango","Uniform","Victor","Whiskey","X-ray",
    "Yankee","Zulu"
    } ;
    tableView.Source = new myViewSource (myData);
}
private class myViewSource : UITableViewSource {
    private List<string>dupPremTeams;
    public myViewSource(List<string>prems) {
        dupPremTeams = prems;
    }
    //设置表中的行数
    public override int RowsInSection(UITableView table,
        int section) {
        return dupPremTeams.Count;
    }
    //每一行的内容
    public override UITableViewCell GetCell(UITableView tableView,
    NSIndexPath index) {
        UITableViewCell theCell = new UITableViewCell();
        theCell.TextLabel.Text = dupPremTeams[index.Row];
        return theCell;
    }
    //添加索引
    public override string[] SectionIndexTitles (UITableView
    tableView)
    {
        return dupPremTeams.Select(s => Convert.ToString(s[0])).
        Distinct().ToArray();
    }
}
……                      //这里省略了视图加载和卸载前后的一些方法
#endregion
}
```

运行效果如图 5.29 所示。

图 5.29　运行效果

5.5 数据的查找

表往往存放了大量的数据，如果用户要查找某一个数据时，一个一个找，既费时又费力。Xamarin.iOS 中提供了 Search Bar with Search Display Controller 对象来解决在表中查找数据的烦恼。Search Bar with Search Display Controller 对象提供了一个搜索框，当用户在此搜索框中输入字符或者字符串后，与此字符或者字符串匹配的内容将会显示在表视图中。

【示例 5-20】以下将使用 Search Bar with Search Display Controller 对象来实现在表中查找数据的功能。具体的操作步骤如下所述。

（1）创建一个 Single View Application 类型的工程。
（2）打开 MainStoryboard.storyboard 文件，对主视图进行设置，效果如图 5.30 所示。

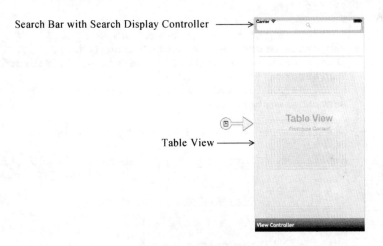

图 5.30 主视图的效果

⚠注意：要将 Search Bar with Search Display Controller 对象拖动到表本身之上，而不是主视图的其他地方。

需要添加的视图以及设置，如表 5-10 所示。

表 5-10 设置视图

视　　图	设　　置
Search Bar with Search Display Controller	
Table View	Name：tableView 位置和大小：(0, 0, 320, 568)

（3）打开 5-12ViewController.cs 文件，编写代码，实现为表视图对象添加内容，以及查找数据的功能。代码如下：

```
using System;
using System.Drawing;
using MonoTouch.Foundation;
using MonoTouch.UIKit;
```

```
using System.Collections.Generic;
using System.Linq;
namespace Application
{
    public partial class __12ViewController : UIViewController
    {
        private List<string> tableData;
        private List<string> filterDataList;
        ……
        #region View lifecycle
        public override void ViewDidLoad ()
        {
            base.ViewDidLoad ();

            tableData = new List<string>() {"Alpha","Bravo","Charlie",
            "Delta","Echo","Foxtrot","Golf","Hotel","India","Juliet",
            "Kilo","Lima","Mike","November","Oscar","Papa","Quebec",
            "Romeo","Sierra","Tango","Uniform","Victor","Whiskey",
            "X-ray","Yankee","Zulu"
            } ;
            filterDataList = new List<string>();
            tableView.Source = new TableSource (this);
            this.searchDisplayController.SearchResultsSource= new
            TableSource (this);
            this.searchDisplayController.Delegate = new
            SearchDelegate(this);
        }
        private class TableSource : UITableViewSource
        {
            public TableSource (__12ViewController controller)
            {
                this.cellID = "cellIdentifier";
                this.parentController = controller;
            }
            private string cellID;
            private __12ViewController parentController;
            //设置行数
            public override int RowsInSection (UITableView tableview, int
            section)
            {
                if (tableview.Equals (this.parentController.tableView))
                {
                    return this.parentController.tableData.Count;
                } else
                {
                    return this.parentController.filterDataList.Count;
                }
            }
            //设置每一行的内容
            public override UITableViewCell GetCell (UITableView tableView,
            NSIndexPath indexPath)
            {
                int rowIndex = indexPath.Row;
                UITableViewCell cell = tableView.DequeueReusableCell
                (this.cellID);
                if (null == cell)
                {
                    cell = new UITableViewCell (UITableViewCellStyle.
                    Default, this.cellID);
```

```csharp
        }
        if (tableView.Equals (this.parentController.tableView))
        {
            cell.TextLabel.Text = this.parentController.
            tableData[rowIndex];
        } else
        {
            cell.TextLabel.Text = this.parentController.
            filterDataList[rowIndex];
        }
        return cell;
    }
}
private class SearchDelegate : UISearchDisplayDelegate
{
    public SearchDelegate (__12ViewController controller)
    {
        this.parentController = controller;
    }
    private __12ViewController parentController;
    //显示与搜索条件匹配的结果
    public override bool ShouldReloadForSearchString
    (UISearchDisplayController controller, string forSearchString)
    {
        //筛选数据
        this.parentController.filterDataList = this.parentController.
        tableData
            .Where (s => s.ToLower ().Contains (forSearchString.
            ToLower ()))
            .ToList ();
        //将结果按字母进行排序
        this.parentController.filterDataList.Sort  (delegate(string
        firstStr, string secondStr) {
            return firstStr.CompareTo (secondStr);
        } );
        return true;
    }
}
……
#endregion
}
```

运行效果如图 5.31 所示。

在此程序中 Search Bar with Search Display Controller 对象，一般包含两个部分：一个是 UISearchBar 类实现的搜索栏，另一个是 UISearchDisplayController 类实现的搜索显示控制器（它本身又包含 UISearchBar 和 UITableView）。要实现数据的查找，必须要实现以下3个步骤。

（1）一个用于存放与搜索条件相匹配的结果的集合。这个集合会在每一次搜索条件时进行更新。在此程序中使用的是 filterDataList 字符串列表。

（2）创建一个 UITableViewSource 类的子类（在此程序中是 TableSource 类），搜索显示控制器创建的 UITableView 会用到这个子类。这个类会负责显示搜索到的结果行。

（3）创建一个 UISearchDisplayDelegate 类的子类（在此程序中是 SearchDelegate 类），它会对来自搜索显示控制器的事件进行响应。必须使用这个类上的方法来响应搜索事件，例如更新搜索结果集合，所有的 UITableView 表能够显示这些搜索的结果。

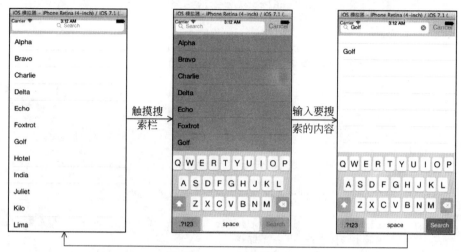

图 5.31 运行效果

> 注意：在此程序中搜索栏的外观和行为也可以改变，常用的属性如表 5-11 所示。

表 5-11 常用的属性

属 性	功 能
Placeholder	当搜索框为空时，会出现占位符，即出现"水印"一样的灰色文本
Prompt	当搜索框处于使用状态时，搜索框上方会出现说明文本
Text	允许预先将搜索文本设定为某一个特定的值
KeyboardType	用来指定键盘类型
AutocorrectionType	搜索输入文本是否应该自动更正
AutocapitalizationType	搜索输入文本是否应该自动大写
BarStyle	搜索栏的颜色

【示例 5-21】 以下以【示例 5-20】为基础，对搜索框进行设置。代码如下：

```
public override void ViewDidLoad ()
{
    base.ViewDidLoad ();
    ……//这里省略了为集合添加数据，设置委托等内容
    searchDisplayController.SearchBar.Placeholder="输入要搜索的内容";     //设置占位符
    //设置键盘类型
    searchDisplayController.SearchBar.KeyboardType = UIKeyboardType.NamePhonePad;
    //搜索输入文本是否应该自动更正
    searchDisplayController.SearchBar.AutocorrectionType = UITextAutocorrectionType.Yes;
}
```

运行效果如图 5.32 所示。

图 5.32 运行效果

5.6 创建简单的网页浏览器

网页浏览器想必对于每一个人来说都不陌生。在 Windows 上有 IE 网页浏览器,在 Mac 机上或者 iPhone 和 iPad 上有 Safari 网页浏览器,用户可以通过这些浏览器获取网页上的信息。网页浏览器的实现主要是通过网页视图来实现的。网页视图对应的类是 UIWebView 类。本节将讲解对网页视图的一些操作。

5.6.1 加载网页视图的内容

在一个网页浏览器中,最基本的功能就是要实现对网页视图的加载。以下将讲解 3 种加载方式。

1. 加载网址中的内容

LoadRequest()方法实现为网页视图加载网址内容的功能。使用 LoadRequest()需要完成 3 个步骤:给出网址、传递网址和加载网址。以下就是这 3 个步骤的具体介绍。

(1) 给出网址。要加载指定网址的网页内容,必须实例化一个 NSUrl 对象,并初始化一个网址,其语法形式如下:

```
NSUrl 对象名= new NSUrl (网址);
```

注意:网址必须是完整的。

(2) 传递网址。给出网址后,需要将网址传递给 NSUrlRequest 对象,其语法形式如下:

```
NSUrlRequest 对象名 = new NSUrlRequest (NSUrl 对象);
```

（3）加载网址。将网址传递到 NSUrlRequest 对象中后，就可以使用加载方法 loadRequest()将 NSUrlRequest 对象进行加载了，其语法形式如下：

网页视图对象.LoadRequest(NSUrlRequest 对象);

【示例 5-22】以下将实现在网页视图中加载网址为"http://www.baidu.com"的网页内容。具体操作步骤如下所述。

（1）创建一个 Single View Application 类型的工程。

（2）打开 MainStoryboard.storyboard 文件，从工具栏中拖动 Web View 网页视图对象到主视图中，将它的位置和大小设置为（0,0,320,568）；将它的 Name 设置为 webView。

（3）打开 5-13ViewController.cs 文件，编写代码，实现一个简单的网页浏览器。代码如下：

```
using System;
using System.Drawing;
using MonoTouch.Foundation;
using MonoTouch.UIKit;
namespace Application
{
    public partial class __13ViewController : UIViewController
    {
        ……
        #region View lifecycle
        public override void ViewDidLoad ()
        {
            base.ViewDidLoad ();
            NSUrl url = new NSUrl ("http://www.baidu.com");
            NSUrlRequest urlRequest = new NSUrlRequest (url);
            this.webView.LoadRequest (urlRequest);         //加载内容
        }
        ……
        #endregion
    }
}
```

运行效果如图 5.33 所示。

2. 加载HTML代码

如果通过网址进行加载的每一个网页都不是开发者想看到的。那么，开发者可以使用 HTML 代码自己编写内容，然后通过使用 LoadHtmlString()方法，将自己编写的 HTML 代码加载到网页视图中。

【示例 5-23】以下将使用 LoadHtmlString()方法在网页视图中加载 HTML 代码。具体操作步骤如下所述。

（1）创建一个 Single View Application 类型的工程，命名为 5-14。

（2）打开 MainStoryboard.storyboard 文件，从工具栏中拖动 Web View 网页视图对象到主视图中，将它的位置和大小设置为（0,0,320,568）；将它的 Name 设置为 webView。

（3）打开 5-14ViewController.cs 文件，编写代码，实现一个简单的网页浏览器。代码如下：

```
using System;
using System.Drawing;
```

第 5 章　显示数据

```
using MonoTouch.Foundation;
using MonoTouch.UIKit;
namespace Application
{
    public partial class __14ViewController : UIViewController
    {
        ……                              //这里省略了视图控制器的构造方法和析构方法
        #region View lifecycle
        public override void ViewDidLoad ()
        {
            base.ViewDidLoad ();
            string html = "This is <font color=blue><font size=35>HTML";
            webView.LoadHtmlString (html, null);
        }
        ……                              //这里省略了视图加载和卸载前后的一些方法
        #endregion
    }
}
```

运行效果如图 5.34 所示。

图 5.33　运行效果

图 5.34　运行效果

3．加载本地文件

除了可以加载网址和 HTML 代码外，在网页视图中还可以使用 LoadData()方法加载一些本地文件，这些文本如表 5-12 所示。

表 5-12　加载的文本

文　　件	扩　展　名
Excel	.xls
Keynote	.key.zip
Numbers	.numbers.zip
Pages	.pages.zip

· 197 ·

续表

文　件	扩　展　名
PDF	.pdf
Powerpoint	.ppt
Word	.doc
Rich Text Format	.rtf
Rich Text Format Directory	.rtfd.zip
Keynote	.key
Numbers	.numbers
Pages	.pages

【示例 5-24】 以下将使用 LoadData()方法为网页视图加载一个 PDF 文件。具体的操作步骤如下所述。

（1）创建一个 Single View Application 类型的工程。

（2）添加 PDF 文件 "VM 安装 MAC10.9 苹果虚拟机.pdf" 到创建工程的 Resources 文件夹中。

（3）打开 MainStoryboard.storyboard 文件，从工具栏中拖动 Web View 网页视图对象到主视图中，将它的位置和大小设置为（0,0,320,568）；将它的 Name 设置为 webView。

（4）打开 5-14ViewController.cs 文件，编写代码，实现一个简单的网页浏览器。代码如下：

```
using System;
using System.Drawing;
using MonoTouch.Foundation;
using MonoTouch.UIKit;
namespace Application
{
    public partial class __22ViewController : UIViewController
    {
        ......
        #region View lifecycle

        public override void ViewDidLoad ()
        {
            base.ViewDidLoad ();

            // Perform any additional setup after loading the view, typically
            from a nib.
            string path = NSBundle.MainBundle.PathForResource ("VM 安装
            MAC10.9 苹果虚拟机","pdf");
            NSUrl url = NSUrl.FromFilename ("VM 安装 MAC10.9 苹果虚拟机.pdf");
            NSData data = NSData.FromFile (path);
            webView.LoadData (data, "application/pdf", "UTF-8", url);
            //加载 PDF 文件
        }
        ......
        #endregion
    }
}
```

运行效果如图 5.35 所示。

第 5 章 显示数据

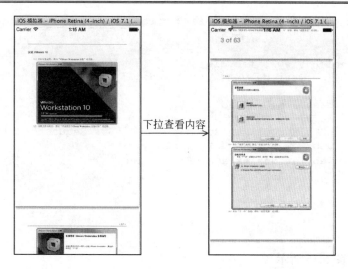

图 5.35 运行效果

5.6.2 设置网页视图

使用网页视图加载的内容，显示出来都是一个风格。当网页上的内容很多时，手机屏幕就变成了滚动的。那么是否可以将网页上所有的内容一次性都显示在手机屏幕上，或者是否可以将网页的内容进行辨别呢。以下就来对这两个问题进行解决。

1. 设置网页视图的自动缩放功能

自动缩放，就是手机的网页中加载的内容相当多时，网页视图不会变为滚动的形式，而是自动将网页中的内容进行合理的缩小，使其在固定的网页视图中显示。此功能的实现需要使用 ScalesPageToFit 属性，其语法形式如下：

网页视图对象.ScalesPageToFit=Bool 类型的值；

其中，当 Bool 值设置为 true 时，实现自动缩放的功能，当 Bool 值为 false 时，不会实现自动缩放。

【示例 5-25】以下将网页视图加载的内容通过 ScalesPageToFit 属性实现自动缩放。具体的操作步骤如下所述。

（1）创建一个 Single View Application 类型的工程。

（2）打开 MainStoryboard.storyboard 文件，从工具栏中拖动 Web View 网页视图对象到主视图中，将它的位置和大小设置为（0,0,320,568）；将它的 Name 设置为 webView。

（3）打开 5-15ViewController.cs 文件，编写代码，实现一个网页视图的自动缩放。代码如下：

```
using System;
using System.Drawing;
using MonoTouch.Foundation;
using MonoTouch.UIKit;
namespace Application
{
```

```
public partial class __15ViewController : UIViewController
{
    ……
    #region View lifecycle
    public override void ViewDidLoad ()
    {
        base.ViewDidLoad ();
        NSUrl url = new NSUrl ("http://www.apple.com");
        NSUrlRequest urlRequest = new NSUrlRequest (url);
        webView.LoadRequest (urlRequest);
        webView.ScalesPageToFit = true;                    //实现自动缩放
    }
    ……
    #endregion
}
```

运行效果如图 5.36 所示。

没有设置此属性值为 false   设置此属性值为 true

图 5.36　运行效果

△注意：如果此网页中的内容是手机版的，那么它的自动缩放就没有任何效果了。

2. 设置网页视图的自动识别功能

使用网页视图加载的网页，可以看到一些特殊信息，如电话号码、邮箱和网址等。这些特殊内容在长按后，除了电话号码会出现一个对话框之外，其他的都不会有反应。这就是网页视图自动识别网页内容的能力。要想实现其他能力的识别，就需要对 DataDetectorTypes 属性进行设置，其语法形式如下：

网页视图对象.DataDetectorTypes = 识别的内容；

其中，识别的内容如表 5-13 所示。

表 5-13 自动识别网页中的内容

内容	功能
Address	识别电子邮件
All	识别网页中的所有内容
CalendarEvent	识别日期
link	识别网址
None	不识别网页中的任何内容
PhoneNumber	识别电话号码

【示例 5-26】以下将实现自动识别网页视图中内容的功能。具体步骤如下：

(1) 创建一个 Single View Application 类型的工程。

(2) 打开 MainStoryboard.storyboard 文件，从工具栏中拖动 Web View 网页视图对象到主视图中，将它的位置和大小设置为（0,0,320,568）；将它的 Name 设置为 webView。

(3) 打开 5-16ViewController.cs 文件，编写代码，实现网页视图中内容的识别。代码如下：

```
using System;
using System.Drawing;
using MonoTouch.Foundation;
using MonoTouch.UIKit;
namespace Application
{
    public partial class __16ViewController : UIViewController
    {
        ……                      //这里省略了视图控制器的构造方法和析构方法
        #region View lifecycle
        public override void ViewDidLoad ()
        {
            base.ViewDidLoad ();
            string phone="电话:10086<br>邮箱:163@qq.com<br>网
            址:http://www.baidu.com";
            webView.LoadHtmlString (phone, null);
            webView.DataDetectorTypes = UIDataDetectorType.All;
                    //识别所有内容
        }
        ……            //这里省略了视图加载和卸载前后的一些方法
        #endregion
    }
}
```

运行效果如图 5.37 所示。

5.6.3 网页视图常用事件

UIWebView 类中有一些非常有用的事件，并且这些事件在制作网页浏览器时常常用到。这些事件如表 5-14 所示。

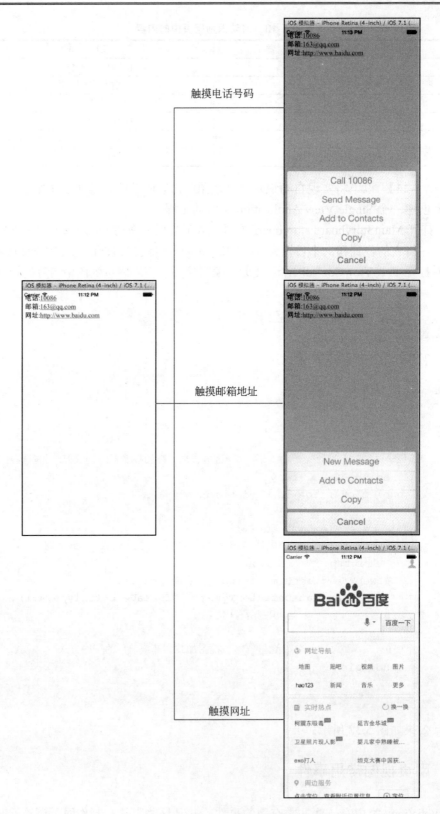

图 5.37 运行效果

表 5-14　网页视图常用事件

事　　件	功　　能
LoadStarted	在控件开始加载内容时触发
LoadFinished	在控件加载内容结束后触发
LoadFailed	在控件加载内容失败后触发
LoadError	在控件加载内容失败后触发

【示例 5-27】以下将使用这些常用事件制作一个网页浏览器。具体操作步骤如下所述。

（1）创建一个 Single View Application 类型的工程。

（2）打开 MainStoryboard.storyboard 文件，对主视图进行设置，效果如图 5.38 所示。

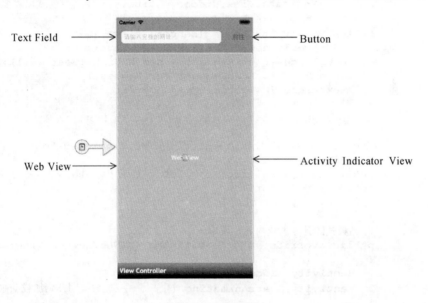

图 5.38　主视图

需要添加的视图以及设置，如表 5-15 所示。

表 5-15　设置视图

视　　图	设　　置
Text Field	Name：textField Text：（空） Placeholder：请输入完整的网址 位置和大小：(10, 30, 237, 30)
Button	Name：button Title：前往 位置和大小：(264, 30, 46, 30)
Web View	Name：webView 位置和大小：(0, 81, 320, 488)
Activity Indicator View	Name：activityIndicatorView 位置和大小：(142, 306, 37, 37)

（3）打开 5-17ViewController.cs 文件，编写代码，实现一个简单的网页浏览器。代码如下：

```csharp
using System;
using System.Drawing;
using MonoTouch.Foundation;
using MonoTouch.UIKit;
namespace Application
{
    public partial class __17ViewController : UIViewController
    {
        ……                    //这里省略了视图控制器的构造方法和析构方法
        #region View lifecycle
        public override void ViewDidLoad ()
        {
            base.ViewDidLoad ();
            activityIndicatorView.Hidden = true;         //隐藏加载视图
            //触摸按钮后触发的事件
            button.TouchUpInside += (sender, e) => {
                NSUrl url= new NSUrl(textField.Text);
                NSUrlRequest urlRequest = new NSUrlRequest (url);
                this.webView.LoadRequest (urlRequest);
                textField.ResignFirstResponder();
            };
            webView.Delegate = new MyWebViewDelegate (activityIndicatorView);
        }
        private class MyWebViewDelegate : UIWebViewDelegate {
            UIActivityIndicatorView activity;
            public MyWebViewDelegate(UIActivityIndicatorView
            activityIndicatorView){
                activity=activityIndicatorView;
            }
            //网页视图开始加载内容时调用
            public override void LoadStarted (UIWebView webView)
            {
                activity.Hidden = false;
                activity.StartAnimating ();      //加载视图开始加载动画的实现
            }
            //网页视图加载内容结束后调用
            public override void LoadingFinished (UIWebView webView)
            {
                activity.Hidden = true;
                activity.StopAnimating ();       //加载视图结束加载动画的实现
            }
            //网页视图加载内容失败后调用
            public override void LoadFailed (UIWebView webView, NSError
            error)
            {
                activity.Hidden = true;
                activity.StopAnimating ();
                UIAlertView a = new UIAlertView ();
                a.Title="内容加载出错";
                a.AddButton("Cancel");
                a.Show ();
            }
        }
        ……                        //这里省略了视图加载和卸载前后的一些方法
        #endregion
    }
}
```

运行效果如图 5.39 所示。

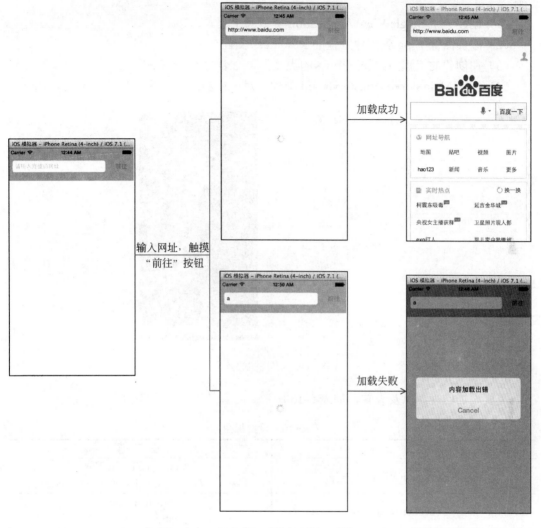

图 5.39　运行效果

5.7　在网格中显示数据

UICollectionView 类提供了一个集合视图,它可以使数据显示在类似网格的布局中。UICollectionView 类的引入是在 iOS 6 中,对于现在的 Xamarin.iOS 也提供了此类来实现数据的显示。本节将讲解如何在集合视图中显示数据。

5.7.1　网格中内容的显示

在网格中该如何实现内容的显示呢?以下将针对这个问题进行介绍。

【示例 5-28】 以下将实现在集合视图中显示图像 1.jpg 的功能。具体的操作步骤如下所述。

（1）创建一个 Single View Application 类型的工程。
（2）添加图像 1.jpg 到创建工程的 Resources 文件夹中。
（3）在创建的工程中添加一个 C#的类文件，命名为 ImageCell。
（4）打开 MainStoryboard.storyboard 文件，对主视图进行设置，效果如图 5.40 所示。

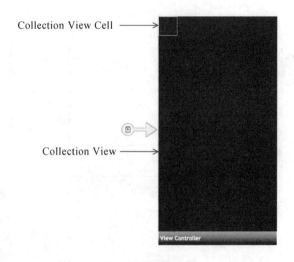

图 5.40　主视图的效果

需要添加的视图以及设置，如表 5-16 所示。

表 5-16　设置视图

视　　图	设　　置
Collection View	Name：collectionView 位置和大小：(0, 0, 320, 568)
Collection View Cell	Class：ImageCell Identifier：ImageCell

注意：选择集合视图对象中的任意一个网格（集合单元格），就可以实现对 ollection View Cell 的设置。

（5）打开 ImageCell.cs 文件，编写代码，实现对集合视图中网格的设置。代码如下：

```
using System;
using MonoTouch.Foundation;
using MonoTouch.UIKit;
using System.CodeDom.Compiler;
namespace Application
{
    partial class ImageCell : UICollectionViewCell
    {
        public const string CELLID = "ImageCell";
        //网格的构造器方法
        public ImageCell (IntPtr handle) : base (handle)
        {
```

第5章 显示数据

```
            this.Initialize();              //调用初始化方法
    }
    //设置属性
    public UIImageView ImageView
    {
        get;
        private set;
    }
    //初始化
    private void Initialize()
    {
        this.ImageView = new UIImageView(this.ContentView.Bounds);
        //实例化对象
        this.ImageView.ContentMode = UIViewContentMode.ScaleAspectFit;
        //设置显示模式
        this.ContentView.AddSubview(this.ImageView);
        //添加
    }
}
```

（6）打开 5-18ViewController.cs 文件，编写代码，实现图像的显示。代码如下：

```
using System;
using System.Drawing;
using MonoTouch.Foundation;
using MonoTouch.UIKit;
using System.Collections.Generic;
namespace Application
{
    public partial class __18ViewController : UIViewController
    {
        private List<UIImage> collectionData;
        ……                         //这里省略了视图控制器的构造方法和析构方法
        #region View lifecycle
        public override void ViewDidLoad ()
        {
            base.ViewDidLoad ();
            collectionData = new List<UIImage>();
            //遍历
            for (int i = 0; i < 50; i++)
            {
                collectionData.Add (UIImage.FromFile ("1.jpg"));//添加图像
            }
            collectionView.Source = new CollectionSource(this);
        }
        private class CollectionSource : UICollectionViewSource
        {
            public CollectionSource (__18ViewController parentController)
            {
                this.parentController = parentController;
            }
            private __18ViewController parentController;
            //设置网格的个数
            public override int GetItemsCount (UICollectionView collectionView, int section)
            {
                return this.parentController.collectionData.Count;
            }
            //设置每个网格中的内容
```

· 207 ·

```
public override UICollectionViewCell GetCell (UICollectionView
collectionView, NSIndexPath indexPath)
{
    int rowIndex = indexPath.Row;
    ImageCell cell = (ImageCell)collectionView.
DequeueReusableCell((NSString)ImageCell.CELLID, indexPath);
    cell.ImageView.Image = this.parentController.
collectionData[rowIndex];
    return cell;
}
......                          //这里省略了视图加载和卸载前后的一些方法
#endregion
```

运行效果如图 5.41 所示。

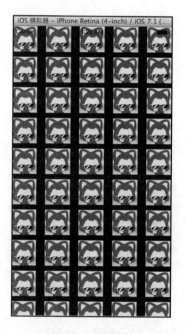

图 5.41　运行效果

UICollectionView 类类似于 UITableView 类，不同的是，它们所实现的布局不同。UITableView 实现的布局是表，UICollectionView 实现的布局是网格。UICollectionView 类中对于数据的显示需要在 UICollectionViewSource 类的子类中重写 GetItemsCount()方法和 GetCell()方法。

注意：CollectionViewCell 类不像 UITableViewCell 类，它为我们提供很多的方法以及属性，我们需要重写这些属性和方法来实现网格的创建。

5.7.2　自定义网格

如果开发者不喜欢集合视图中默认的网格，还可以对网格进行一些设置，形成自定义的网格，这样可以让开发的应用程序更加独特。

【示例 5-29】 以下将集合视图中的网格变为自定义的网格。具体的操作步骤如下所述。

（1）创建一个 Single View Application 类型的工程。

（2）添加图像 1.jpg 到创建工程的 Resources 文件夹中。

（3）在创建的工程中添加一个 C#的类文件，命名为 ImageCell。

（4）打开 MainStoryboard.storyboard 文件，对主视图进行设置，效果如图 5.42 所示。

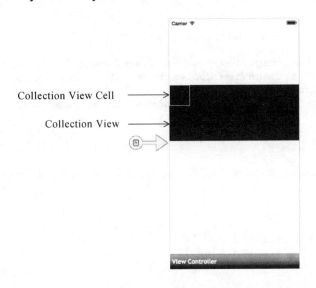

图 5.42 主视图的效果

需要添加的视图以及设置，如表 5-17 所示。

表 5-17 设置视图

视 图	设 置
Collection View	Name：collectionView 位置和大小：(0, 160, 320, 135)
Collection View Cell	Class：ImageCell Identifier：ImageCell

注意：选择集合视图对象中的任意一个集合单元，就可以实现对 ollection View Cell 的设置。

（5）打开 ImageCell.cs 文件，编写代码，实现对集合视图中网格的设置。代码如下：

```
using System;
using MonoTouch.Foundation;
using MonoTouch.UIKit;
using System.CodeDom.Compiler;
namespace Application
{
    partial class ImageCell : UICollectionViewCell
    {
        public const string CELLID = "ImageCell";
        //网格的构造器方法
```

```
public ImageCell (IntPtr handle) : base (handle)
{
    this.Initialize();
}
//属性设置
public UIImageView ImageView
{
    get;
    private set;
}
//初始化
private void Initialize()
{
    this.ImageView = new UIImageView(this.ContentView.Bounds);
    this.ImageView.ContentMode = UIViewContentMode.ScaleAspectFit;
    this.ContentView.AddSubview(this.ImageView);
}
}
}
```

（6）打开 5-19ViewController.cs 文件，编写代码，实现自定义的网格。代码如下：

```
using System;
using System.Drawing;
using MonoTouch.Foundation;
using MonoTouch.UIKit;
using System.Collections.Generic;
namespace Application
{
    public partial class __19ViewController : UIViewController
    {
        private List<UIImage> collectionData;
        ……
        #region View lifecycle
        public override void ViewDidLoad ()
        {
            base.ViewDidLoad ();
            this.collectionData = new List<UIImage>();
            //添加图像
            for (int i = 0; i < 30; i++)
            {
                collectionData.Add(UIImage.FromFile("1.jpg"));
            }
            collectionView.Source = new CollectionSource(this);
            UICollectionViewFlowLayout flowLayout = new
            UICollectionViewFlowLayout();
            flowLayout.MinimumLineSpacing = 30f;              //设置行距
            flowLayout.MinimumInteritemSpacing = 10f;         //设置列距
            flowLayout.SectionInset = new UIEdgeInsets(4f, 4f, 4f, 4f);
                                                              //设置边界
            flowLayout.ItemSize = new SizeF(20f, 20f);        //设置网格的大小
            this.collectionView.CollectionViewLayout = flowLayout;
        }
        private class CollectionSource : UICollectionViewSource
        {
            public CollectionSource (__19ViewController parentController)
            {
                this.parentController = parentController;
            }
```

第 5 章 显示数据

```
        private __19ViewController parentController;
        //设置网格个数
        public override int GetItemsCount (UICollectionView
        collectionView, int section)
        {
            return this.parentController.collectionData.Count;
        }
        //设置网格中的内容
        public override UICollectionViewCell GetCell (UICollectionView
        collectionView, NSIndexPath indexPath){
            int rowIndex = indexPath.Row;
            ImageCell cell = (ImageCell)collectionView.
        DequeueReusableCell((NSString)ImageCell.CELLID, indexPath);
            cell.ImageView.Image = this.parentController.
        collectionData[rowIndex];
            return cell;
        }
    }
    ……
    #endregion
}
```

运行效果如图 5.43 所示。

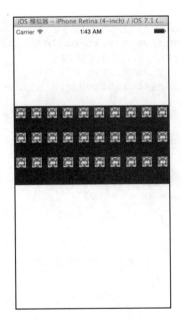

图 5.43 运行效果

在程序中使用到了几个属性对网格进行设置，这些属性的功能如表 5-18 所示。

表 5-18 常用属性

属　　性	功　　能
MinimumLineSpacing	网格的行之间的最小距离
MinimumInteritemSpacing	网格的列之间的最小距离
SectionInset	设置边界
ItemSize	集合视图中网格的尺寸

这些属性对应的设置如图 5.44 所示。

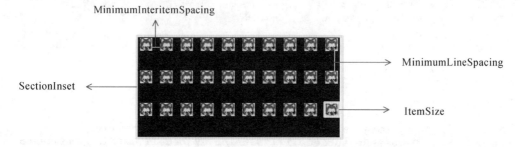

图 5.44　属性对应的设置

注意：在图 5.43 所示的运行效果中，可以看到 ItemSize 属性在设置之后，改变了所有网格的大小，那么如果想要让集合视图中某一些网格大小有所改变，该如何实现呢？以下将为开发者实现这一功能。

【示例 5-30】 以下将以【示例 5-29】为基础，实现将某一些网格变大的功能。代码如下：

```
//导出 Objective-C 中的 collectionView:layout:sizeForItemAtIndexPath:
[Export("collectionView:layout:sizeForItemAtIndexPath:")]
//要导出的方法实现 GetSizeForItem 方法中的内容
public     SizeF     GetSizeForItem(UICollectionView     collectionView,
UICollectionViewLayout layout, NSIndexPath indexPath) {
    //判断
    if (indexPath.Item > 10 && indexPath.Item < 16)
    {
        return new SizeF(40f, 40f);
    } else
    {
        return new SizeF(20f, 20f);
    }
}
```

运行效果如图 5.45 所示。

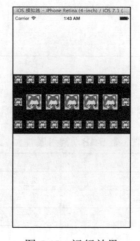

图 5.45　运行效果

5.7.3 网格的响应

集合视图中的网格也是可以进行响应的,需要重写 ItemSelected()方法。

【示例 5-31】 以下以【示例 5-28】为基础,实现网格的响应。代码如下:

```
public override void ItemSelected (UICollectionView collectionView,
NSIndexPath indexPath)
{
    int i = indexPath.Row;
    UIAlertView alartView = new UIAlertView ();
    alartView.Title = string.Format ("你选择的是{ 0}个网格", i+1);
    alartView.AddButton ("Cancel");
    alartView.Show ();
}
```

运行效果如图 5.46 所示。

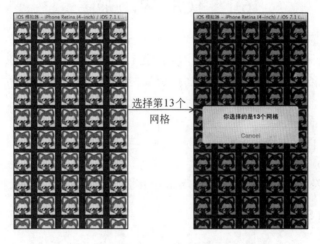

图 5.46 运行效果

第6章 网络服务

网络服务是指一些在网络上运行的、面向服务的、基于分布式程序的软件模块。网络服务采用 HTTP 和 XML（标准通用标记语言的子集）等互联网通用标准，使人们可以在不同的地方通过不同的终端设备访问 Web 服务器上的数据，如网上订票、查看订座情况。本章将讲解如何使用常用的网络服务。

6.1 使用 Web 服务

Web 服务是一种服务导向架构的技术，通过标准的 Web 协议提供服务，目的是保证不同平台的应用服务可以互操作。本节将讲解如何构建一个 Web 服务，以及 Web 服务的使用。

6.1.1 构建一个 Web 服务

要构建一个 Web 服务，需要实现以下 3 个步骤。

1. 使用Visual Studio创建Web服务

（1）打开 Visual Studio，选择"新建项目"命令，弹出"新建项目"对话框，如图 6.1 所示。

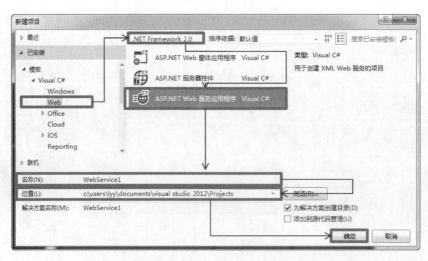

图 6.1 操作步骤 1

（2）选择 Web 模版，将基于的.Net 框架设置为.NET Framework 2.0，选择 ASP.NET Web 服务应用程序。输入名称，并选择位置后，单击"确定"按钮。这样就创建好一个项目名为 WebService1 的项目了。

> 注意：名称以及位置都有默认的内容。

2. 发布Web服务

（1）选择菜单栏中的"生成"|"发布选定的内容"命令，或者右击 Visual Studio 右边的项目名，在弹出的下拉菜单中选择"发布(B)…"选项，弹出"发布 Web"对话框，其第一项内容为"配置文件"，如图 6.2 所示。

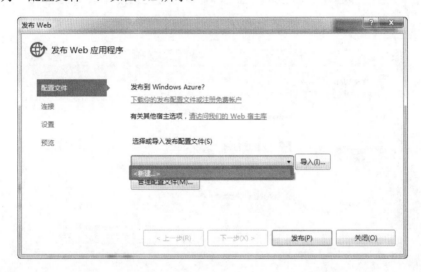

图 6.2 操作步骤 2

（2）选择"选择或导入发布配置文件(S)"下方的下拉列表，在显示的下拉列表中选择"<新建…>"选项，弹出"新建配置文件"对话框，如图 6.3 所示。

图 6.3 操作步骤 3

（3）输入配置文件名称后，单击"确定"按钮，弹出"发布 Web"对话框，该对话框的第二项内容为"连接"，如图 6.4 所示。

（4）将"发布方法"设置为 Web Deploy 包，选择程序包的位置后，单击"下一步"按钮，弹出"发布 Web"对话框，该对话框的第三项内容为"设置"，如图 6.5 所示。

（5）最后单击"发布"按钮，创建的 Web 就被发布了。

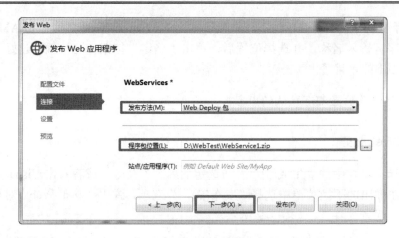

图 6.4　操作步骤 4

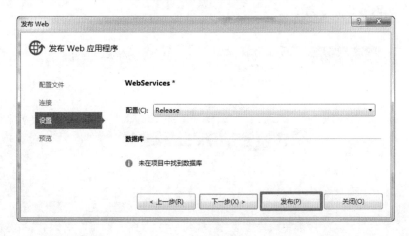

图 6.5　操作步骤 5

3. 搭建Web服务

（1）找到生成 Web 服务应用程序的文件夹，解压压缩文件。由于在创建项目时，将项目名称设置为 WebService1，所以压缩文件名为 WebService1。在解压的文件夹中找到如图 6.6 所示的内容。

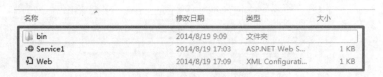

图 6.6　操作步骤 1

（2）创建一个新的文件夹（这里将文件夹命名为 WebServiceTest），将这 3 个内容复制到此文件夹中。

（3）单击"开始"按钮，选择"控制面板"菜单，弹出"控制面板"窗口，如图 6.7 所示。

第 6 章 网络服务

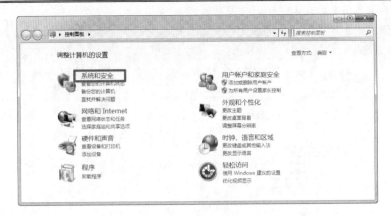

图 6.7 操作步骤 2

（4）选择"系统和安全"选项，弹出"系统和安全"窗口，如图 6.8 所示。

图 6.8 操作步骤 3

（5）选择"管理工具"选项，弹出"管理工具"窗口，如图 6.9 所示。

图 6.9 操作步骤 4

(6) 找到"Internet 信息服务（IIS）管理器"选项。双击它，将其打开，如图 6.10 所示。

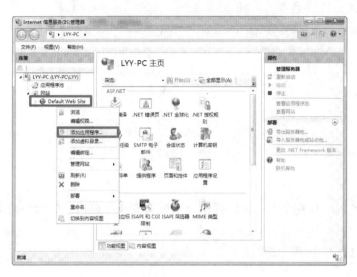

图 6.10 操作步骤 5

(7) 选择"网站"文件夹下面的 Default Web Site 选项，右击，在弹出的下拉菜单中选择"添加应用程序…"命令，弹出"添加应用程序"对话框，如图 6.11 所示。

(8) 输入别名以及物理路径后，单击"确定"按钮，退出"添加应用程序"对话框。然后打开浏览器，在网址搜索栏中输入网址，再按回车键，就可以看到如图 6.12 所示的效果。

注意：在网址搜索栏中输入的网址构成如图 6.13 所示。

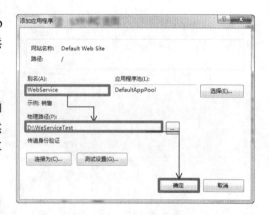

图 6.11 操作步骤 6

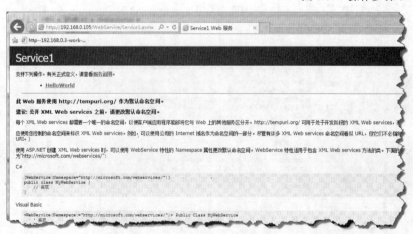

图 6.12 效果

第 6 章 网络服务

图 6.13 网址构成

其中，别名是在"添加应用程序"对话框中输入的别名；.asmx 服务文件是"添加应用程序"对话框中设置的物理路径中的服务文件。

6.1.2 Web 服务的使用

本小节将会讲解如何在 Xamarin.iOS 中使用 Web 服务。

【示例 6-1】以下将触摸按钮，获取 Web 服务中的字符串。具体的操作步骤如下所述。

（1）创建一个 Single View Application 类型的工程。

（2）在创建的工程中添加 6.1.1 节中的 Web 服务的引用到 Web References 文件夹中。添加 Web 引用的具体步骤如下：

首先，右击引用文件夹上的 6-1，显示下拉菜单，如图 6.14 所示。

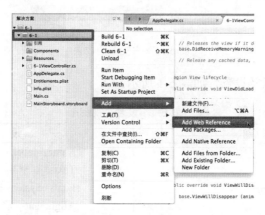

图 6.14 操作步骤 1

然后，选择下拉菜单中的 Add|Add Web Reference 命令，弹出 Add Web Reference 对话框，如图 6.15 所示。

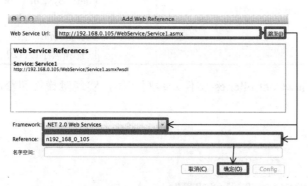

图 6.15 操作步骤 2

• 219 •

在此对话框中输入 Web 服务的 Url，再单击"跳至"按钮，将 Framework 设置为.NET 2.0 Web Service，在 Reference 文本框中输入应用名称，如果不输入名称，将会有默认的名称。然后单击"确定"按钮，此时会在左侧的导航栏中添加一个 Web Reference 的文件夹，在此文件夹中有一个名为 n192_168_0_105 的 Web 应用。

（3）打开 MainStoryboard.storyboard 文件，对主视图进行设置，效果如图 6.16 所示。

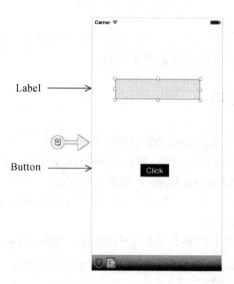

图 6.16　主视图的效果

需要添加的视图以及设置，如表 6-1 所示。

表 6-1　设置主视图

视　　图	设　　置
Label	Name：label Text：（空） Font：System 30 pt Alignment：居中 位置和大小：(60, 149, 205, 46)
Button	Name：btnShow Title：Click Font：System 19 pt Text Color：白色 Background：黑色 位置和大小：(120, 350, 70, 30)

（4）打开 6-1ViewController.cs 文件，编写代码，实现触摸按钮获取 Web 服务中的字符串。代码如下：

```
using System;
using System.Drawing;
using MonoTouch.Foundation;
using MonoTouch.UIKit;
using Application.n192_168_0_105;
```

```
namespace Application
{
    public partial class __1ViewController : UIViewController
    {
        ……
        #region View lifecycle
        public override void ViewDidLoad ()
        {
            base.ViewDidLoad ();
            //触摸按钮获取 Web 服务中的字符串
            btnShow.TouchUpInside += (s, e) => {
                using (Service1 webService = new Service1())
                {
                    label.Text = webService.HelloWorld ();        //调用方法
                }
            };
        }
        ……
        #endregion
    }
}
```

运行效果如图 6.17 所示。

图 6.17　运行效果

注意：在使用 Web 服务时，必须要将此服务开启，6.1.1 节在构建一个 Web 服务时就将此服务开启了。

在 Web 服务中有很多异步方法。

【示例 6-2】 以下将使用 Web 服务中的异步方法来获取 Web 服务中的字符串。具体的操作步骤如下所述。

（1）创建一个 Single View Application 类型的工程。

（2）在创建的工程中添加 6.1.1 节中的 Web 服务的引用到 Web References 文件夹中。

（3）打开 MainStoryboard.storyboard 文件，拖动 Label 标签对象到主视图中，将此对象的 Name 设置为 label；将 Font 设置为 System 26 pt；将 Alignment 设置为居中。

（4）打开 6-2ViewController.cs 文件，编写代码，实现获取 Web 服务中的字符串。代码如下：

```csharp
using System;
using System.Drawing;
using MonoTouch.Foundation;
using MonoTouch.UIKit;
using Application.n192_168_0_105;
namespace Application
{
    public partial class __2ViewController : UIViewController
    {
        ……
        #region View lifecycle
        public override void ViewDidLoad ()
        {
            base.ViewDidLoad ();
            Service1 webService = new Service1 ();
            webService.HelloWorldCompleted += (sender, args) => this.
            InvokeOnMainThread (() => label.Text = args.Result);
                        //当 HelloWorld()方法调用完成后调用
            webService.HelloWorldAsync ();                          //调用异步方法
        }
        ……
        #endregion
    }
}
```

运行效果如图 6.18 所示。

图 6.18　运行效果

6.2　使用 REST 服务

代表性状态传输（Representational State Transfer，REST）在 Web 领域已经得到了广泛的应用，是基于 SOAP 和 Web 服务描述语言（Web Services Description Language，WSDL）

的 Web 服务的更为简单的替代方法。本节将讲解如何在 Xamarin.iOS 中使用 REST 服务。

【示例 6-3】 以下将触摸按钮，获取 REST 服务中的内容。具体的操作步骤如下所述。

（1）创建一个 Single View Application 类型的工程。

（2）添加 System.Json 和 System.Net.Http 引用到创建工程的"引用"文件夹中。

（3）打开 MainStoryboard.storyboard 文件，对主视图进行设置，效果如图 6.19 所示。

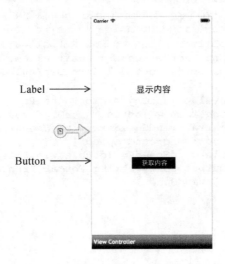

图 6.19 主视图的效果

需要添加的视图以及设置，如表 6-2 所示。

表 6-2 设置主视图

视　图	设　置
Label	Name：lblOutput Text：显示内容 Font：System 23 pt Alignment：居中 位置和大小：(24, 141, 272, 99)
Button	Name：btnForecast Title：获取内容 Font：System 17 pt Text Color：白色 Background：黑色 位置和大小：(106, 366, 116, 30)

（4）打开 6-3ViewController.cs 文件，编写代码，实现获取 REST 服务中的内容。代码如下：

```
using System;
using System.Drawing;
using MonoTouch.Foundation;
using MonoTouch.UIKit;
using System.Net.Http;
using System.Json;
namespace Application
```

```
{
    public partial class __3ViewController : UIViewController
    {
        ……
        #region View lifecycle
        public override void ViewDidLoad ()
        {
            base.ViewDidLoad ();
            //触摸按钮后调用，TouchUpInside 事件需要被标记为异步
            btnForecast.TouchUpInside += async (sender, e) => {
                HttpClient client = new HttpClient();         //实例化对象
                string jsonResponse = await client.GetStringAsync("http:
//api.ometfn.net/0.1/forecast/eu12/46.5,6.32/now.json");
                //解析字符串
                JsonValue jsonObj = JsonValue.Parse(jsonResponse);
                string temp = (string)jsonObj["run"];
                JsonArray windSpeedArray = (JsonArray)jsonObj
["wind_10m_ground_speed"];
                double windSpeed = (double)windSpeedArray[0];
                lblOutput.Text = string.Format("run: {0}\nWind speed: {1}",
                temp, windSpeed);
            } ;
        }
        ……
        #endregion
    }
}
```

运行效果如图 6.20 所示。

图 6.20　运行效果

> 注意：在此程序中引入了两个新的命名空间：一个是 System.Net.Http，另一个是 System.Json。其中，System.Net.Http 命名空间提供用于 HTTP 应用程序的编程接口，System.Json 命名空间提供了对序列化 JavaScript 对象 JSON 的支持。

6.3 使用原生的 API 进行通信

在 Xamarin.iOS 中，本身自带 API 进行网络服务之间的通信。

【示例 6-4】以下将使用 Xamarin.iOS 自带的 API 对 REST 服务进行连接和使用。具体的操作步骤如下所述。

（1）创建一个 Single View Application 类型的工程。

（2）添加 System.Json 和 System.Net.Http 引用到创建工程的"引用"文件夹中。

（3）打开 MainStoryboard.storyboard 文件，对主视图进行设置，效果如图 6.21 所示。

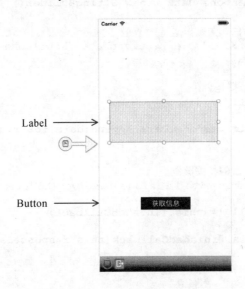

图 6.21　主视图的效果

需要添加的视图以及设置，如表 6-3 所示。

表 6-3　设置主视图

视　　图	设　　置
Label	Name：lblOutput Text：（空） Font：System 25 pt 位置和大小：(27, 200, 262, 95)
Button	Name：lblOutput Title：获取信息 Font：System 17 pt Text Color：白色 Background：黑色 位置和大小：(100, 427, 121, 30)

（4）在创建的工程中添加一个 C#的类文件，并命名为 ConnectionDelegate。

（5）打开 ConnectionDelegate.cs 文件，编写代码，实现代码如下：

```csharp
using System;
using MonoTouch.Foundation;
using System.Text;
namespace Application
{
    public class ConnectionDelegate:NSUrlConnectionDelegate
    {
        ……
        private Action<string> finishedCallback;
        private StringBuilder responseData;
        //构造方法
        public ConnectionDelegate(Action<string> callback)
        {
            this.finishedCallback = callback;
            this.responseData = new StringBuilder();
        }
        //结束数据后的相关操作
        public override void ReceivedData (NSUrlConnection connection, NSData data)
        {
            //判断数据是否存储
            if (null != data)
            {
                this.responseData.Append(data.ToString());
            }
        }
        //当请求结束后的相关操作
        public override void FinishedLoading (NSUrlConnection connection)
        {
            if (null != this.finishedCallback)
            {
                this.finishedCallback(this.responseData.ToString());
            }
            this.responseData.Clear();
        }
        //当请求失败时的相关操作
        public override void FailedWithError (NSUrlConnection connection, NSError error)
        {
            //判断是否有错误
            if (null != error)
            {
                Console.WriteLine("Error in connection! { 0}", error.LocalizedDescription);
            }
        }
    }
}
```

（6）打开 6-4ViewController.cs 文件，编写代码，实现使用自带的 API 进行 REST 的连接和使用。代码如下：

```csharp
using System;
using System.Drawing;
using MonoTouch.Foundation;
using MonoTouch.UIKit;
using System.Json;
namespace Application
{
```

第 6 章 网络服务

```
public partial class __4ViewController : UIViewController
{
    ......
    #region View lifecycle
    public override void ViewDidLoad ()
    {
        base.ViewDidLoad ();
        this.btnForecast.TouchUpInside += (sender, e) => {
            NSUrlRequest request = new NSUrlRequest(new NSUrl("http:
            //api.ometfn.net/0.1/forecast/eu12/46.5,6.32/now.json"));
            NSUrlConnection connection = new NSUrlConnection(request,
            new ConnectionDelegate((response) => {
                //解析数据
                JsonValue jsonObj = JsonValue.Parse(response);
                JsonArray timesArray = (JsonArray)jsonObj["times"];
                double times = (double)timesArray[0];
                string status = (string)jsonObj["status"];
                this.lblOutput.Text = string.Format("Times: {0}\
                nStatus {1}", times,status);
            } ));
        } ;
    }
    ......
    #endregion
}
```

运行效果如图 6.22 所示。

图 6.22 运行效果

> **注意**：在此程序中引入了一个新的命名空间 System.Text。此命名空间包含表示 ASCII、Unicode、UTF-7 和 UTF-8 字符编码的类，用于将字符块转换为字节块和将字节块转换为字符块的抽象基类，以及操作和格式化 String 对象而不创建 String 的中间实例的 Helper 类。

第 7 章　多媒体资源

多媒体是在计算机系统中，组合两种或两种以上媒体的一种人机交互式信息交流和传播媒体，即多种信息载体的表现形式和传递方式。在 iOS 开发中，使用多媒体，可以使应用程序变得与众不同。多媒体中的资源包括图形图像、音频和视频等。本章将讲解这些多媒体资源的使用。

7.1　选择图像和视频

在 iOS 的真实设备或者模拟的设备中，都有一个 Photos 应用程序。它是 iOS 设备的内置应用程序，一般称之为照片库。在此应用程序中可以存放一些图像以及自己录制的视频。如果获取图像或者某一视频作为一个应用程序的背景，该如何做到呢？此时需要使用 UIImagePickerController 类。此类用来管理一个系统，此系统提供了一系列选择照片的用户接口，即图像选择控制器。此类用来负责提供与用户互动交流的界面，以及选择图像和视频的功能。本节将讲解如何选择图像和视频。

7.1.1　选择图像

选择照片库中的图像，需要用到 UIImagePickerController 类中的 FinishedPickingMedia 事件。

【示例 7-1】以下将在照片库中选择图像，作为当前应用程序的背景。具体的操作步骤如下所述。

（1）创建一个 Single View Application 类型的工程。
（2）打开 MainStoryboard.storyboard 文件，对主视图进行设置，效果如图 7.1 所示。

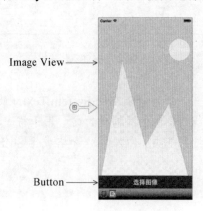

图 7.1　主视图的效果

需要添加的视图以及设置，如表 7-1 所示。

表 7-1 设置主视图

视　　图	设　　置
Image View	Name：imgView 位置和大小：(0, 0, 320, 568)
Button	Name：btnSelect Title：选择图像 Font：System 21 pt Text Color：白色 Background：深灰色 位置和大小：(0, 525, 320, 43)

（3）打开 7-1ViewController.cs 文件，编写代码，实现图像的选择功能。代码如下：

```
using System;
using System.Drawing;
using MonoTouch.Foundation;
using MonoTouch.UIKit;
namespace Application
{
    public partial class __1ViewController : UIViewController
    {
        UIImagePickerController imagePicker;
        ……                    //这里省略了视图控制器的构造方法和析构方法
        #region View lifecycle
        public override void ViewDidLoad ()
        {
            base.ViewDidLoad ();
            imagePicker = new UIImagePickerController();
            imagePicker.FinishedPickingMedia += this.ImagePicker_
            FinishedPickingMedia;
            imagePicker.Canceled += this.ImagePicker_Cancelled;
            //设置图像来源
            imagePicker.SourceType = UIImagePickerControllerSourceType.
            PhotoLibrary;
            //触摸按钮
            btnSelect.TouchUpInside += async delegate {
                await this.PresentViewControllerAsync(this.imagePicker, true);
                //进入照片库应用程序
            } ;
        }
        //图像选择后调用的方法
        private async void ImagePicker_FinishedPickingMedia (object sender,
        UIImagePickerMediaPickedEventArgs e)
        {
            //获取选择后的图像
            UIImage pickedImage = e.Info[UIImagePickerController.
            OriginalImage] as UIImage;
            .imgView.Image = pickedImage;        //将选择的图像显示在图像视图中
            await imagePicker.DismissViewControllerAsync(true);
                                                //退出照片库应用程序
        }
        //取消选择后调用的方法
        private async void ImagePicker_Cancelled (object sender, EventArgs e)
        {
```

第 2 篇　资源使用篇

```
                await this.imagePicker.DismissViewControllerAsync(true);
        }
        ……                              //这里省略了视图加载和卸载前后的一些方法
        #endregion
    }
}
```

运行效果如图 7.2 所示。

图 7.2　运行效果

7.1.2　向模拟器中添加图像

在此程序中，使用到了 iOS Simulator 中的 Photos 应用程序，如何向 iOS Simulator 内的 Photos 应用程序中添加图像是一直困惑着开发者的问题。以下是向 Photos 应用程序中添加图像的具体步骤。

· 230 ·

（1）打开 iOS Simulator。（这里有两种方法可以打开 iOS Simulator，一种是在创建好工程之后，单击运行按钮；另一种是启动 Xcode，选择 Xcode|Open Developer Tool|iOS Simulator 命令）。

（2）拖动图像到 iOS Simulator 上，如图 7.3 所示。

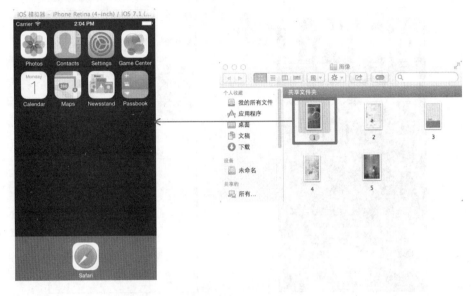

图 7.3　操作步骤 1

（3）这时 iOS Simulator 打开 Safari 浏览器，并在浏览器上显示拖动到 iOS Simulator 的图像，如图 7.4 所示。

（4）长按该照片，就会出现一个动作表单，如图 7.5 所示。

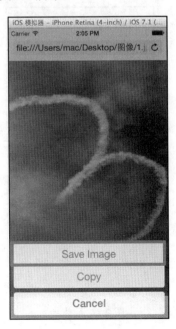

图 7.4　操作步骤 2　　　　　　　　　　图 7.5　操作步骤 3

(5) 选择 Save Image 按钮，照片就添加到了 Photos 应用程序中。

当然，添加在 Photos 应用程序中的图像也可以删除，具体的删除步骤如下：

(1) 打开 iOS Simulator 应用程序的首页，如图 7.6 所示。

(2) 单击 Photos 应用程序，将其打开，如图 7.7 所示。

图 7.6　操作步骤 1

图 7.7　操作步骤 2

(3) 选择要进行删除的照片，打开此图像，如图 7.8 所示。

(4) 触摸删除图标，弹出动作表单，如图 7.9 所示。

图 7.8　操作步骤

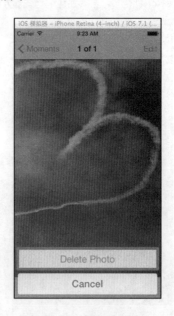
图 7.9　操作步骤

(5) 选择 Delete Photo 按钮，删除图像。此时，图像就从 Photos 应用程序中被删除了。

7.1.3 设置图像显示来源

在此应用程序中，使用了 SourceTypes 属性对图像的显示来源进行了设置。图像的显示来源可以设置为 3 种，如表 7-2 所示。

表 7-2 图像的来源

图像的来源	解　释
PhotoLibrary	照相库
SavedPhotosAlbum	相册
Camera	相机

【示例 7-2】以下将实现当触摸按钮后，会打开相应的图像来源显示界面。具体的操作步骤如下所述。

（1）创建一个 Single View Application 类型的工程。
（2）添加图像 1.png 到创建工程的 Resources 文件夹中。
（3）打开 MainStoryboard.storyboard 文件，对主视图进行设置，效果如图 7.10 所示。

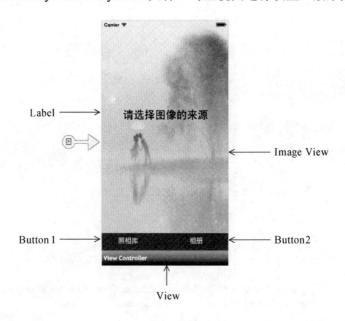

图 7.10 主视图的效果

需要添加的视图以及设置，如表 7-3 所示。

表 7-3 设置主视图

视　图	设　置
Label	Text：请选择图像的来源 Font：System Bold 27 pt Alignment：居中 位置和大小：(50, 208, 224, 48)

续表

视 图	设 置
Image View	Image：1.png 位置和大小：(0, 0, 320, 525)
Button1	Name：pbtn Title：照相库 Font：System 17 pt Text Color：白色 位置和大小：(28, 9, 77, 30)
Button2	Name：abtn Title：相册 Font：System 17 pt Text Color：白色 位置和大小：(200, 9, 77, 30)
View	Background：深灰色 位置和大小：(0, 520, 320, 48)

（4）打开 7-14ViewController.cs 文件，编写代码，实现不同来源的图像显示。代码如下：

```csharp
using System;
using System.Drawing;
using MonoTouch.Foundation;
using MonoTouch.UIKit;
namespace Application
{
    public partial class __14ViewController : UIViewController
    {
        UIImagePickerController picker;
        ……                              //这里省略了视图控制器的构造方法和析构方法
        #region View lifecycle
        public override void ViewDidLoad ()
        {
            base.ViewDidLoad ();
            picker = new UIImagePickerController ();
            picker.Canceled += this.ImagePicker_Cancelled;
            //触摸"照相库"按钮，打开图像来源为照相库的界面
            pbtn.TouchUpInside += (sender, e) => {
                picker.SourceType=UIImagePickerControllerSourceType.PhotoLibrary;
                this.PresentViewControllerAsync(picker,true);
            };
            //触摸"相册"按钮，打开图像来源为相册的界面
            abtn.TouchUpInside += (sender, e) => {
                picker.SourceType=UIImagePickerControllerSourceType.SavedPhotosAlbum;
                this.PresentViewControllerAsync(picker,true);
            };
        }
        //关闭图像来源的界面
        private async void ImagePicker_Cancelled (object sender, EventArgs e)
        {
            await this.picker.DismissViewControllerAsync(true);
        }
        ……                              //这里省略了视图加载和卸载前后的一些方法
        #endregion
```

第 7 章 多媒体资源

```
    }
}
```

运行效果如图 7.11 所示。

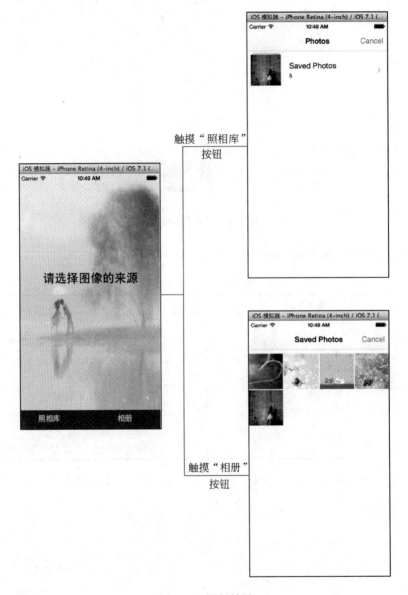

图 7.11 运行效果

7.1.4 选择视频

视频的选择，也需要使用到 UIImagePickerController 类中的 FinishedPickingMedia 事件。

【示例 7-3】 以下将实现选择视频的功能。具体的操作步骤如下所述。

（1）创建一个 Single View Application 类型的工程。

（2）打开 MainStoryboard.storyboard 文件，对主视图进行设置，效果如图 7.12 所示。

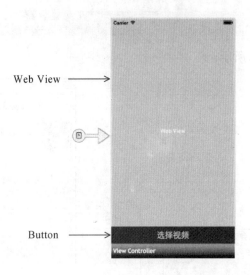

图 7.12 主视图的效果

需要添加的视图以及设置，如表 7-4 所示。

表 7-4 设置主视图

视 图	设 置
Web View	Name：webView 位置和大小：(0, 0, 320, 568)
Button	Name：btnSelect Title：选择视频 Font：System 21 pt Text Color：白色 Background：深灰色 位置和大小：(0, 525, 320, 43)

（3）打开 7-2ViewController.cs 文件，编写代码，实现视频选择的功能。代码如下：

```
using System;
using System.Drawing;
using MonoTouch.Foundation;
using MonoTouch.UIKit;
namespace Application
{
    public partial class __2ViewController : UIViewController
    {
        UIImagePickerController imagePicker;
        ……                          //这里省略了视图控制器的构造方法和析构方法
        #region View lifecycle
        public override void ViewDidLoad ()
        {
            base.ViewDidLoad ();
            imagePicker = new UIImagePickerController();
            imagePicker.FinishedPickingMedia += this.ImagePicker_
                FinishedPickingMedia;
            imagePicker.Canceled += this.ImagePicker_Cancelled;
            imagePicker.MediaTypes = new string[] {"public.movie" };
            btnSelect.TouchUpInside += async delegate {
```

```
            await this.PresentViewControllerAsync(this.imagePicker, true);
        };
    }
    //视频选择后调用的方法
    private async void ImagePicker_FinishedPickingMedia (object sender,
    UIImagePickerMediaPickedEventArgs e)
    {
        NSUrl mediaUrl = e.Info[UIImagePickerController.MediaURL] as
            NSUrl;
        NSUrlRequest aa = new NSUrlRequest (mediaUrl);
        webView.LoadRequest (aa);
        await imagePicker.DismissViewControllerAsync(true);
    }
    //取消选择后调用的方法
    private async void ImagePicker_Cancelled (object sender, EventArgs e)
    {
        await imagePicker.DismissViewControllerAsync(true);
    }
    ......                      //这里省略了视图加载和卸载前后的一些方法
    #endregion
    }
}
```

运行效果如图 7.13 所示。

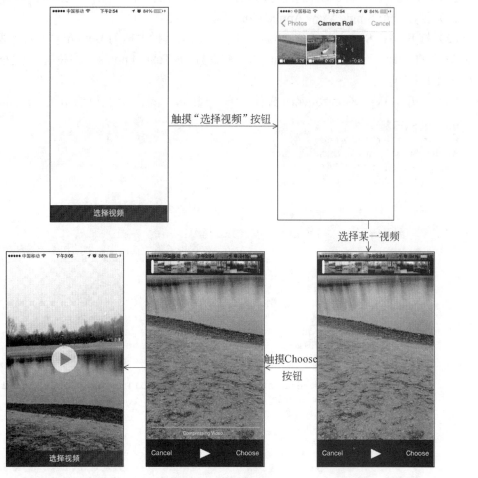

图 7.13　运行效果

> **注意**：在此程序中使用了 MediaTypes 属性设置媒体的类型，这样在出现图像选择控制器时，只会出现视频的内容，不会出现图像的内容。

7.2 使用相机捕获媒体

21 世纪是一个风靡自拍的世纪，打开微博或者空间，你都会看到一张张熟悉的自拍脸，不仅如此，有的还会录成视频。自从手机上有了相机功能后，很多的开发商看到了相机的商机，做出了很多与相机有关的应用，如美颜相机、魔幻相机等，充分满足了广大用户的需求。本节将讲解如何使用相机进行媒体的捕获，通俗一点讲就是如何使用相机进行拍照。

7.2.1 打开相机

在 7.1.1 节中提供了对图像显示来源的设置，在此设置中有一个 Camera。如果将 SourceTypes 属性设置为 Camera，那么打开图像选择器时，打开的就是相机。

【示例 7-4】 以下将实现在触摸按钮后，打开相机的功能。具体的操作步骤如下所述。

（1）创建一个 Single View Application 类型的工程。

（2）打开 MainStoryboard.storyboard 文件，从工具栏中拖动 Button 对象对主视图中。将此对象的 Name 设置为 Openbtn，Ttile 设置为 Take Photos，将位置和大小设置为（102,311,116,30）。

（3）打开 7-3ViewController.cs 文件，编写代码，实现打开相机的功能。代码如下：

```
using System;
using System.Drawing;
using MonoTouch.Foundation;
using MonoTouch.UIKit;
namespace Application
{
    public partial class __3ViewController : UIViewController
    {
        UIImagePickerController imagePicker;
        ……                       //这里省略了视图控制器的构造方法和析构方法
        #region View lifecycle
        public override void ViewDidLoad ()
        {
            base.ViewDidLoad ();
            imagePicker = new UIImagePickerController();
            Openbtn.TouchUpInside +=async delegate {
                if (UIImagePickerController.IsSourceTypeAvailable
                (UIImagePickerControllerSourceType.Camera))
                {
                    imagePicker.SourceType = UIImagePickerController
                    SourceType.Camera;
                    await this.PresentViewControllerAsync(this.imagePicker,
                    true);
                } else
                {
                    UIAlertView alertView = new UIAlertView ();
                    alertView.Title = "对不起,此设备没有相机功能";
```

```
                alertView.AddButton ("Cancel");
                alertView.Show ();
            }
        };
    }
    ……                    //这里省略了视图加载和卸载前后的一些方法
    #endregion
}
```

运行效果如图 7.14 所示。

图 7.14　运行效果

> 注意：在使用相机时，需要对设备上的相机是否可用进行检查，拍照时需要使用真机。

7.2.2　设置相机

开发者可以对相机的摄像头和闪光灯等进行设置，可以在打开相机后，就可以看到自己想要的效果,而不是在打开相机之后用手进行设置。

1. 设置摄像头

在 iOS 设备中通常都两个摄像头像,一个在设备的前面,另一个在设备的后面,那么如何实现这两个摄像头的切换呢?这时只要对属性 CameraDevice 进行设置就可以了,其语法形式如下:

图像选择控制器对象.CameraDevice=前置摄像头/后置摄像头

其中,前置摄像头用 Front 表示;后置摄像头用 Rear 表示,一般默认为后置摄像头。

2. 设置闪光灯

在昏暗条件下用手机拍照,闪光灯的使用是必不可少的。它的功能是实现光亮的加强。如果想要实现闪光灯的打开或关闭,需要对 cameraFlashMode 属性进行设置,其语法形式如下:

图像选择控制器对象.CameraFlashMode=闪光灯的状态;

其中,闪光灯的状态有 3 种,如表 7-5 所示。

表 7-5 闪光灯的状态

状 态	功 能
On	闪光灯不管在什么环境光线下都是打开的
Off	闪光灯不管在什么环境光线下都是关闭的
Auto	在不同的环境光线下,设备自己判断是否打开闪光灯

注意:一般不对此属性进行设置,默认闪光灯是 Auto 状态。

3. 设置相机模式

相机的作用有两个:一个是拍摄静止的图像(照片),另一种是拍摄动的事物(视频)。要想实现摄像头的切换,就要使用到 cameraDevice 属性,其语法形式如下:

图像选择控制器对象.CameraDevice=相机的模式;

其中,相机的模式有两种,一种是 Photo,它表示相机拍摄的是静止的图像;另一种是 Video,它表示相机拍摄的是视频。一般不对此属性进行设置,默认相机模式为 Photo。

【示例 7-5】 以下将改变默认相机的摄像头和闪光灯。具体的操作步骤如下所述。

(1)创建一个 Single View Application 类型的工程。

(2)打开 MainStoryboard.storyboard 文件,从工具栏中拖动 Button 对象对主视图中,将此对象的 Name 设置为 btn,Ttile 设置为打开相机,将位置和大小设置为(115,255,90,30)。

(3)打开 7-4ViewController.cs 文件,编写代码,实现打开相机的功能。代码如下:

```
using System;
using System.Drawing;
using MonoTouch.Foundation;
using MonoTouch.UIKit;
namespace Application
{
```

```
public partial class __4ViewController : UIViewController
{
    ......                          //这里省略了视图控制器的构造方法和析构方法
    #region View lifecycle
    public override void ViewDidLoad ()
    {
        base.ViewDidLoad ();
        UIImagePickerController imagePicker = new UIImagePickerController();

        btn.TouchUpInside +=async delegate  {
            imagePicker.SourceType = UIImagePickerControllerSource
            Type.Camera;
            //将相机的摄像头改为前置摄像头
            imagePicker.CameraDevice=UIImagePickerControllerCamera
            Device.Front;
            //将相机的闪光灯改为Off，即关闭相机的闪光灯
            imagePicker.CameraFlashMode=UIImagePickerController
            CameraFlashMode.Off;
            await this.PresentViewControllerAsync(imagePicker, true);
        };
    }
    ......                          //这里省略了视图加载和卸载前后的一些方法
    #endregion
}
```

运行效果如图 7.15 所示。

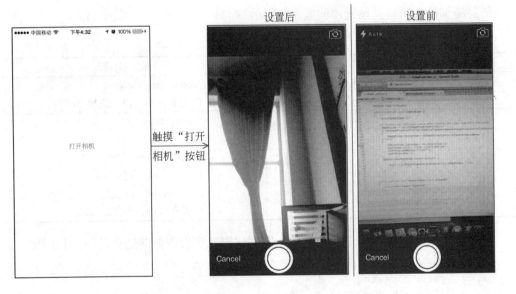

图 7.15 运行效果

7.2.3 捕获媒体

相机最重要的功能就是进行拍照和录制视频，即捕获媒体。

【示例 7-6】 以下将实现相机的拍照功能。具体的操作步骤如下所述。

（1）创建一个 Single View Application 类型的工程。

（2）打开 MainStoryboard.storyboard 文件，对主视图进行设置，效果如图 7.16 所示。

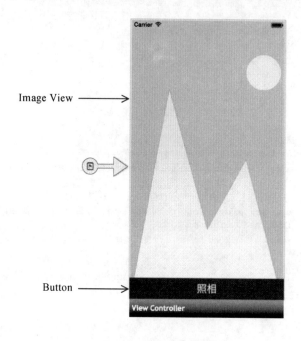

图 7.16 主视图的效果

需要添加的视图以及设置，如表 7-6 所示。

表 7-6 设置主视图

视　　图	设　　置
Image View	Name：imgView 位置和大小：(0, 0, 320, 568)
Button	Name：take Title：照相 Font：System 21 pt Text Color：白色 Background：深灰色 位置和大小：(0, 525, 320, 43)

（3）打开 7-5ViewController.cs 文件，编写代码，实现图像的选择功能。代码如下：

```
using System;
using System.Drawing;
using MonoTouch.Foundation;
using MonoTouch.UIKit;
namespace Application
{
    public partial class __5ViewController : UIViewController
    {
        UIImagePickerController imagePicker;
        ……
        #region View lifecycle
        public override void ViewDidLoad ()
        {
```

```csharp
        base.ViewDidLoad ();
        imagePicker = new UIImagePickerController();
        imagePicker.FinishedPickingMedia += this.ImagePicker_
        FinishedPickingMedia;
        imagePicker.Canceled += this.ImagePicker_Cancelled;
        //判断是否有相机
        if (UIImagePickerController.IsSourceTypeAvailable
        (UIImagePickerControllerSourceType.Camera))
        {
            this.imagePicker.SourceType = UIImagePickerController
            SourceType.Camera;
        } else
        {
            UIAlertView alertView = new UIAlertView ();
            alertView.Title = "对不起，此设备没有相机功能";
            alertView.AddButton ("Cancel");
            alertView.Show ();

        }
        //触摸按钮打开相机
        takebtn.TouchUpInside += async delegate {
            await this.PresentViewControllerAsync(this.imagePicker, true);
        } ;
    }
    //获取照片，并将其永久的保存在相册中
    private async void ImagePicker_FinishedPickingMedia (object sender,
    UIImagePickerMediaPickedEventArgs e)
    {
        UIImage pickedImage = e.Info[UIImagePickerController.
        OriginalImage] as UIImage;
        imgView.Image = pickedImage;
        //保存图像到相册中
        pickedImage.SaveToPhotosAlbum((s, error) => {
            if (null != error)
            {
                Console.WriteLine("Image not saved! Message: { 0}",
                error.LocalizedDescription);
            }
        } );
        await imagePicker.DismissViewControllerAsync(true);
    }
    //取消选择后调用的方法
    private async void ImagePicker_Cancelled (object sender, EventArgs e)
    {
        await this.imagePicker.DismissViewControllerAsync(true);
    }
    ……
    #endregion
}
}
```

运行效果如图 7.17 所示。

第 2 篇　资源使用篇

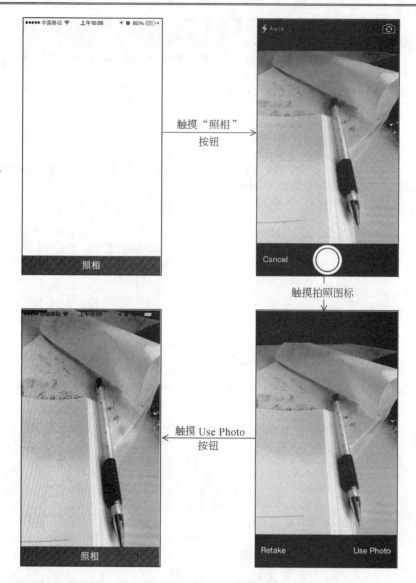

图 7.17　运行效果

> 注意：在此程序中，使用了 SaveToPhotosAlbum()方法将制定的图像即拍摄的照相保存在相册中。

7.2.4　自定义相机

如果开发者不希望使用 iOS 设备内置的相机界面，还可以使用 CameraOverlayView 属性对此界面进行自定义。其语法形式如下：

图像选择控制器对象.CameraOverlayView=自定义的视图；

【示例 7-7】以下将实现自定义的相机。具体的操作步骤如下所述。

（1）创建一个 Single View Application 类型的工程。

（2）添加图像 1.png 和 2.png 到创建工程的 Resources 文件夹中。
（3）打开 MainStoryboard.storyboard 文件，对主视图进行设置，效果如图 7.18 所示。

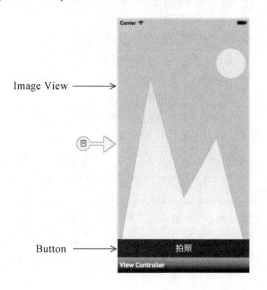

图 7.18　主视图的效果

需要添加的视图以及设置，如表 7-7 所示。

表 7-7　设置主视图

视　图	设　置
Image View	Name：imgView 位置和大小：(0, 0, 320, 568)
Button	Name：takebtn Title：拍照 Font：System 20 pt Text Color：白色 Background：深灰色 位置和大小：(0, 525, 320, 43)

（4）打开 7-6ViewController.cs 文件，编写代码，实现自定义的相机。代码如下：

```
using System;
using System.Drawing;
using MonoTouch.Foundation;
using MonoTouch.UIKit;
namespace Application
{
    public partial class __6ViewController : UIViewController
    {
        UIImagePickerController imagePicker;
        ……                              //这里省略了视图控制器的构造方法和析构方法
        #region View lifecycle
        public override void ViewDidLoad ()
        {
            base.ViewDidLoad ();
            imagePicker = new UIImagePickerController();
```

```csharp
imagePicker.FinishedPickingMedia += this.ImagePicker_
FinishedPickingMedia;
this.imagePicker.SourceType = UIImagePickerControllerSource
Type.Camera;
//实例化并设置视图对象vv
UIView vv = new UIView ();
vv.Frame = imagePicker.View.Frame;
//实例化并设置工具栏对象controlView
UIToolbar controlView = new UIToolbar ();
controlView.BackgroundColor = UIColor.Black;
controlView.Frame = new RectangleF (0, this.View.Frame.
Size.Height - 88, this.View.Frame.Size.Width ,60);
//实例化并设置按钮对象camerabtn
UIButton camerabtn = new UIButton ();
camerabtn.Frame = new RectangleF (0, 0, 50, 50);
camerabtn.SetImage (UIImage.FromFile ("2.png"), UIControl
State.Normal);
camerabtn.ShowsTouchWhenHighlighted = true;
//触摸camerabtn按钮实现摄像头的切换
camerabtn.TouchUpInside += (sender, e) => {
    //判断按钮的摄像头是否为后置摄像头
    if(imagePicker.CameraDevice==UIImagePickerController
    CameraDevice.Rear)
    {
        //判断相机是否使用
        if(UIImagePickerController.IsCameraDeviceAvailable
        (UIImagePickerControllerCameraDevice.Front)){
            imagePicker.CameraDevice=UIImagePickerController
            CameraDevice.Front;
        }
    } else
    {
        imagePicker.CameraDevice=UIImagePickerController
        CameraDevice.Rear;
    }
};
UIBarButtonItem takePickItem = new UIBarButtonItem (camerabtn);
//实例化并设置按钮对象spbtn
UIButton spbtn = new UIButton ();
spbtn.Frame = new RectangleF (0, 0, 10, 10);
UIBarButtonItem spItem = new UIBarButtonItem (spbtn);
//实例化并设置按钮对象takebutton
UIButton takebutton = new UIButton ();
takebutton.Frame = new RectangleF (0, 0, 150, 150);
takebutton.ShowsTouchWhenHighlighted = true;
takebutton.SetBackgroundImage(UIImage.FromFile ("1.png"),
UIControlState.Normal);
//触摸takebutton按钮实现拍照的功能
takebutton.TouchUpInside += (sender, e) => {
    **imagePicker.TakePicture ();**
};
UIBarButtonItem takeButtonItem=new UIBarButtonItem(takebutton);
//实例化并设置按钮对象cancelbtn
UIButton cancelbtn = new UIButton ();
cancelbtn.Frame = new RectangleF (0, 0, 40, 40);
cancelbtn.ShowsTouchWhenHighlighted = true;
cancelbtn.SetTitle ("取消", UIControlState.Normal);
cancelbtn.SetTitleColor (UIColor.Black, UIControlState.Normal);
//触摸cancelbtn按钮实现关闭图像选择控制器的功能
```

```
            cancelbtn.TouchUpInside += (sender, e) => {
                imagePicker.DismissViewControllerAsync(true);
            };
            UIBarButtonItem cancelItem =new UIBarButtonItem(cancelbtn);
            //设置工具栏对象controlView的条目
            controlView.Items=new UIBarButtonItem[] {takePickItem,spItem,
            takeButtonItem,cancelItem};
            vv.AddSubview (controlView);
            imagePicker.ShowsCameraControls=false;
            imagePicker.CameraOverlayView=vv;     //设置图像选择控制器的界面
            //触摸"拍照"按钮,打开图像选择控制器
            takebtn.TouchUpInside += async delegate {
                await this.PresentViewControllerAsync(this.imagePicker, true);
            };
        }
        //获取照片,并将其永久的保存在相册中
        private async void ImagePicker_FinishedPickingMedia (object sender,
        UIImagePickerMediaPickedEventArgs e)
        {
            UIImage pickedImage = e.Info[UIImagePickerController.
            OriginalImage] as UIImage;
            imgView.Image = pickedImage;
            pickedImage.SaveToPhotosAlbum((s, error) => {
                if (null != error)
                {
                    Console.WriteLine("Image not saved! Message: { 0}",
                    error.LocalizedDescription);
                }
            } );
            await imagePicker.DismissViewControllerAsync(true);
        }
        ……                       //这里省略了视图加载和卸载前后的一些方法
        #endregion
    }
}
```

运行效果如图 7.19 所示。

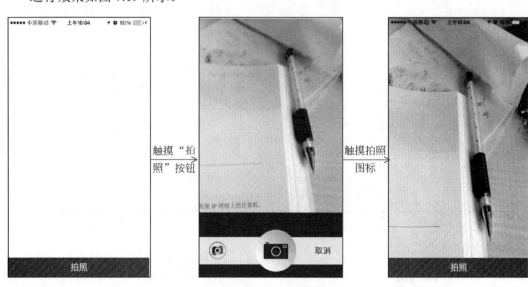

图 7.19 运行效果

⚠注意：当用户触摸图像选择控制器中的拍照图标时，就会实现拍照功能，此功能的实现，使用了 TakePicture()方法。

7.3 播 放 视 频

在很多的应用程序中都有视频的播放功能，如暴风客户端和优酷客户端等。本节将讲解如何在自己开发的应用程序中播放视频。

7.3.1 播放视频文件

MPMoviePlayerController 类提供了一个视频播放控制器，可以用来播放视频。
【示例 7-8】 以下将实现对视频的播放。具体的操作步骤如下所述。
（1）创建一个 Single View Application 类型的工程。
（2）添加视频 Movie.m4v 到创建工程的 Resources 文件夹中。
（3）打开 MainStoryboard.storyboard 文件，对主视图进行设置，效果如图 7.20 所示。

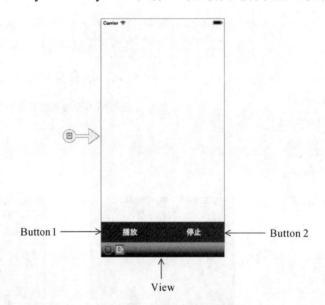

图 7.20 主视图的效果

需要添加的视图以及设置，如表 7-8 所示。

表 7-8 设置主视图

视　　图	设　　置
Button1	Name：playbtn Title：播放 Font：System Bold 19 pt Text Color：白色 位置和大小：(51, 11, 46, 30)

续表

视　图	设　置
Button2	Name：stopbtn Title：停止 Font：System Bold 19 pt Text Color：白色 位置和大小：(217, 11, 46, 30)
View	Background：深灰色 位置和大小：(0, 516, 320, 52)

（4）打开 7-7ViewController.cs 文件，编写代码，实现视频的播放。代码如下：

```
using System;
using System.Drawing;
using MonoTouch.Foundation;
using MonoTouch.UIKit;
using MonoTouch.MediaPlayer;          //引入命名控件 MonoTouch.MediaPlayer
namespace Application
{
    public partial class __7ViewController : UIViewController
    {
        MPMoviePlayerController moviePlayer;
        ……                            //这里省略了视图控制器的构造方法和析构方法
        #region View lifecycle
        public override void ViewDidLoad ()
        {
            base.ViewDidLoad ();
            moviePlayer = new MPMoviePlayerController(NSUrl.FromFilename
            ("Movie.m4v"));
            moviePlayer.View.Frame = new RectangleF(0f, 20f, this.View.
            Frame.Width, 500f);
            this.View.AddSubview(moviePlayer.View);
            //触摸"播放"按钮，开始播放视频
            playbtn.TouchUpInside += delegate {
                moviePlayer.Play();
            } ;
            //触摸"停止"按钮，播放的视频将会停止
            stopbtn.TouchUpInside += delegate {
                moviePlayer.Stop();
            };
        }
        ……            //这里省略了视图加载和卸载前后的一些方法
        #endregion
    }
}
```

运行效果如图 7.21 所示。

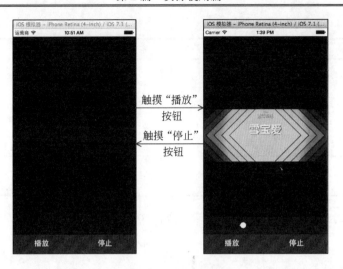

图 7.21　运行效果

在此程序中需要注意以下 3 点：

（1）新的命名空间

在此程序中引入了新的命名空间 MonoTouch.MediaPlayer。MPMoviePlayerController 类就位于此命名空间中。它提供了播放影片、准备播放影片和停止播放影片的功能。

（2）视频的播放控制

当触摸"播放"按钮后，视频就会播放，当触摸"停止"按钮后，视频就会停止播放。以下就是它们的详细介绍：

① 播放视频。使用 Play()方法进行视频的播放，其语法形式如下：

视频播放控制器.Play();

② 停止播放视频。Stop()方法可以将正在播放的视频关闭，其语法形式如下：

视频播放控制器.Stop();

（3）播放的文件

在此程序中，播放了一个扩展名为.m4v 的视频文件，iOS 除了此文件以外，还支持扩展名为.mov、.mp4、mpv 和.3GP 的文件。

7.3.2　设置视频控制器

本小节将讲解视频播放控制器常用的一些设置。

1．设置播放控件的风格

每一个视频中都会有一个播放控件，里面有开始和暂停按钮，以及进度条、全屏切换等控件。播放控件的显示风格是可以有变化的，使用 ControlStyle 属性可以实现风格的变化，其语法形式如下：

视频播放控制器.ControlStyle=播放控件的风格;

其中,播放控件的风格有 4 种,如表 7-9 所示。

表 7-9 播放控件的风格

风　　格	特　　点
Default	显示播放控件
Embedded	内嵌的播放控件(和 Default 的效果一样)
Fullscreen	全屏播放显示播放控件
None	不显示播放控件

【示例 7-9】 以下以【示例 7-8】为基础,改变视频控制器中的播放控件。代码如下:

```
public override void ViewDidLoad ()
{
    base.ViewDidLoad ();
    moviePlayer = new MPMoviePlayerController(NSUrl.FromFilename
    ("Movie.m4v"));
    moviePlayer.View.Frame = new RectangleF(0f, 20f, this.View.Frame.Width,
    500f);
    this.View.AddSubview(moviePlayer.View);
    moviePlayer.ControlStyle = MPMovieControlStyle.Fullscreen;
    //播放视频
    playbtn.TouchUpInside += delegate {
        moviePlayer.Play();
    };
    //停止播放视频
    stopbtn.TouchUpInside += delegate {
        moviePlayer.Stop()
    };
}
```

运行效果如图 7.22 所示。

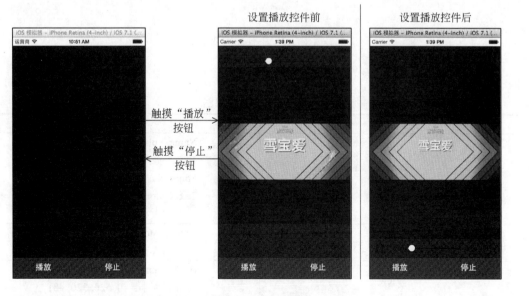

图 7.22 运行效果

2. 设置全屏模式

Fullscreen 属性可以设置视频是否以全屏模式进行播放,其语法形式如下:

`视频播放控制器.Fullscreen=Bool 值;`

其中,当布尔值为 true 时,表示以全屏模式播放;当布尔值为 false 时,表示不会以全屏模式播放。

【**示例 7-10**】 以下将实现对视频全屏模式的设置。具体的操作步骤如下所述。

(1)创建一个 Single View Application 类型的工程。

(2)添加视频 Movie.m4v 到创建工程的 Resources 文件夹中。

(3)打开 MainStoryboard.storyboard 文件,对主视图进行设置,效果如图 7.23 所示。

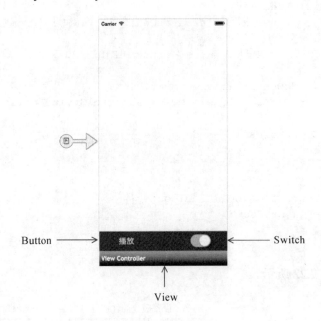

图 7.23 主视图的效果

需要添加的视图以及设置,如表 7-10 所示。

表 7-10 设置主视图

视 图	设 置
Button	Name:pbtn Title:播放 Font:System 19 pt Text Color:白色 位置和大小:(44, 8, 46, 30)
Switch	Name:stopbtn 位置和大小:(228, 8, 51, 31)
View	Background:深灰色 位置和大小:(0, 520, 320, 48)

(4)打开 7-15ViewController.cs 文件，编写代码，实现视频以全屏模式播放。代码如下：

```csharp
using System;
using System.Drawing;
using MonoTouch.Foundation;
using MonoTouch.UIKit;
using MonoTouch.MediaPlayer;
namespace Application
{
    public partial class __15ViewController : UIViewController
    {
        MPMoviePlayerController moviePlayer;
        ……                          //这里省略了视图控制器的构造方法和析构方法
        #region View lifecycle
        public override void ViewDidLoad ()
        {
            base.ViewDidLoad ();
            moviePlayer = new MPMoviePlayerController(NSUrl.FromFilename
            ("Movie.m4v"));
            moviePlayer.View.Frame = new RectangleF(0f, 20f, this.View.
            Frame.Width, 500f);
            this.View.AddSubview(moviePlayer.View);
            fswich.On = false;
            //播放视频
            pbtn.TouchUpInside += delegate {
                moviePlayer.Play();
            } ;
            //使用开关控制屏幕模式
            fswich.ValueChanged+=(sender, e) =>{
                //判断开关是否为开的状态
                if(fswich.On==true){
                    moviePlayer.Fullscreen=true;
                }else{
                    moviePlayer.Fullscreen=false;
                }
            };
        }
        ……                          //这里省略了视图加载和卸载前后的一些方法
        #endregion
    }
}
```

运行效果如图 7.24 所示。

7.3.3 视频播放控制器常用的监听事件

视频播放控制器有一些常用的监听事件，如表 7-11 所示。它们可以用来监听视频播放器的一些变化。

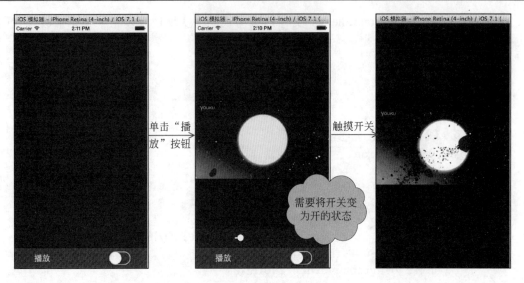

图 7.24　运行效果

表 7-11　视频播放控制器常用的监听事件

事　　件	功　　能
ObservePlaybackStateDidChange	监听视频播放的状态
ObservePlaybackDidFinish	此事件的视频播放结束后调用
ObserveDidEnterFullscreen	监听用户轻拍全屏控件并且控制器已进入全屏模式
ObserveDidExitFullscreen	监听控制器已离开全屏模式
ObserveDurationAvailable	监听视频的持续时间
ObserveLoadStateDidChange	此事件在播放网络视频时调用的，当控制器在缓冲器中已经完成预加载的媒体被触发
ObserveNaturalSizeAvailable	当电影框架的尺寸被制成时被触发
ObserveNowPlayingMovieDidChange	视频播放器的内容已经改变时触发

【示例 7-11】以下将使用 ObservePlaybackStateDidChange、ObservePlaybackDidFinish 和 ObserveDidEnterFullscreen 监听事件实现对视频播放控制器的监听。具体的操作步骤如下所述。

（1）创建一个 Single View Application 类型的工程。

（2）添加视频 Movie.m4v 到创建工程的 Resources 文件夹中。

（3）打开 MainStoryboard.storyboard 文件，从工具栏中拖动 Button 对象对主视图中，将此对象的 Name 设置为 pbtn，Ttile 设置为播放，System 设置为 System 20 pt，Text Color 设置为白色，Background 设置为深灰色，位置和大小设置为（0,523,320,45）。

（4）打开 7-16ViewController.cs 文件，编写代码，实现对视频播放控制器的监听。代码如下：

```
using System;
using System.Drawing;
using MonoTouch.Foundation;
using MonoTouch.UIKit;
using MonoTouch.MediaPlayer;
namespace Application
```

第7章 多媒体资源

```csharp
{
    public partial class __16ViewController : UIViewController
    {
        NSObject playbackStateChanged;
        NSObject finishedPlaying;
        NSObject fullScreenChanged;
        MPMoviePlayerController moviePlayer;
        ……                        //这里省略了视图控制器的构造方法和析构方法
        #region View lifecycle
        public override void ViewDidLoad ()
        {
            base.ViewDidLoad ();
            moviePlayer = new MPMoviePlayerController(NSUrl.FromFilename
            ("Movie.m4v"));
            moviePlayer.View.Frame = new RectangleF(0f, 20f, this.View.
            Frame.Width, 320f);
            View.AddSubview(this.moviePlayer.View);
            //实现监听事件
            playbackStateChanged = MPMoviePlayerController.Notifications.
            ObservePlaybackStateDidChange(this.MoviePlayer_PlaybackStateChanged);
            finishedPlaying = MPMoviePlayerController.Notifications.
            ObservePlaybackDidFinish(this.MoviePlayer_FinishedPlayback);
            fullScreenChanged = MPMoviePlayerController.Notifications.
            ObserveDidEnterFullscreen (MoviePlayer_FullScreenChanged);
            //触摸按钮，播放视频
            pbtn.TouchUpInside += delegate {
                moviePlayer.Play();
            } ;
        }
        //监听视频播放的状态
        private void MoviePlayer_PlaybackStateChanged(object sender,
        NSNotificationEventArgs e)
        {
            Console.WriteLine("Movie player load state changed: {0}",
            this.moviePlayer.PlaybackState);
        }
        //此事件在视频播放后调用
        private void MoviePlayer_FinishedPlayback(object sender,
        NSNotificationEventArgs e)
        {
            Console.WriteLine("Movie player finished playing.");
        }
        //监听用户轻拍全屏控件并且控制器已进入全屏模式
        private void MoviePlayer_FullScreenChanged(object sender,
        NSNotificationEventArgs e)
        {
            Console.WriteLine("Full Screen state changed:{0}",
            this.moviePlayer.Fullscreen);
        }
        ……                        //这里省略了视图加载和卸载前后的一些方法
        #endregion
    }
}
```

运行效果如图 7.25 所示。

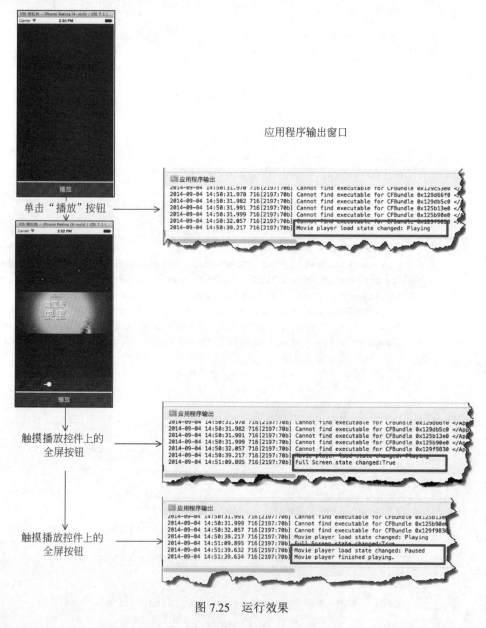

图 7.25 运行效果

7.4 播放音频

在 iOS 的应用或者游戏开发中对音频的支持也是必不可少的。本节将讲解在开发的应用中如何播放音频。

7.4.1 播放较短的音频文件

较短的音频文件专指一些系统声音，如短信铃声和闹铃等。系统声音的类 SystemSound

第 7 章 多媒体资源

包含在 MonoTouch.AudioToolbox 命名空间中。

【示例 7-12】 以下将使用一些较短的音频文件制作一个小钢琴。具体的操作步骤如下所述。

（1）创建一个 Single View Application 类型的工程。

（2）添加音频文件 1.mp3、2.mp3、3.mp3、4.mp3、5.mp3、6.mp3、7.mp3、C.mp3、D.mp3、E.mp3 和 F.mp3 到创建工程的 Resources 文件夹中。

（3）添加图像 keyBackground1.png、keyBackground2.png 和 pianoBackground.png 到创建工程的 Resources 文件夹中。

（4）打开 MainStoryboard.storyboard 文件，对主视图进行设置，效果如图 7.26 所示。

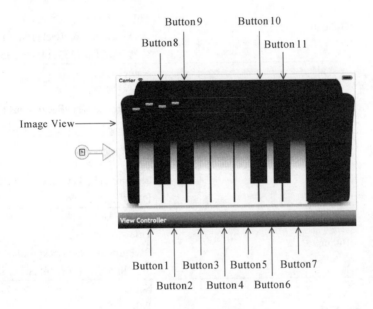

图 7.26　主视图的效果

需要添加的视图以及设置，如表 7-12 所示。

表 7-12　设置主视图

视　图	设　置
Button1	Name：Do Title：（空） Background：keyBackground1.png 位置和大小：(48, 145, 61, 153)
Button2	Name：Re Title：（空） Background：keyBackground1.png 位置和大小：(105, 145, 61, 153)
Button3	Name：Mi Title：（空） Background：keyBackground1.png 位置和大小：(162, 145, 61, 153)

续表

视图	设置
Button4	Name：Fa Title：（空） Background：keyBackground1.png 位置和大小：(219, 145, 61, 153)
Button5	Name：So Title：（空） Background：keyBackground1.png 位置和大小：(276, 145, 61, 153)
Button6	Name：La Title：（空） Background：keyBackground1.png 位置和大小：(333, 145, 61, 153)
Button7	Name：Si Title：（空） Background：keyBackground1.png 位置和大小：(391, 145, 61, 153)
Button8	Name：C Title：（空） Background：keyBackground2.png 位置和大小：(86, 145, 61, 153)
Button9	Name：D Title：（空） Background：keyBackground2.png 位置和大小：(142, 145, 61, 153)
Button10	Name：E Title：（空） Background：keyBackground2.png 位置和大小：(316, 145, 61, 153)
Button11	Name：F Title：（空） Background：keyBackground2.png 位置和大小：(372, 145, 61, 153)

（5）打开 7-10ViewController.cs 文件，编写代码，实现小钢琴的功能。代码如下：

```csharp
using System;
using System.Drawing;
using MonoTouch.Foundation;
using MonoTouch.UIKit;
using MonoTouch.AudioToolbox;
namespace Application
{
    public partial class __10ViewController : UIViewController
    {
        ……                          //这里省略了视图控制器的构造方法和析构方法
        #region View lifecycle
        public override void ViewDidLoad ()
        {
```

第 7 章 多媒体资源

```
        base.ViewDidLoad ();
        //触摸按钮播放 1.mp3 的音频文件
        Do.TouchUpInside += delegate {
            string path = NSBundle.MainBundle.PathForResource ("1", "mp3");
            NSUrl url = NSUrl.FromFilename (path);
            SystemSound aa = new SystemSound (url);
                                                //创建 SystemSound 类的对象
            aa.PlaySystemSound ();              //播放声音
        };
        //触摸按钮播放 2.mp3 的音频文件
        Re.TouchUpInside += delegate {
            string path = NSBundle.MainBundle.PathForResource ("2", "mp3");
            NSUrl url = NSUrl.FromFilename (path);
            SystemSound aa = new SystemSound (url);
            aa.PlaySystemSound ();
        };
        ……                    //这里省略了 3.mp3~E.mp3 音频文件的播放
        //触摸按钮播放 F.mp3 的音频文件
        F.TouchUpInside += delegate {
            string path = NSBundle.MainBundle.PathForResource ("F", "mp3");
            NSUrl url = NSUrl.FromFilename (path);
            SystemSound aa = new SystemSound (url);
            aa.PlaySystemSound ();
        };

    }
    ……                       //这里省略了视图加载和卸载前后的一些方法
    #endregion
  }
}
```

运行效果如图 7.27 所示。

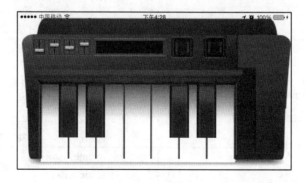

图 7.27 运行效果

注意：在程序中，PlaySystemSound()方法实现的功能就是让音频文件开始播放。

7.4.2 播放较长的音频文件

MonoTouch.AVFoundation 命名空间中的 AVAudioPlayer 类可以用来实现对较长音频文件的播放。

【示例 7-13】 以下将在应用程序中实现一首完整歌曲的播放。具体的操作步骤如下：
（1）创建一个 Single View Application 类型的工程。

（2）添加音乐文件 Music.mp3 到创建工程的 Resources 文件夹中。

（3）打开 MainStoryboard.storyboard 文件，对主视图进行设置，效果如图 7.28 所示。

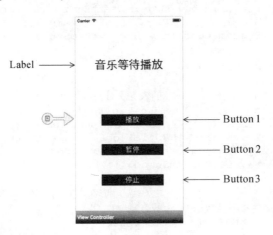

图 7.28　主视图的效果

需要添加的视图以及设置，如表 7-13 所示。

表 7-13　设置主视图

视　　图	设　　置
Button1	Name：playbtn Title：播放 Font：System 21 pt Text Color：白色 Background：黑色 位置和大小：(78, 285, 184, 32)
Button2	Name：pasuebtn Title：暂停 Font：System 21 pt Text Color：白色 Background：黑色 位置和大小：(78, 373, 184, 32)
Button3	Name：stopbtn Title：停止 Font：System 21 pt Text Color：白色 Background：黑色 位置和大小：(78, 462, 184, 32)
Label	Name：label Text：音乐等待播放 Font：System 34 pt Alignment：居中 Lines：2 位置和大小：(20, 81, 280, 119)

第 7 章 多媒体资源

（4）打开 7-8ViewController.cs 文件，编写代码，实现播放歌曲的功能。代码如下：

```csharp
using System;
using System.Drawing;
using MonoTouch.Foundation;
using MonoTouch.UIKit;
using MonoTouch.AVFoundation;
namespace Application
{
    public partial class __8ViewController : UIViewController
    {
        AVAudioPlayer audioplayer;
        ……                              //这里省略了视图控制器的构造方法和析构方法
        #region View lifecycle
        public override void ViewDidLoad ()
        {
            base.ViewDidLoad ();
            string path = NSBundle.MainBundle.PathForResource ("Music", "mp3");
            NSUrl url = NSUrl.FromFilename (path);
            audioplayer = new AVAudioPlayer (url, null);
                                                      //实例化对象audioplayer
            audioplayer.PrepareToPlay ();
            //触摸按钮，播放歌曲
            playbtn.TouchUpInside += (sender, e) => {
                audioplayer.Play ();
                label.Text="音乐正在播放";
            } ;
            //触摸按钮，暂停播放的歌曲
            pasuebtn.TouchUpInside += (sender, e) => {
                audioplayer.Pause ();
                label.Text="音乐暂停中，正在等待音乐播放";
            } ;
            //触摸按钮，停止播放的歌曲
            stopbtn.TouchUpInside += (sender, e) => {
                audioplayer.Stop ();
                label.Text="音乐停止，重新开始播放音乐";
            } ;
        }
        ……                              //这里省略了视图加载和卸载前后的一些方法
        #endregion
    }
}
```

运行效果如图 7.29 所示。

注意：在此程序中对于歌曲的播放使用了 Play()方法；对于播放歌曲的暂停使用了 Pause() 方法；对于播放歌曲的停止使用了 Stop()方法。

在一般的音乐播放器中，会使用到音量大小的调节，以及显示音乐播放的进度。其中，对音量的调节需要使用 Volume 属性实现，对于播放进度的显示需要使用 CurrentTime 属性实现。

第 2 篇　资源使用篇

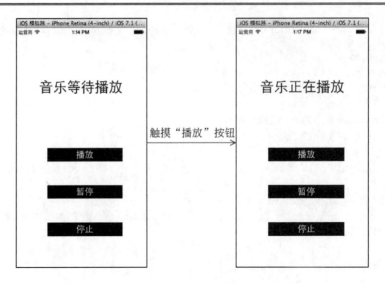

图 7.29　运行效果

【示例 7-14】以下将使用 Volume 属性实现对音量的调节，以及使用 CurrentTime 属性实现播放进度的显示。具体的操作步骤如下所述。

（1）创建一个 Single View Application 类型的工程。
（2）添加音乐文件 Music.mp3 和图像 1.jpg 到创建工程的 Resources 文件夹中。
（3）打开 MainStoryboard.storyboard 文件，对主视图进行设置，效果如图 7.30 所示。

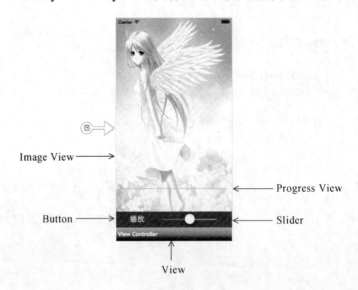

图 7.30　主视图的效果

需要添加的视图以及设置，如表 7-14 所示。

表 7-14　设置主视图

视　　图	设　　置
Image View	Image：1.jpg 位置和大小：(0, 0, 320, 525)

续表

视图	设置
Progress View	Name：progressView Progress：0 位置和大小：(10, 461, 296, 2)
Button	Name：pbtn Title：播放 Font：System 20 pt Text Color：白色 位置和大小：(34, 6, 46, 30)
Slider	Name：slider 位置和大小：(124, 6, 156, 31)
View	Background：深灰色 位置和大小：(0, 525, 320, 43)

（4）打开 7-17ViewController.cs 文件，编写代码，实现声音的调节以及显示播放的进度。代码如下：

```
using System;
using System.Drawing;
using MonoTouch.Foundation;
using MonoTouch.UIKit;
using MonoTouch.AVFoundation;
using System.Threading;
using System.Threading.Tasks;
namespace Application
{
    public partial class __17ViewController : UIViewController
    {
        AVAudioPlayer audioplayer;
        ……                      //这里省略了视图控制器的构造方法和析构方法
        #region View lifecycle
        public override void ViewDidLoad ()
        {
            base.ViewDidLoad ();
            string path = NSBundle.MainBundle.PathForResource ("Music", "mp3");
            NSUrl url = NSUrl.FromFilename (path);
            audioplayer = new AVAudioPlayer (url, null);
                                                    //实例化对象audioplayer
            audioplayer.PrepareToPlay ();
            pbtn.TouchUpInside += (sender, e) => {
                audioplayer.Play ();                //播放音乐
                Task.Factory.StartNew(this.StartProgress);
            } ;
            //调节音量
            slider.ValueChanged += (sender, e) => {
                audioplayer.Volume = slider.Value;
            };
        }
        //实现音乐播放进度的显示
        public void StartProgress ()
```

```
        {
            float currentProgress = 0f;
            //判断当前的进度，并循环
            while (currentProgress < 1f)
            {
                Thread.Sleep(1000);                    //1000毫秒后暂停当前线程
                this.InvokeOnMainThread(delegate {
                    double a= audioplayer.CurrentTime / audioplayer.Duration;
                    progressView.Progress = (float)a;
                } );
            }
        }
        ……                              //这里省略了视图加载和卸载前后的一些方法
        #endregion
    }
}
```

运行效果如图 7.31 所示。

图 7.31　运行效果

7.4.3　访问音乐库

开发者如果不想在应用程序中播放添加到 Resources 文件中的音频文件，可以播放音乐库中的音频。在每一个 iOS 设备中都有一个音乐库，在其中保存音频文件，并可以对这些音频文件进行播放和删除等操作。如果想要在其他的应用程序中访问音乐库，并且播放其中的音频需要使用到 MonoTouch.MediaPlayer 命名空间中的 MPMediaPickerController 类。

【示例 7-15】以下将实现对音乐库的访问。具体的操作步骤如下所述。

（1）创建一个 Single View Application 类型的工程。

（2）添加图像 1.png 到创建工程的 Resources 文件夹中。

（3）打开 MainStoryboard.storyboard 文件，对主视图进行设置，效果如图 7.32 所示。

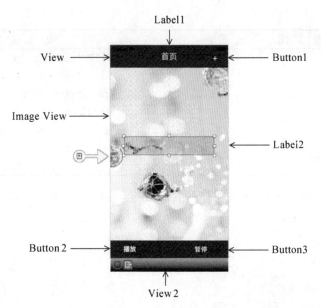

图 7.32 主视图的效果

需要添加的视图以及设置,如表 7-15 所示。

表 7-15 设置主视图

视　　图	设　　置
View1	Background:深灰色 位置和大小:(0, 0, 320, 61)
Image View	Image:1.png 位置和大小:(0, 60, 320, 462)
Label1	Text:首页 Color:白色 Font:System 22 pt Alignment:居中 位置和大小:(106, 14, 108, 33)
Label2	Name:label Text:(空) Font:System Bold 23 pt Alignment:居中 位置和大小:(39, 246, 242, 47)
Button1	Name:showbtn Title:+ Font:System 22 pt Text Color:白色 位置和大小:(262, 20, 46, 26)
Button2	Name:playbtn Title:播放 Font:System Bold 17 pt Text Color:白色 位置和大小:(29, 8, 46, 30)

视 图	设 置
Button3	Name：pausebtn Title：暂停 Font：System Bold 17 pt Text Color：白色 位置和大小：(223, 8, 46, 30)
View	Background：深灰色 位置和大小：(0, 522, 320, 46)

（4）打开 7-9ViewController.cs 文件，编写代码，实现音乐库的访问，以及选择音乐库中的音乐进行播放。代码如下：

```
using System;
using System.Drawing;
using MonoTouch.Foundation;
using MonoTouch.UIKit;
using MonoTouch.MediaPlayer;
namespace Application
{
    public partial class __9ViewController : UIViewController
    {
        private MPMediaPickerController mediaPicker;
        private MPMusicPlayerController musicPlayer;
        ……                          //这里省略了视图控制器的构造方法和析构方法
        #region View lifecycle
        public override void ViewDidLoad ()
        {
            base.ViewDidLoad ();
            mediaPicker = new MPMediaPickerController(MPMediaType.Music);
            mediaPicker.ItemsPicked += MediaPicker_ItemsPicked;
            mediaPicker.DidCancel += MediaPicker_DidCancel;
            musicPlayer= MPMusicPlayerController.ApplicationMusicPlayer;
            mediaPicker.ItemsPicked+=MediaPicker_ItemsPicked;  //选择歌曲
            //触摸按钮，打开音乐库
            showbtn.TouchUpInside += async delegate {
                await this.PresentViewControllerAsync(mediaPicker,true);
            };
            //播放歌曲
            playbtn.TouchUpInside += delegate {
                musicPlayer.Play();
                label.Text="所选择的音乐处于播放状态";
            };
            //暂停歌曲
            pasuebtn.TouchUpInside += delegate {
                musicPlayer.Pause();
                label.Text="所选择的音乐处于暂停状态";
            } ;
        }
        //选择歌曲
        private async void MediaPicker_ItemsPicked (object sender,
        ItemsPickedEventArgs e)
        {
```

```
            this.musicPlayer.SetQueue(e.MediaItemCollection);
            await this.DismissViewControllerAsync(true);
        }
        //取消选择的歌曲
        private async void MediaPicker_DidCancel (object sender, EventArgs e)
        {
            await this.mediaPicker.DismissViewControllerAsync(true);
        }
        ……                          //这里省略了视图加载和卸载前后的一些方法
        #endregion
    }
}
```

运行效果如图 7.33 所示。

图 7.33 运行效果

> **注意**：在使用 MPMediaPickerController 类时，首先要对这个类进行实例化，在此程序中实例化的代码如下：

```
mediaPicker = new MPMediaPickerController(MPMediaType.Music);
```

其中，括号内的内容 MPMediaType.Music 用来指定一个媒体项目。媒体项目如表 7-16 所示。

表 7-16 媒体项目

名 称	功 能
音频媒体项目	
Music	媒体项目包含音乐
Podcast	媒体项目包含博客
AudioBook	媒体项目包含有声书
AudioITunesU	媒体项目包含了 iTunes U 音频
AnyAudio	媒体项目包含一个未指定类型的音频内容
视频媒体项目	
Movie	媒体项目包含影片
TVShow	媒体项目包含一个电视节目
VideoPodcast	媒体项目包含一个视频播客
MusicVideo	媒体项目包含了一个音乐视频
VideoITunesU	媒体项目包含了 iTunes U 视频
AnyVideo	媒体项目包含一个未指定类型的视频内容
HomeVideo	媒体项目包含一个家庭录像
TypeAnyVideo	体项目包含一个视频
通用媒体项目	
Any	媒体项目包含一个未指定类型的音频

7.5 使用麦克风录音

麦克风录音在很多的应用程序中都会使用到，如会说话的 Tom 猫、会说话的狗狗本等。本节将讲解如何使用麦克风录音。

【示例 7-16】以下将使用麦克风录音，并将录制的声音进行播放。具体的操作步骤如下所述。

（1）创建一个 Single View Application 类型的工程。

（2）添加图像 1.jpg 和 2.png 到创建工程的 Resources 文件夹中。

（3）打开 MainStoryboard.storyboard 文件，对主视图进行设置，效果如图 7.34 所示。

第 7 章 多媒体资源

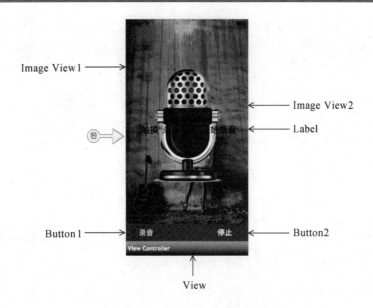

图 7.34 主视图的效果

需要添加的视图以及设置，如表 7-17 所示。

表 7-17 设置主视图

视　　图	设　　置
Image View1	Name：imageView1 Image：1.jpg 位置和大小：(0, 0, 320, 523)
Image View2	Name：imageView2 Image：2.png 位置和大小：(54, 113, 212, 312)
Button1	Name：btnStar Title：录音 Font：System Bold 19 pt Text Color：白色 位置和大小：(30, 8, 46, 30)
Button2	Name：btnStop Title：停止 Font：System Bold 19 pt Text Color：白色 位置和大小：(236, 8, 46, 30)
Label	Name：label Text：触摸"录音"按钮开始录音 Font：System 23 pt Alignment：居中 位置和大小：(38, 261, 254, 43)
View	Background：深灰色 位置和大小：(0, 523, 320, 45)

（4）打开 7-11ViewController.cs 文件，编写代码，实现录音以及播放录音的功能。代码如下：

```csharp
using System;
using System.Drawing;
using MonoTouch.Foundation;
using MonoTouch.UIKit;
using MonoTouch.AVFoundation;
using MonoTouch.AudioToolbox;
using System.IO;
namespace Application
{
    public partial class __11ViewController : UIViewController
    {
        private AVAudioRecorder audioRecorder;
        ……                          //这里省略了视图控制器的构造方法和析构方法
        #region View lifecycle
        public override void ViewDidLoad ()
        {
            base.ViewDidLoad ();
            imageView1.Hidden = true;
            imageView2.Hidden = true;
            NSUrl soundFileUrl = null;
            NSError error;
            //实例化对设置音频会话对象session
            AVAudioSession session = AVAudioSession.SharedInstance();
            session.SetCategory(AVAudioSession.CategoryPlayAndRecord, out error);
            session.SetActive(true, out error);
            bool grantedPermission = false;
            //用来查询用户是否允许应用程序使用麦克风
            session.RequestRecordPermission((granted) => {
                Console.WriteLine("Permission granted: {0}", granted);
                if (granted)
                {
                    grantedPermission = true;
                    //设置用来保存录制声音的文件
                    string soundFile = Path.Combine(Environment.GetFolderPath
                        (Environment.SpecialFolder.Personal), "sound.wav");
                    soundFileUrl = new NSUrl(soundFile);
                    //对录音的设置
                    NSDictionary recordingSettings =
                        NSDictionary.FromObjectAndKey(AVAudioSettings.
                        AVFormatIDKey, NSNumber.FromInt32((int)AudioFile
                        Type.WAVE));
                    audioRecorder = AVAudioRecorder.ToUrl(soundFileUrl,
                        recordingSettings, out error);
                }
            } );
            //录音
            btnStar.TouchUpInside += delegate {
                if (grantedPermission)
                {
                    imageView1.Hidden = true;
                    imageView2.Hidden = false;
                    label.Hidden=true;
                    this.audioRecorder.Record();
                }
            } ;
            //停止录音，并播放录制的声音
```

```
            btnStop.TouchUpInside += delegate {
                if (grantedPermission)
                {
                    this.audioRecorder.Stop();
                    imageView2.Hidden = true;
                    imageView1.Hidden = false;
                    AVAudioPlayer player = AVAudioPlayer.FromUrl(soundFileUrl);
                    player.Play();
                }
            };
        }
        ……                              //这里省略了视图加载和卸载前后的一些方法
        #endregion
    }
}
```

运行效果如图 7.35 所示。

图 7.35　运行效果

> **注意**：在此程序中使用了两个关键的类，一个是 AVAudioSession 类，另一个是 AVAudioRecorder 类。其中，AVAudioSession 类用来指定关于声音的策略，这个策略将解决当屏幕被锁定时，声音使用停止等问题。AVAudioRecorder 类提供了录音的功能。

7.6　直接管理相册

本节将采用编程的方式对相册进行访问和管理。

7.6.1　获取相册中内容的路径

ALAsset 的 DefaultRepresentation 属性可以用来获取相册中每一个资源的路径。

【示例7-17】 以下将实现获取相册中照片的路径。具体的操作步骤如下所述。

（1）创建一个 Single View Application 类型的工程。

（2）打开 MainStoryboard.storyboard 文件，从工具栏中拖动 Button 对象对主视图中，将此对象的 Name 设置为 btnEnumerate，Ttile 设置为直接访问相册，System 设置为 System 22 pt，Text Color 设置为白色，Background 设置为深灰色，位置和大小设置为（0,250,320,43）。

（3）打开 7-12ViewController.cs 文件，编写代码，实现获取相册中照片的路径。代码如下：

```
using System;
using System.Drawing;
using MonoTouch.Foundation;
using MonoTouch.UIKit;
using MonoTouch.AssetsLibrary;
namespace Application
{
    public partial class __12ViewController : UIViewController
    {
        private ALAssetsLibrary assetsLibrary;
        ……                              //这里省略了视图控制器的构造方法和析构方法
        #region View lifecycle
        public override void ViewDidLoad ()
        {
            base.ViewDidLoad ();
            //触摸按钮，获取路径
            this.btnEnumerate.TouchUpInside += delegate {
                if (ALAssetsLibrary.AuthorizationStatus == ALAuthorization
                    Status.Authorized ||
                    ALAssetsLibrary.AuthorizationStatus == ALAuthorization
                    Status.NotDetermined)
                {
                    this.assetsLibrary = new ALAssetsLibrary();
                    //遍历相册中的资源
                    this.assetsLibrary.Enumerate(ALAssetsGroupType.All,
                    this.GroupsEnumeration, this.GroupsEnumerationFailure);
                }
            } ;
        }
        //遍历相册组
        private void GroupsEnumeration(ALAssetsGroup assetGroup, ref bool stop)
        {
            if (null != assetGroup)
            {
                stop = false;
                assetGroup.SetAssetsFilter(ALAssetsFilter.AllPhotos);
                                                        //设置组的过滤器
                assetGroup.Enumerate(this.AssetEnumeration);
                                                        //遍历在组中所有的照片
            }
        }
        //遍历单个照片，并输出
        private void AssetEnumeration(ALAsset asset, int index, ref bool stop)
        {
```

```
            if (null != asset)
            {
                stop = false;
                //输出单个照片信息
                Console.WriteLine("Asset url: {0}", asset.
                DefaultRepresentation.Url.AbsoluteString);
            }
        }
        //遍历相册中的资源失败后调用
        private void GroupsEnumerationFailure(NSError error)
        {
            if (null != error)
            {
                Console.WriteLine("Error enumerating asset groups! Message:
                {0}", error.LocalizedDescription);
            }
        }
        ……                      //这里省略了视图加载和卸载前后的一些方法
        #endregion
    }
}
```

运行效果如图 7.36 所示。

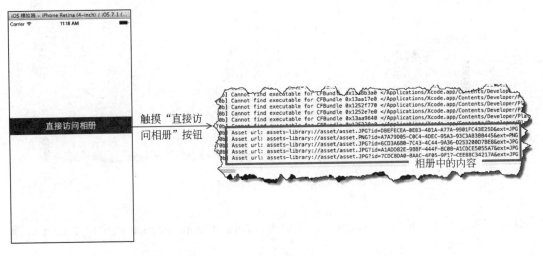

图 7.36 运行效果

7.6.2 读取相册中 EXIF 数据

EXIF 是一种图像文件格式，它的数据存储与 JPEG 格式是完全相同的。实际上 EXIF 格式就是在 JPEG 格式头部插入了数码照片的信息，包括拍摄时的光圈、快门、白平衡、ISO、焦距、日期和时间以及相机品牌、型号、色彩编码、拍摄时录制的声音、全球定位系统（GPS）和缩略图等。本小节将讲解如何获取相册中的 EXIF 数据。

【示例 7-18】 以下以【示例 7-17】为基础，获取相册中照片的 EXIT 数据。代码如下：

```
private void AssetEnumeration(ALAsset asset, int index, ref bool stop)
```

```
{
    if (null != asset)
    {
        stop = false;
        NSDictionary metaData = asset.DefaultRepresentation.Metadata;
        if (null != metaData)
        {
            NSDictionary exifData = (NSDictionary)metaData[new NSString("{Exif}")];
            Console.WriteLine("Asset url: {0}", exifData);          //输出
        }
    }
}
```

运行效果如图 7.37 所示。

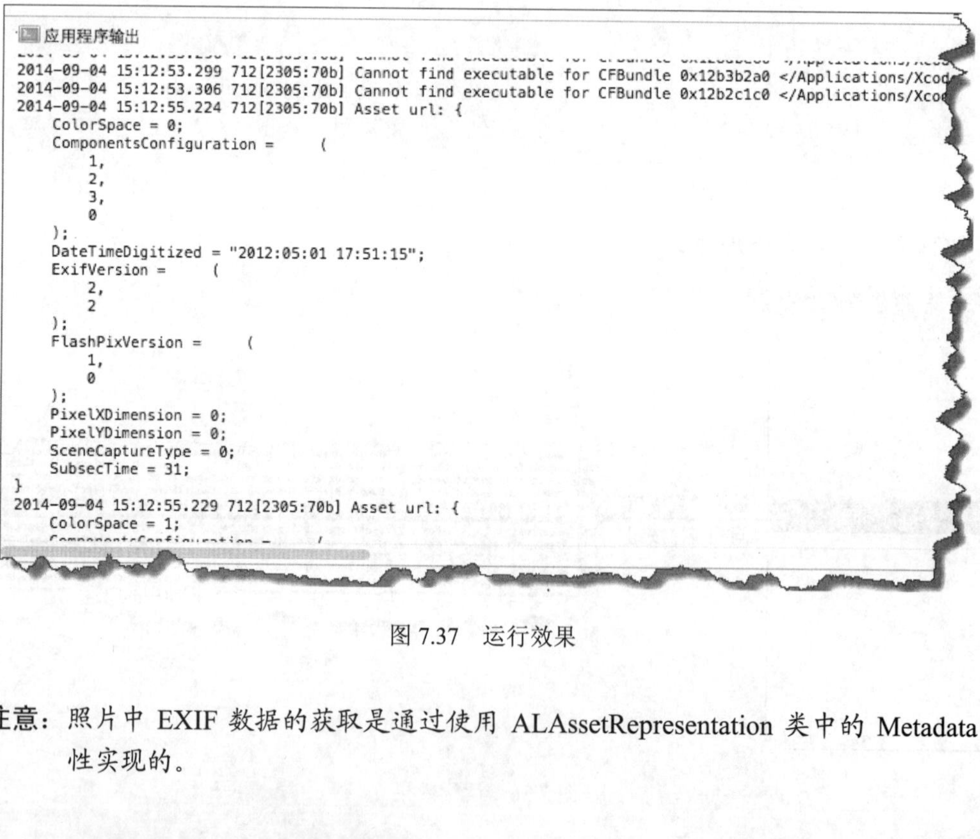

图 7.37　运行效果

> 注意：照片中 EXIF 数据的获取是通过使用 ALAssetRepresentation 类中的 Metadata 属性实现的。

7.6.3　获取相册中的实际的照片

使用编程的方式不仅可以获取相册中照片的路径和 EXIT 数据，还可以通过 DefaultRepresentation 属性获取实际的照片。

【示例 7-19】 以下将使用 DefaultRepresentation 属性获取相册中实际的照片。具体的操作步骤如下所述。

（1）创建一个 Single View Application 类型的工程。

（2）打开 MainStoryboard.storyboard 文件，对主视图进行设置，效果如图 7.38 所示。

第 7 章 多媒体资源

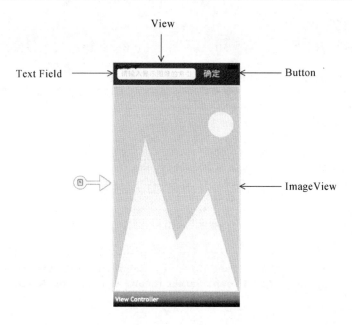

图 7.38 主视图的效果

需要添加的视图以及设置，如表 7-18 所示。

表 7-18 设置主视图

视　　图	设　　置
View	Background：深灰色 位置和大小：(0, 0, 320, 55)
Text Field	Name：textField Text：（空） Font：System 28 pt Alignment：居中 Placeholder：请输入显示图像的索引 Keyboard：Number Pad 位置和大小：(9, 12, 200, 30)
Button	Name：btn Title：确定 Font：System 21 pt Text Color：白色 位置和大小：(226, 12, 46, 30)
Image View	Name：imageView 位置和大小：(0, 56, 320, 512)

（3）打开 7-13ViewController.cs 文件，编写代码，实现照片的获取。代码如下：

```
using System;
using System.Drawing;
using MonoTouch.Foundation;
using MonoTouch.UIKit;
using MonoTouch.AssetsLibrary;
using MonoTouch.CoreGraphics;
```

```csharp
namespace Application
{
    public partial class __13ViewController : UIViewController
    {
        private ALAssetsLibrary assetsLibrary;
        ……                              //这里省略了视图控制器的构造方法和析构方法
        #region View lifecycle
        public override void ViewDidLoad ()
        {
            base.ViewDidLoad ();
            btn.TouchUpInside += delegate {
                if (ALAssetsLibrary.AuthorizationStatus ==
                ALAuthorizationStatus.Authorized ||
                    ALAssetsLibrary.AuthorizationStatus ==
                    ALAuthorizationStatus.NotDetermined)
                {
                    this.assetsLibrary = new ALAssetsLibrary();
                    this.assetsLibrary.Enumerate(ALAssetsGroupType.All,
                    this.GroupsEnumeration, this.GroupsEnumerationFailure);
                    textField.ResignFirstResponder();
                }
            };
        }
        //遍历相册组
        private void GroupsEnumeration(ALAssetsGroup assetGroup, ref bool stop)
        {
            if (null != assetGroup)
            {
                stop = false;
                assetGroup.SetAssetsFilter(ALAssetsFilter.AllPhotos);
                assetGroup.Enumerate(this.AssetEnumeration);
            }
        }
        //遍历单个照片
        private void AssetEnumeration(ALAsset asset, int index, ref bool stop)
        {
            if (null != asset)
            {
                stop = false;
                int myselect=int.Parse(textField.Text);  //将字符串转换为整型
                if (index == myselect) {
                    //获取照片
                    CGImage aa = asset.DefaultRepresentation.GetImage();
                    imageView.Image = UIImage.FromImage (aa);
                }
            }
        }
        //遍历相册中的资源失败后调用
        private void GroupsEnumerationFailure(NSError error)
        {
            if (null != error)
            {
                Console.WriteLine("Error enumerating asset groups! Message:
                {0}", error.LocalizedDescription);
            }
        }
```

```
       ......                    //这里省略了视图加载和卸载前后的一些方法
        #endregion
    }
}
```

运行效果如图 7.39 所示。

图 7.39 运行效果

第 8 章 内置应用程序

iOS 移动设备提供了很多内置的应用程序给用户，其中包括 Mail、Phone、Safari 和 SMS 等。这使得它长期以来都是最受欢迎的移动设备之一。作为一名 iOS 开发人员，可以使用编程的方式使用这些内置的应用程序。本章将讲解这些应用程序的使用。

8.1 打 电 话

在 iOS 移动设备中都有打电话的功能，即 Phone 应用程序，如果想要调用此应用程序，需要使用如下的代码：

```
UIApplication.SharedApplication.OpenUrl (NSUrl 对象);
```

其中，NSUrl 对象是由字符串初始化的，此字符串必须以"tel:"作为前缀。

【示例 8-1】 以下将实现打电话的功能，当用户在文本框中输入电话号码后，触摸具有电话图标的按钮，就会自动调用内置的 Phone 应用程序。具体的操作步骤如下所述。

（1）创建一个 Single View Application 类型的工程。
（2）添加图像 1.jpg 和 2.png 到创建工程的 Resources 文件夹中。
（3）打开 MainStoryboard.storyboard 文件，对主视图进行设置，效果如图 8.1 所示。

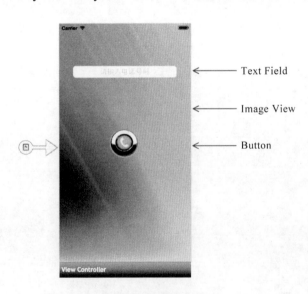

图 8.1 主视图的效果

需要添加的视图以及设置，如表 8-1 所示。

第 8 章 内置应用程序

表 8-1 设置主视图

视图	设置
Text Field	Name：textField Text：（空） Font：System 17 pt Alignment：居中 Placeholder：请输入电话号码 Keyboard：Number Pad 位置和大小：(33, 100, 254, 30)
Image View	Image：1.jpg 位置和大小：(0, 0, 320, 568)
Button	Name：btn Title：（空） Background：2.png 位置和大小：(122, 250, 75, 75)

（4）打开 8-1ViewController.cs 文件，编写代码，实现打电话的功能。代码如下：

```
using System;
using System.Drawing;
using MonoTouch.Foundation;
using MonoTouch.UIKit;
namespace Application
{
    public partial class __1ViewController : UIViewController
    {
        ……                              //这里省略了视图控制器的构造方法和析构方法
        #region View lifecycle
        public override void ViewDidLoad ()
        {
            base.ViewDidLoad ();
            //触摸按钮，调用 Phone 应用程序
            btn.TouchUpInside += (sender, e) => {
                textField.ResignFirstResponder();
                string number = string.Format ("tel:{0}", textField.Text);
                NSUrl url = new NSUrl (number);        //创建并初始化 url 对象
                //判断 Phone 应用程序是否可用
                if (UIApplication.SharedApplication.CanOpenUrl (url)) {
                    UIApplication.SharedApplication.OpenUrl (url);
                                                        //调用 Phone 应用程序
                } else {
                    UIAlertView alertView=new UIAlertView ();
                    alertView.Title="对不起,你没有打电话的软件";
                    alertView.AddButton("Cancel");
                    alertView.Show();
                }
            };
        }
        ……                              //这里省略了视图加载和卸载前后的一些方法
        #endregion
    }
}
```

运行效果如图 8.2 所示。

第 2 篇 资源使用篇

图 8.2 运行效果

注意：此应用程序必须使用真机进行测试，否则不会调用 Phone 应用程序，而是弹出"对不起，你没有打电话的软件"的警告视图。

8.2 使用 Safari

很多的开发者为了方便，在制作网页浏览器时，都会调用内置的 Safari 应用程序。它的调用需要使用以下的代码：

```
UIApplication.SharedApplication.OpenUrl (NSUrl 对象);
```

其中，NSUrl 对象是由字符串初始化的，这个字符串一定要是一个完整的网址。

【示例 8-2】以下将通过对 Safari 应用程序的调用实现一个简单的网页浏览器。具体的操作步骤如下所述。

（1）创建一个 Single View Application 类型的工程。
（2）添加图像 1.jpg 到创建工程的 Resources 文件夹中。
（3）打开 MainStoryboard.storyboard 文件，对主视图进行设置，效果如图 8.3 所示。

第 8 章 内置应用程序

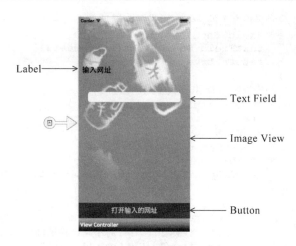

图 8.3 主视图的效果

需要添加的视图以及设置,如表 8-2 所示。

表 8-2 设置主视图

视 图	设 置
Label	Text:输入网址 Font:System Bold 21 pt 位置和大小:(10, 134, 199, 32)
Text Field	Name:textField Text:(空) Keyboard:URL 位置和大小:(26, 211, 268, 30)
Image View	Image:1.jpg 位置和大小:(0, 0, 320, 568)
Button	Name:btn Title:打开输入的网址 Font:System Bold 21 pt Text Color:白色 Background:深灰色 位置和大小:(0, 525, 320, 43)

(4)打开 8-2ViewController.cs 文件,编写代码,实现内置 Safari 应用程序的调用。代码如下:

```
using System;
using System.Drawing;
using MonoTouch.Foundation;
using MonoTouch.UIKit;
namespace Application
{
    public partial class __2ViewController : UIViewController
    {
        ……                              //这里省略了视图控制器的构造方法和析构方法
        #region View lifecycle
        public override void ViewDidLoad ()
        {
            base.ViewDidLoad ();
```

```
            //触摸键盘上的 return 键关闭键盘
            textField.ShouldReturn=delegate{
                textField.ResignFirstResponder();
                return true;
            };
            //触摸按钮, 调用 Safari 应用程序
            btn.TouchUpInside += (sender, e) => {
                string number = string.Format ("{0}", textField.Text);
                NSUrl url = new NSUrl (number);
                //判断 Safari 应用程序是否可用
                if (UIApplication.SharedApplication.CanOpenUrl (url)) {
                    UIApplication.SharedApplication.OpenUrl (url);
                                                //打开 Safari 应用程序
                } else {
                    UIAlertView alertView=new UIAlertView ();
                    alertView.Title="对不起，你的网址有误";
                    alertView.AddButton("Cancel");
                    alertView.Show();
                }
            };
        }
        ……                        //这里省略了视图加载和卸载前后的一些方法
        #endregion
    }
}
```

运行效果如图 8.4 所示。

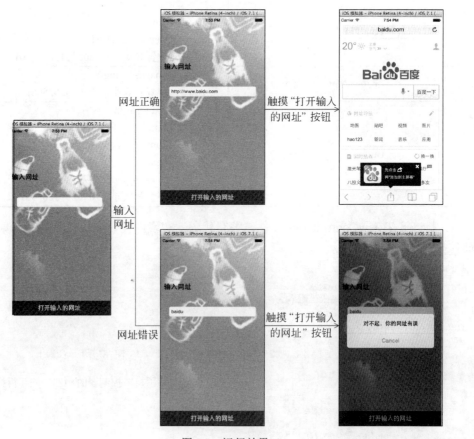

图 8.4　运行效果

8.3 发送短信和电子邮件

在 iOS 的移动设备中,有发送短信和电子邮件的内置应用程序,分别为 SMS 应用程序和 Email 应用程序。本节将讲解这两个内置应用程序的调用。

8.3.1 发送短信

发送短信需要调用 SMS 内置应用程序。它的调用需要使用以下的代码:

```
UIApplication.SharedApplication.OpenUrl (NSUrl 对象);
```

其中,NSUrl 对象是由字符串初始化的,此字符串必须要以"sms:"作为前缀。

【示例 8-3】 以下将实现当用户输入电话号码,并触摸"发送短信"按钮后,实现对 SMS 应用程序的调用。具体的操作步骤如下所述。

(1)创建一个 Single View Application 类型的工程。
(2)添加图像 1.jpg 到创建工程的 Resources 文件夹中。
(3)打开 MainStoryboard.storyboard 文件,对主视图进行设置,效果如图 8.5 所示。

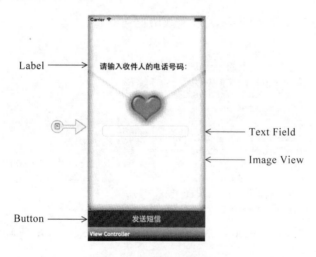

图 8.5 主视图的效果

需要添加的视图以及设置,如表 8-3 所示。

表 8-3 设置主视图

视 图	设 置
Label	Text:请输入收件人的电话号码: Font:System Bold 21 pt Alignment:居中 位置和大小:(20, 118, 274, 38)
Text Field	Name:tf Text:(空) 位置和大小:(39, 295, 236, 30)

续表

视　　图	设　　置
Image View	Image：1.jpg 位置和大小：(0, 0, 320, 524)
Button	Name：btn Title：发送短信 Font：System 20 pt Text Color：白色 Background：深灰色 位置和大小：(0, 524, 320, 44)

（4）打开 8-3ViewController.cs 文件，编写代码，实现对 SMS 应用程序的调用。代码如下：

```
using System;
using System.Drawing;
using MonoTouch.Foundation;
using MonoTouch.UIKit;
namespace Application
{
    public partial class __3ViewController : UIViewController
    {
        ……                               //这里省略了视图控制器的构造方法和析构方法
        #region View lifecycle
        public override void ViewDidLoad ()
        {
            base.ViewDidLoad ();
            //触摸键盘上的 return 键关闭键盘
            tf.ShouldReturn=delegate{
                tf.ResignFirstResponder();
                return true;
            };
            //触摸按钮，实现 SMS 应用程序的调用
            btn.TouchUpInside += (s, e) => {
                string number = string.Format ("sms:{0}", tf.Text);
                NSUrl textUrl = new NSUrl (number);
                //判断 SMS 应用程序是否可用
                if (UIApplication.SharedApplication.CanOpenUrl(textUrl))
                {
                    UIApplication.SharedApplication.OpenUrl(textUrl);
                                                       //调用 SMS 应用程序
                } else
                {
                    UIAlertView alertView=new UIAlertView ();
                    alertView.Title="对不起，不能发送短信";
                    alertView.AddButton("Cancel");
                    alertView.Show();
                }
            } ;
        }
        ……                               //这里省略了视图加载和卸载前后的一些方法
        #endregion
    }
}
```

运行效果如图 8.6 所示。

第 8 章 　内置应用程序

图 8.6　运行效果

注意：此程序需要在真机上运行。在 iOS Simulator 上没有提供内置的 SMS 应用程序。

8.3.2　发送电子邮件

电子邮件的发送需要使用 Email 内置应用程序。此程序的调用代码如下：

UIApplication.SharedApplication.OpenUrl (NSUrl 对象);

其中，NSUrl 对象是由字符串初始化的，此字符串必须以"mailto:"作为前缀。

【示例 8-4】　以下将实现当用户输入电话号码，并触摸"发送电子邮件"按钮后，实现对 Email 应用程序的调用。具体的操作步骤如下所述。

（1）创建一个 Single View Application 类型的工程。
（2）添加图像 1.jpg 到创建工程的 Resources 文件夹中。
（3）打开 MainStoryboard.storyboard 文件，对主视图进行设置，效果如图 8.7 所示。

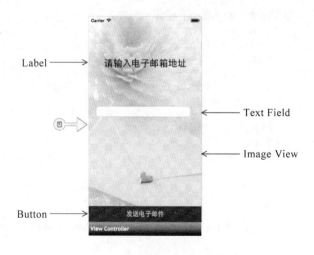

图 8.7　主视图的效果

需要添加的视图以及设置，如表 8-4 所示。

表 8-4 设置主视图

视　　图	设　　置
Label	Text：请输入电子邮箱地址: Font：System Bold 21 pt Alignment：居中 位置和大小：(20, 118, 274, 38)
Text Field	Name：tf Text：（空） 位置和大小：(39, 295, 236, 30)
Image View	Image：1.jpg 位置和大小：(0, 0, 320, 524)
Button	Name：btn Title：发送电子邮件 Font：System 20 pt Text Color：白色 Background：深灰色 位置和大小：(0, 524, 320, 44)

（4）打开 8-4ViewController.cs 文件，编写代码，实现对 Email 应用程序的调用。代码如下：

```
using System;
using System.Drawing;
using MonoTouch.Foundation;
using MonoTouch.UIKit;
namespace Application
{
    public partial class __4ViewController : UIViewController
    {
        ……                              //这里省略了视图控制器的构造方法和析构方法
        #region View lifecycle
        public override void ViewDidLoad ()
        {
            base.ViewDidLoad ();
            //触摸键盘上的 return 键，关闭键盘
            tf.ShouldReturn=delegate{
                tf.ResignFirstResponder();
                return true;
            };
            //触摸按钮，调用 Email 应用程序
            btn.TouchUpInside += (s, e) => {
                string eamil= string.Format ("mailto:{0}", tf.Text);
                //判断在字符串中是否包含"@"
                if(eamil.IndexOf("@")>-1){
                    NSUrl emailUrl = new NSUrl(eamil);
                    //判断 Eail 应用程序是否可用
                    if (UIApplication.SharedApplication.CanOpenUrl
                    (emailUrl))
                    {
                        UIApplication.SharedApplication.OpenUrl(emailUrl);
                                                     //调用 Email 应用程序
                    } else
```

```
                    {
                         UIAlertView alertView=new UIAlertView ();
                         alertView.Title="对不起，不能发送邮件";
                         alertView.AddButton("Cancel");
                         alertView.Show();
                    }
               }else
               {
                    UIAlertView alertView=new UIAlertView ();
                    alertView.Title="对不起，邮件地址不对";
                    alertView.AddButton("Cancel");
                    alertView.Show();
               }
          };
     }
     ……                          //这里省略了视图加载和卸载前后的一些方法
     #endregion
}
```

运行效果如图 8.8 所示。

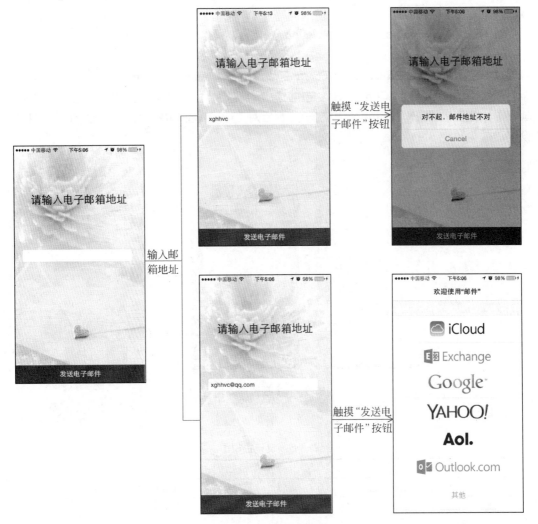

图 8.8　运行效果

> **注意**：此程序需要在真机上运行。在 iOS Simulator 上没有提供内置的 Email 应用程序。

8.4 在应用程序中使用短信

在 8.3.1 节中，调用 SMS 应用程序有一个缺点，就是在触摸"发送短信"时，SMS 应用程序将获取控制权，当前的程序则被推入后台。当发送短信时，需要手动把应用程序调入前台，否则应用程序将不会显示。为了在应用程序中编写短信，然后让 SMS 应用程序代替发送，可以使用 MFMessageComposeViewController 类来实现。MFMessageComposeViewController 类显示 SMS 编辑器模态窗口，而不会使当前的应用程序进入后台。如果想要在发送短信后恢复当前的应用程序，使用此类是十分有用的。

【示例 8-5】 以下将制作一个用来发送短信的应用程序。当用户在文本框和文本视图中输入内容后，触摸"发送短信"按钮，可以打开 SMS 编辑器模态窗口。触摸此窗口中的"发送"按钮，就会将编写的短信发送给指定的联系人，并退出 SMS 编辑器模态窗口。具体的操作步骤如下所述。

（1）创建一个 Single View Application 类型的工程。
（2）添加图像 1.png 到创建工程的 Resources 文件夹中。
（3）打开 MainStoryboard.storyboard 文件，拖动 Navigation Controller 导航控制器到画布中，将 Is Initial View Controller 复选框选中，将导航控制器关联的根视图控制器设置为画布中原先有的 View Controller 控制器，然后对 View Controller 视图控制器的主视图进行设置，效果如图 8.9 所示。

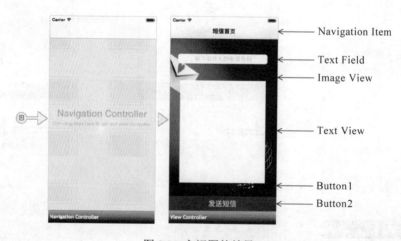

图 8.9 主视图的效果

需要添加的视图以及设置，如表 8-5 所示。

表 8-5 设置主视图

视 图	设 置
Navigation Item	Title：短信首页
Text Field	Name：tf Text：（空）

视　图	设　置
Text Field	Font：System 17 pt Alignment：居中 Placeholder：输入收件人的电话号码 Keyboard：Number Pad 位置和大小：(22,108, 269, 30)
Image View	Image：1.png 位置和大小：(0, 64, 320, 461)
Text View	Name：tv Text：（空） 位置和大小：(30, 186, 256, 300)
Button1	Name：cbtn Title：（空） 位置和大小：(0, 64, 320, 461)
Button2	Name：btn Title：发送短信 Font：System 21 pt Text Color：白色 Background：深灰色 位置和大小：(0, 525, 320, 43)

（4）打开 8-5ViewController.cs 文件，编写代码，实现在应用程序中使用短信的功能。代码如下：

```
using System;
using System.Drawing;
using MonoTouch.Foundation;
using MonoTouch.UIKit;
using MonoTouch.MessageUI;
namespace Application
{
    public partial class __5ViewController : UIViewController
    {
        private MFMessageComposeViewController messageController;
        ……                          //这里省略了视图控制器的构造方法和析构方法
        #region View lifecycle
        public override void ViewDidLoad ()
        {
            base.ViewDidLoad ();
            //触摸按钮，关闭键盘
            cbtn.TouchUpInside += (sender, e) => {
                tf.ResignFirstResponder();
                tv.ResignFirstResponder();
            };
            //触摸"发送短信"按钮，打开 SMS 编辑器模态窗口
            btn.TouchUpInside += async (s, e) => {
                //判断当前的设备是否可以发送文本消息
                if (MFMessageComposeViewController.CanSendText)
                {
                    messageController = new MFMessageComposeViewController();
                                                     //实例化对象
```

```
            messageController.Recipients = new string[] {tf.Text};
                                                  //设置收件人
            messageController.Body =tv.Text;      //设置短信内容
            //发送成功或者失败后的处理
            messageController.Finished += MessageController_
            Finished;
            await this.PresentViewControllerAsync
            (messageController, true);
        }
        else
        {
            UIAlertView alertView=new UIAlertView();
            alertView.Title="对不起,短信不可以发送";
            alertView.AddButton("Cancel");
            alertView.Show();
        }
    };
}
//短信发送成功或者失败后的处理
private async void MessageController_Finished (object sender,
MFMessageComposeResultEventArgs e) {
    switch (e.Result)
    {
    //如果短信成功发送
    case MessageComposeResult.Sent:
        UIAlertView alertView1=new UIAlertView();
        alertView1.Title="信息已发送";
        alertView1.AddButton("Cancel");
        alertView1.Show();
        break;
    //如果用户触摸了"取消"按钮,消息没有被发送
    case MessageComposeResult.Cancelled:
        UIAlertView alertView2=new UIAlertView();
        alertView2.Title="信息已取消";
        alertView2.AddButton("Cancel");
        alertView2.Show();
        break;
    //消息发送失败
    default:
        UIAlertView alertView3=new UIAlertView();
        alertView3.Title="信息发送失败";
        alertView3.AddButton("Cancel");
        alertView3.Show();
        break;
    }
    e.Controller.Finished -= MessageController_Finished;
    await e.Controller.DismissViewControllerAsync(true);
}
……                            //这里省略了视图加载和卸载前后的一些方法
#endregion
    }
}
```

在此程序中,引入了新的命名空间 MonoTouch.MessageUI,此命名空间包含一些重要的允许开发者执行文本信息的 UI 元素。运行效果如图 8.10 所示。

⚠️ **注意**:此程序需要在真机上运行。在 iOS Simulator 上没有提供 SMS 发送功能。

第 8 章　内置应用程序

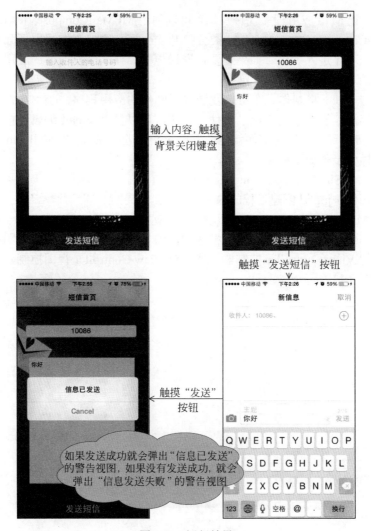

图 8.10　运行效果

在 Finished 方法中，提供了 MessageComposeResult 的 3 个参数。这 3 个参数的功能如表 8-6 所示。

表 8-6　MessageComposeResult参数

参　　数	功　　能
Sent	信息发送成功
Cancelled	用户触摸"取消"按钮
Failed	信息发送失败

8.5　在应用程序中使用电子邮件

和 8.3.1 节一样，8.3.2 节中调用 Email 应用程序也有相同的一个缺点，就是在触摸"发送电子邮件"时，Mail 应用程序将获取控制权，当前的程序则被推入后台。当发送电子邮件时，需要手动把应用程序调入前台，否则应用程序将不会显示。为了在应用程序中编写电子邮件，然后让 Mail 应用程序代替发送，可以使用 MFMailComposeViewController 类来实现。MFMailComposeViewController 类会打开一个模态窗口来编写电子邮件消息，而不会使当前的应用程序进入后台。如果想要在发送电子邮件后恢复当前的应用程序，使用此类是十分有用的。

【示例 8-6】 以下将制作一个用来发送电子邮件的应用程序。当用户在文本框和文本视图中输入内容后，触摸"发送电子邮件"按钮，可以打开 Email 编辑器模态窗口，触摸此窗口中的 Send 按钮，编写的邮件就会被发送给指定的联系人，并退出 Email 编辑器模态窗口。具体的操作步骤如下所示。

（1）创建一个 Single View Application 类型的工程。

（2）添加图像 1.jpg 到创建工程的 Resources 文件夹中。

（3）打开 MainStoryboard.storyboard 文件，拖动 Navigation Controller 导航控制器到画布中，将 Is Initial View Controller 复选框选中，将导航控制器关联的根视图控制器设置为画布中原先有的 View Controller 控制器，然后对 View Controller 视图控制器的主视图进行设置，效果如图 8.11 所示。

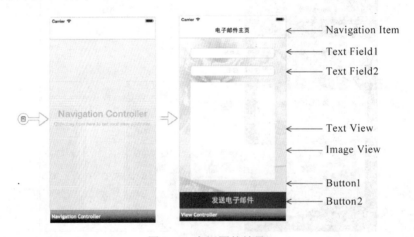

图 8.11 主视图的效果

需要添加的视图以及设置，如表 8-7 所示。

表 8-7 设置主视图

视 图	设 置
Navigation Item	Title：电子邮件主页
Text Field1	Name：rtf Text：（空） Font：System 16 pt Alignment：居中 Placeholder：收件人 位置和大小：(34, 95, 252, 30)
Text Field2	Name：stf Text：（空） Font：System 16 pt Alignment：居中 Placeholder：主题 位置和大小：(34, 147, 252, 30)

续表

视　　图	设　　置
Image View	Image：1.jpg 位置和大小：(0, 64, 320, 456)
Text View	Name：tv Text：（空） 位置和大小：(34, 196, 252, 287)
Button1	Name：cbtn Title：（空） 位置和大小：(0, 64 320, 456)
Button2	Name：btn Title：发送电子邮件 Font：System 21 pt Text Color：白色 Background：深灰色 位置和大小：(0, 520, 320, 48)

（4）打开 8-6ViewController.cs 文件，编写代码，实现在应用程序中使用电子邮件的功能。代码如下：

```
using System;
using System.Drawing;
using MonoTouch.Foundation;
using MonoTouch.UIKit;
using MonoTouch.MessageUI;
namespace Application
{
    public partial class __6ViewController : UIViewController
    {
        private MFMailComposeViewController mailController;
        ……                                  //这里省略了视图控制器的构造方法和析构方法
        #region View lifecycle
        public override void ViewDidLoad ()
        {
            base.ViewDidLoad ();
            //触摸按钮，关闭键盘
            cbtn.TouchUpInside += (sender, e) => {
                rtf.ResignFirstResponder();
                stf.ResignFirstResponder();
                tv.ResignFirstResponder();
            };
            //触摸按钮，打开 Email 编辑器模态窗口
            btn.TouchUpInside += async (s, e) => {
                //判断当前的设备是否可以发送电子邮件
                if (MFMailComposeViewController.CanSendMail)
                {
                    mailController = new MFMailComposeViewController();
                    mailController.SetToRecipients(new string[] {rtf.Text});
                                                                //设置收件人
                    mailController.SetSubject(stf.Text);
                                                                //设置邮件主题
                    mailController.SetMessageBody(tv.Text, false);
```

```csharp
                    //设置邮件内容
            //响应用户操作
            mailController.Finished += this.MailController_Finished;
            await this.PresentViewControllerAsync(mailController,
            true);
        }
        else
        {
            UIAlertView alertView=new UIAlertView();
            alertView.Title="对不起,不能发送电子邮件";
            alertView.AddButton("Cancel");
            alertView.Show();
        }
    };
}
//响应用户操作
private async void MailController_Finished (object sender,
MFComposeResultEventArgs e)
{
    switch (e.Result)
    {
    //电子邮件排到等待发送
    case MFMailComposeResult.Sent:
        UIAlertView alertView1=new UIAlertView();
        alertView1.Title="邮件已发送";
        alertView1.AddButton("Cancel");
        alertView1.Show();
        break;
    //用户点击 Save Draft 按钮
    case MFMailComposeResult.Saved:
        UIAlertView alertView2=new UIAlertView();
        alertView2.Title="信息已保存";
        alertView2.AddButton("Cancel");
        alertView2.Show();
        break;
    //用户点击 Delete Draft 按钮
    case MFMailComposeResult.Cancelled:
        UIAlertView alertView3=new UIAlertView();
        alertView3.Title="信息发送已取消";
        alertView3.AddButton("Cancel");
        alertView3.Show();
        break;
    //电子邮件发送失败
    case MFMailComposeResult.Failed:
        UIAlertView alertView4=new UIAlertView();
        alertView4.Title="信息发送失败";
        if (null != e.Error)
        {
            alertView4.Message = string.Format ("原因是: {0}",
            e.Error.LocalizedDescription);
        }
        alertView4.AddButton("Cancel");
        alertView4.Show();
        Console.WriteLine("Email sending failed!");
        break;
    }
    e.Controller.Finished -= MailController_Finished;
    await e.Controller.DismissViewControllerAsync(true);
```

```
        }
        ……
#endregion                              //这里省略了视图加载和卸载前后的一些方法
    }
}
```

运行效果如图 8.12 所示。

在 MFMailComposeViewController 的 Finished 方法中，提供了 MessageComposeResult 的 4 个参数。这 4 个参数的功能介绍如下。

- Sent：用户触摸了 Send 按钮，这意味着电子邮件在等待发送。
- Save：用户触摸 Cancel 按钮后，在弹出的动作表单中又触摸了 Save Draft 按钮。如图 8.13 所示，就是触摸 Cancel 按钮后出现的动作表单。
- Cancelled：用户触摸了 Cancel 按钮后，在弹出的动作表单中又触摸了 Delete Draft 按钮。
- Failed：电子邮件发送失败。

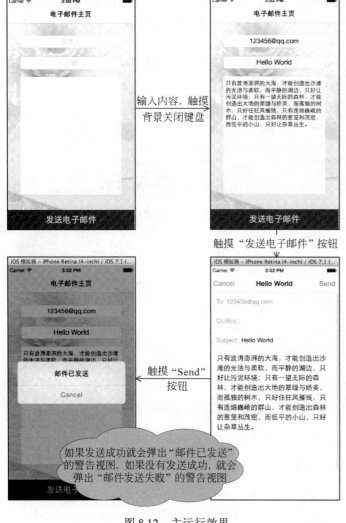

图 8.12　主运行效果

图 8.13　动作表单

8.6 管理地址簿

地址簿是 iOS 的基本功能。它可以让用户存储和编辑联系人的电话和邮件资料。通过地址簿，用户可以更轻松快捷地给选定的联系人发送短信或者打电话。本节将讲解对地址簿的一些操作。

8.6.1 访问地址簿

要想访问地址簿，需要使用到 ABAddressBook 类中的 GetPeople 方法。

【示例 8-7】以下将实现在触摸按钮后，获取地址簿中的联系人名称。具体的操作步骤如下所述。

（1）创建一个 Single View Application 类型的工程。

（2）打开 MainStoryboard.storyboard 文件，从工具栏中拖动 Button 按钮对象到主视图中，将此按钮的 Name 设置为 btn，将 Title 设置为访问地址簿，将 Font 设置为 System 18 pt，将位置和大小设置为（75, 229, 153, 30）。

（3）打开 8-7ViewController.cs 文件，编写代码，实现地址簿的访问，以及在应用程序输出窗口输出联系人的名称。代码如下：

```
using System;
using System.Drawing;
using MonoTouch.Foundation;
using MonoTouch.UIKit;
using MonoTouch.AddressBook;
using System.Threading;
namespace Application
{
    public partial class __7ViewController : UIViewController
    {
        ……
        #region View lifecycle
        public override void ViewDidLoad ()
        {
            base.ViewDidLoad ();
            btn.TouchUpInside += (s, e) => {
                //获取授权状态
                ABAuthorizationStatus abStatus = ABAddressBook.GetAuthorizationStatus();
                NSError error;
                ABAddressBook addressBook = ABAddressBook.Create(out error);
                //判断授权状态
                if (abStatus == ABAuthorizationStatus.NotDetermined)
                {
                    //请求访问地址簿事件
                    addressBook.RequestAccess((g, err) => {
                        if (!g)
                        {
                            Console.WriteLine("User denied address book access!");
                        } else
```

```
                    {
                        this.InvokeOnMainThread(() => this.ReadContacts
                        (addressBook));
                    }
                } );
            } else if (abStatus == ABAuthorizationStatus.Authorized)
            {
                this.ReadContacts(addressBook);
            } else
            {
                Console.WriteLine("App does not have access to the
                address book!");
            }
        } ;
    }
    //获取地址簿中的联系人信息
    private void ReadContacts(ABAddressBook addressBook)
    {
        ABPerson[] contacts = addressBook.GetPeople();
        //遍历
        foreach (ABPerson eachPerson in contacts)
        {
            Console.WriteLine("{0} {1}", eachPerson.LastName, eachPerson.
            FirstName);
        }
    }
    ……
    #endregion
    }
}
```

在此程序中，引入了新的命名空间 MonoTouch.AddressBook，此命名空间提供了对地址簿进行管理的类。运行效果如图 8.14 所示。

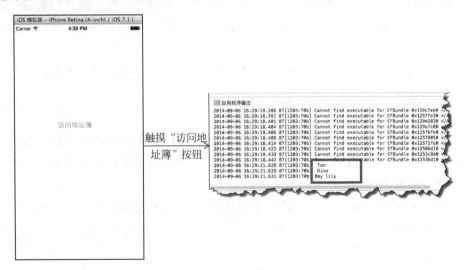

图 8.14　运行效果

在此程序需要注意以下两个问题。
（1）授权状态
在此程序中使用 GetAuthorizationStatus()方法实现了对授权状态的获取。其授权状态有

4 种，如表 8-8 所示。

表 8-8 授权状态

状 态	功 能
NotDetermined	用户还没有决定是否授权你的程序进行访问
Restricted	iOS 设备上的一些许可配置阻止了你的程序与地址簿数据库进行交互
Denied	用户明确的拒绝了你的程序对地址簿的访问
Authorized	用户已经授权给你的程序对地址簿进行访问

（2）真机测试

如果开发者使用真机进行测试，在触摸"访问地址簿"按钮后，会弹出一个警告视图，如图 8.15 所示。

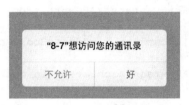

图 8.15 效果

这个警告视图就是让用户决定是否授权你的程序进行访问。如果触摸"不允许"按钮，表明你的程序不可以对地址簿进行访问，如果触摸"好"按钮，表明你的程序可以对地址簿进行访问。

8.6.2 打开地址簿

在 MonoTouch.AddressBookUI 命名空间中提供了和联系人显示信息相关的一些控制器。ABPeoplePickerNavigationController 控制器，它显示了一个导航界面，类似于地址簿应用程序。

【示例 8-8】 以下将使用 ABPeoplePickerNavigationController 控制器实现打开地址簿的功能。具体的操作步骤如下所述。

（1）创建一个 Single View Application 类型的工程。

（2）打开 MainStoryboard.storyboard 文件，从工具栏中拖动 Button 按钮对象到主视图中，将此按钮的 Name 设置为 btn，将 Title 设置为打开地址簿，将 Font 设置为 System 18 pt，将位置和大小设置为（106，283，108，30）。

（3）打开 8-8ViewController.cs 文件，编写代码，实现触摸按钮后打开地址簿。代码如下：

```
using System;
using System.Drawing;
using MonoTouch.Foundation;
using MonoTouch.UIKit;
using MonoTouch.AddressBookUI;
namespace Application
{
    public partial class __8ViewController : UIViewController
    {
```

第 8 章 内置应用程序

```
        ABPeoplePickerNavigationController pn;
        ……              //这里省略了视图控制器的构造方法和析构方法
        #region View lifecycle
        public override void ViewDidLoad ()
        {
            base.ViewDidLoad ();
            //触摸按钮，打开地址簿
            btn.TouchUpInside += (sender, e) => {
                pn = new ABPeoplePickerNavigationController ();
                this.PresentViewControllerAsync (pn, true);
                pn.Cancelled += this.cancelled;
                                //取消事件，触摸地址簿中的 Cancel 按钮后调用
            };
        }
        //关闭地址簿
        public void cancelled(object sender, EventArgs e){
            this.pn.DismissViewControllerAsync (true);
        }
        ……            //这里省略了视图加载和卸载前后的一些方法
        #endregion
    }
}
```

运行效果如图 8.16 所示。

图 8.16 运行效果

8.6.3 添加联系人

在 iOS 中，添加联系人都会打开一个添加联系人界面，此界面需要使用 ABNewPersonViewController 控制器，它显示的是一个新添加的联系人的可编辑的属性界面。

【示例 8-9】 以下将实现为地址簿中添加新的联系人。具体的操作步骤如下所述。

（1）创建一个 Single View Application 类型的工程。

（2）打开 MainStoryboard.storyboard 文件，拖动 Navigation Controller 导航控制器到画布中，将 Is Initial View Controller 复选框选中，将导航控制器关联的根视图控制器设置为画布中原先有的 View Controller 控制器，然后对 View Controller 视图控制器的主视图进行

设置，效果如图 8.17 所示。

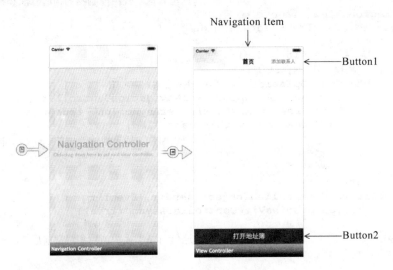

图 8.17　主视图的效果

需要添加的视图以及设置，如表 8-9 所示。

表 8-9　设置主视图

视　　图	设　　置
Navigation Item	Title：首页
Button1	Name：addPerson Title：添加联系人 Font：System 14 pt 位置和大小：(0, 0, 83, 30)
Button2	Name：openaddress Title：打开地址簿 Font：System 19 pt Text Color：白色 Background：深灰色 位置和大小：(0, 524, 320, 44)

（3）打开 8-9ViewController.cs 文件，编写代码，实现联系人的添加。代码如下：

```
using System;
using System.Drawing;
using MonoTouch.Foundation;
using MonoTouch.UIKit;
using MonoTouch.AddressBookUI;
namespace Application
{
    public partial class __9ViewController : UIViewController
    {
        ABPeoplePickerNavigationController pn;
        ABNewPersonViewController pv;
        ……                          //这里省略了视图控制器的构造方法和析构方法
        #region View lifecycle
        public override void ViewDidLoad ()
        {
            base.ViewDidLoad ();
```

```csharp
            //触摸"打开地址簿"按钮，打开地址簿
            openaddress.TouchUpInside += (sender, e) => {
                pn = new ABPeoplePickerNavigationController ();
                this.PresentViewControllerAsync (pn, true);
                pn.Cancelled += this.cancelled;
                               //取消事件，触摸地址簿中的Cancel按钮后调用
            };
            addPerson.TouchUpInside += (sender, e) => {
                pv=new ABNewPersonViewController();
                //完成事件，在触摸Done按钮后调用
                pv.NewPersonComplete+=this.new_PersonComple;
                this.NavigationController.PushViewController(pv,true);
            };
        }
        //退出地址簿
        public void cancelled(object sender, EventArgs e){
            this.pn.DismissViewControllerAsync (true);
        }
        //退出添加联系人界面
        public void new_PersonComple(object sender,EventArgs e){
            this.NavigationController.PopViewControllerAnimated (true);
        }
        ……                   //这里省略了视图加载和卸载前后的一些方法
        #endregion
    }
}
```

运行效果如图 8.18 所示。

图 8.18　运行效果

8.6.4　显示联系人信息

ABPersonViewController 控制器，它显示了一个指定联系人的属性界面，可以用来查看此联系人的个人信息。

【示例 8-10】 以下将实现显示指定联系人的个人信息。具体操作步骤如下所述。

（1）创建一个 Single View Application 类型的工程。

（2）打开 MainStoryboard.storyboard 文件，拖动 Navigation Controller 导航控制器到画布中，将 Is Initial View Controller 复选框选中，将导航控制器关联的根视图控制器设置为画布中原先有的 View Controller 控制器，然后对 View Controller 视图控制器的主视图进行设置，效果如图 8.19 所示。

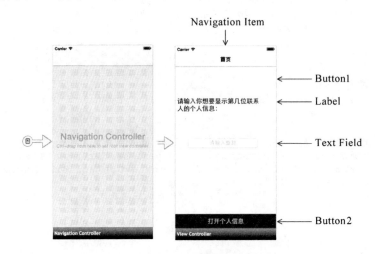

图 8.19 主视图的效果

需要添加的视图以及设置，如表 8-10 所示。

表 8-10 设置主视图

视 图	设 置
Navigation Item	Title：首页
Button1	Name：cbtn Title：（空） 位置和大小：(0, 64, 320, 460)
Label	Text：请输入你想要显示第几位联系人的个人信息: Font：System 20 pt Lines：2 位置和大小：(6, 158, 268, 49)
Text Field	Name：tf Text：（空） Font：System 17 pt Alignment：居中 Placeholder：请输入整数 Keyboard：Number Pad 位置和大小：(40, 288, 222, 30)
Button2	Name：btn Title：打开个人信息 Font：System 19 pt Text Color：白色 Background：深灰色 位置和大小：(0, 524, 320, 44)

(3) 打开 8-10ViewController.cs 文件，编写代码，实现个人信息的显示。代码如下：

```csharp
using System;
using System.Drawing;
using MonoTouch.Foundation;
using MonoTouch.UIKit;
using MonoTouch.AddressBook;
using MonoTouch.AddressBookUI;
namespace Application
{
    public partial class __10ViewController : UIViewController
    {
        ……                                    //这里省略了视图控制器的构造方法和析构方法
        #region View lifecycle
        public override void ViewDidLoad ()
        {
            base.ViewDidLoad ();
            //触摸按钮，关闭键盘
            cbtn.TouchUpInside += (sender, e) => {
                tf.ResignFirstResponder();
            };
            //触摸"打开个人信息"按钮，打开个人信息
            btn.TouchUpInside += (sender, e) => {
                //获取授权状态
                ABAuthorizationStatus status = ABAddressBook.
                GetAuthorizationStatus();
                NSError error;
                ABAddressBook addressBook = ABAddressBook.Create(out error);
                //判断授权状态
                if (status == ABAuthorizationStatus.NotDetermined)
                {
                    //请求访问地址簿事件
                    addressBook.RequestAccess((g, err) => {
                        if (g)
                        {
                            this.InvokeOnMainThread(() =>this.Display-
                            ContactCard(addressBook));
                        } else
                        {
                            Console.WriteLine("User denied access to the
                            address book!");
                        }
                    } );
                } else if (status == ABAuthorizationStatus.Authorized)
                {
                    this.DisplayContactCard(addressBook);
                } else
                {
                    Console.WriteLine("App does not have access to the
                    address book!");
                }
            } ;
        }
        //显示个人信息
        private void DisplayContactCard(ABAddressBook addressBook)
        {
            ABPerson[] contacts = addressBook.GetPeople();
            ABPersonViewController personController = new
            ABPersonViewController();
            int num = int.Parse (tf.Text);
```

```
            if (num <=4) {
                personController.DisplayedPerson = contacts [num];
                this.NavigationController.PushViewController
                (personController, true);
            } else {
                UIAlertView alert = new UIAlertView ();
                alert.Title = "对不起，此数字对应的联系人不存在";
                alert.AddButton ("cancel");
                alert.Show ();

            }
        }
        ……                              //这里省略了视图加载和卸载前后的一些方法
        #endregion
    }
}
```

运行效果如图 8.20 所示。

图 8.20　运行效果

8.7 管理日历

日历就是一个数据库，可以使用 MonoTouch.EventKit 命名空间中的类管理它，也可以使用 MonoTouch.EventKitUI 命名空间中的类访问用户与日历交互的界面。本节将讲解使用这两个命名空间中的类对日历的管理。

8.7.1 访问日历

本小节将讲解如何访问日历事件。

【示例 8-11】 以下将实现触摸按钮，在应用程序输出窗口输出日历中的事件的功能。具体的操作步骤如下所述。

（1）创建一个 Single View Application 类型的工程。

（2）打开 MainStoryboard.storyboard 文件，从工具栏中拖动 Button 按钮对象到主视图中，将此按钮的 Name 设置为 btn，将 Title 设置为访问日历，将位置和大小设置为（120, 282, 79, 30）。

（3）打开 8-11ViewController.cs 文件，编写代码，实现日历事件的访问，以及在应用程序输出窗口输出日历事件。代码如下：

```
using System;
using System.Drawing;
using MonoTouch.Foundation;
using MonoTouch.UIKit;
using MonoTouch.EventKit;
namespace Application
{
    public partial class __11ViewController : UIViewController
    {
        ……
        #region View lifecycle
        public override void ViewDidLoad ()
        {
            base.ViewDidLoad ();
            btn.TouchUpInside += async (sender, e) => {
                //获取授权状态
                EKAuthorizationStatus status = EKEventStore.
                GetAuthorizationStatus(EKEntityType.Event);
                EKEventStore evStore = new EKEventStore();
                //判断授权状态
                if (status == EKAuthorizationStatus.NotDetermined)
                {
                    //请求访问日历事件
                    if (await evStore.RequestAccessAsync(EKEntityType.Event))
                    {
                        this.DisplayEvents(evStore);
                    } else
                    {
                        Console.WriteLine("User denied access to the calendar!");
```

```
                } else if (status == EKAuthorizationStatus.Authorized)
                {
                    this.DisplayEvents(evStore);
                } else
                {
                    Console.WriteLine("App does not have access to the
                    calendar!");
                }
            } ;
        }
        //显示日历事件
        private void DisplayEvents (EKEventStore evStore)
        {
            //查询从现在开始经过 365 天的事件
            NSPredicate evPredicate =
                evStore.PredicateForEvents(DateTime.Now, DateTime.Now.
                AddDays(365), evStore.GetCalendars(EKEntityType.Event));
            //枚举日历事件
            evStore.EnumerateEvents(evPredicate, delegate(EKEvent calEvent,
            ref bool stop) {
                if (null != calEvent) {
                    stop = false;
                    //输出
                    Console.WriteLine("Event title: {0}\n Event start date:
                    {1}", calEvent.Title, calEvent.StartDate);
                }
            } );
        }
        ......
        #endregion
    }
}
```

运行效果如图 8.21 所示。

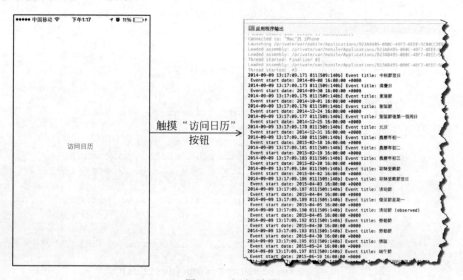

图 8.21　运行效果

> 注意：在此程序中使用到了 NSPredicate 类，它用于查询，和 where 类似。

8.7.2 打开日历事件界面

EKEventViewController 控制器显示了一个单一日历事件的描述界面。

【示例 8-12】 以下将实现触摸按钮打开日历事件界面的功能。具体的操作步骤如下所述。

（1）创建一个 Single View Application 类型的工程。

（2）打开 MainStoryboard.storyboard 文件，拖动 Navigation Controller 导航控制器到画布中，将 Is Initial View Controller 复选框选中，将导航控制器关联的根视图控制器设置为画布中原先有的 View Controller 控制器。从工具栏中拖动 Button 按钮对象到主视图中，将此按钮的 Name 设置为 btn，将 Title 设置为打开日历，将位置和大小设置为（104, 283, 112, 30）。

（3）打开 8-12ViewController.cs 文件，编写代码，实现触摸按钮后打开日历。代码如下：

```
using System;
using System.Drawing;
using MonoTouch.Foundation;
using MonoTouch.UIKit;
using MonoTouch.EventKitUI;
using MonoTouch.EventKit;
namespace Application
{
    public partial class __12ViewController : UIViewController
    {
        ……
        #region View lifecycle
        public override void ViewDidLoad ()
        {
            base.ViewDidLoad ();
            btn.TouchUpInside += async (sender, e) => {
                EKAuthorizationStatus status = EKEventStore.
                GetAuthorizationStatus(EKEntityType.Event);
                EKEventStore evStore = new EKEventStore();
                if (status == EKAuthorizationStatus.NotDetermined)
                {
                    if (await evStore.RequestAccessAsync(EKEntityType.
                    Event))
                    {
                        //打开日历事件界面
                        EKEventViewController evc=new EKEventView-
                        Controller();
                        this.NavigationController.PushViewController(evc,
                        true);
                    } else
                    {
                        Console.WriteLine("User denied access to the
                        calendar!");
                    }
                } else if (status == EKAuthorizationStatus.Authorized)
```

```
                    {
                        EKEventViewController evc=new EKEventViewController();
                        this.NavigationController.PushViewController(evc,true);
                    } else
                    {
                        Console.WriteLine("App does not have access to the
                        calendar!");
                    }
                };
            }
            ……
            #endregion
        }
}
```

运行效果如图 8.22 所示。

图 8.22　运行效果

8.7.3　添加日历事件

在图 8.22 所示的运行结果中看到的这个事件是一个默认的事件。开发者也可以添加一些事件到日历中。以下将讲解添加事件的方式。

1．使用代码添加日历事件

EKEvent 类提供了很多的属性来实现对日历事件的设置，如标题、开始时间、结束事件等。

【示例 8-13】　以下将使用 EKEvent 类实现对日历事件的添加。具体步骤如下所述。

（1）创建一个 Single View Application 类型的工程。

（2）打开 MainStoryboard.storyboard 文件，拖动 Navigation Controller 导航控制器到画布中，将 Is Initial View Controller 复选框选中，将导航控制器关联的根视图控制器设置为画布中原先有的 View Controller 控制器，然后对 View Controller 视图控制器的主视图进行设置，效果如图 8.23 所示。

第 8 章 内置应用程序

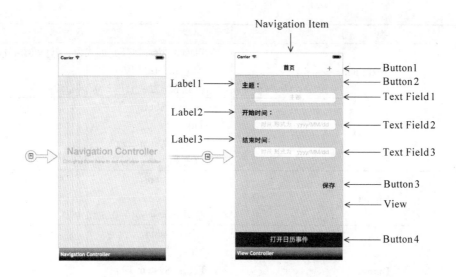

图 8.23 主视图的效果

需要添加的视图以及设置，如表 8-11 所示。

表 8-11 设置主视图

视　　图	设　　置
Navigation Item	Title：首页
Label1	Text：主题： Font：System Bold 19 pt 位置和大小：(20, 20, 68, 21)
Label2	Text：开始时间： Font：System Bold 19 pt 位置和大小：(20, 100, 115, 21)
Label3	Text：结束时间： Font：System Bold 19 pt 位置和大小：(20, 180, 98, 21)
Text Field1	Name：tf1 Text：（空） Font：System 17 pt Alignment：居中 Placeholder：主题 位置和大小：(56, 50, 244, 30)
Text Field2	Name：tf2 Text：（空） Font：System 17 pt Alignment：居中 Placeholder：时间 形式为 yyyy/MM/dd 位置和大小：(56, 130, 244, 30)
Text Field3	Name：tf3 Text：（空） Font：System 17 pt

• 309 •

续表

视 图	设 置
Text Field3	Placeholder：时间 形式为：yyyy/MM/dd Alignment：居中 位置和大小：(56, 210, 244, 30)
Button1	Name：abtn Title：+ Font：System 23 pt 位置和大小：(0, 0, 46, 30)
Button2	Name：cbtn Title：（空） 位置和大小：(0, 0, 320, 467)
Button3	Name：sbtn Title：保存 Font：System 19 pt Text Color：黑色 位置和大小：(247, 308, 63, 30)
Button4	Name：btn Title：打开日历事件 Font：System 19 pt Text Color：白色 Background：深灰色 位置和大小：(0, 524, 320, 44)
View	Name：vv 位置和大小：(0, 64, 320, 460)

（3）打开 8-13ViewController.cs 文件，编写代码，实现对日历事件的添加。代码如下：

```
using System;
using System.Drawing;
using MonoTouch.Foundation;
using MonoTouch.UIKit;
using MonoTouch.EventKitUI;
using MonoTouch.EventKit;
namespace Application
{
    public partial class __13ViewController : UIViewController
    {
        EKEventStore evStore;
        EKEvent newEvent;
        ……
        #region View lifecycle
        public override void ViewDidLoad ()
        {
            base.ViewDidLoad ();
            EKEventStore evStore = new EKEventStore();
            vv.Hidden = true;
            //触摸按钮，关闭键盘
            cbtn.TouchUpInside += (sender1, e1) => {
```

```
                    tf1.ResignFirstResponder();
                    tf2.ResignFirstResponder();
                    tf3.ResignFirstResponder();
                };
                //触摸按钮,隐藏 vv 视图对象
                abtn.TouchUpInside += (sender, e) => {
                    vv.Hidden = false;
                };
                //触摸按钮,保存添加的日历事件
                sbtn.TouchUpInside += (sender, e) => {
                    NSDateFormatter df = new NSDateFormatter ();
                    df.DateFormat="yyyy/MM/dd";
                    //字符串转换为日期
                    NSDate sdate = df.Parse (tf2.Text);
                    NSDate edate = df.Parse (tf3.Text);
                    vv.Hidden=true;
                    newEvent=EKEvent.FromStore(evStore);
                    newEvent.StartDate = sdate;          //设置日历事件的开始时间
                    newEvent.EndDate = edate;            //设置日历事件的结束时间
                    newEvent.Title = tf1.Text;           //设置日历事件的标题
                    newEvent.Calendar = evStore.DefaultCalendarForNewEvents;
                                                         //设置日历
                    NSError error = null;
                    evStore.SaveEvent(newEvent, EKSpan.ThisEvent, out error);
                                                         //保存日历事件
                };
                //触摸按钮,打开日历事件界面
                btn.TouchUpInside += (sender, e) => {
                    EKEventViewController evc=new EKEventViewController();
                    evc.Event=newEvent;
                    this.NavigationController.PushViewController(evc,true);
                };
            }
            ……
            #endregion
    }
}
```

运行效果如图 8.24 所示。

2. 使用添加事件界面添加事件

EKEventEditViewController 控制器提供了一个允许用户添加或者编辑日历的事件界面。

【示例 8-14】以下将使用添加事件界面实现对事件的添加。具体的操作步骤如下所述。

（1）创建一个 Single View Application 类型的工程。

（2）打开 MainStoryboard.storyboard 文件,拖动 Navigation Controller 导航控制器到画布中,将 Is Initial View Controller 复选框选中,将导航控制器关联的根视图控制器设置为画布中原先有的 View Controller 控制器,然后对 View Controller 视图控制器的主视图进行设置,效果如图 8.25 所示。

第 2 篇 资源使用篇

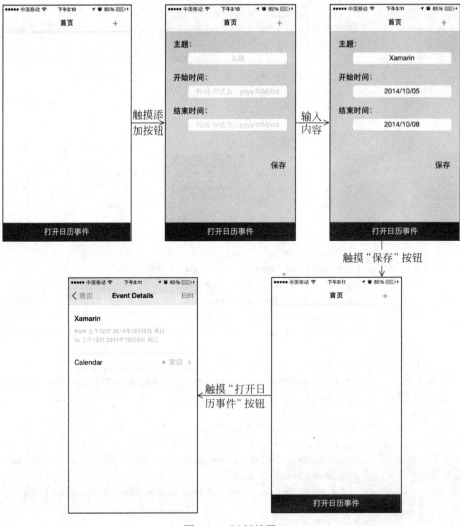

图 8.24 运行效果

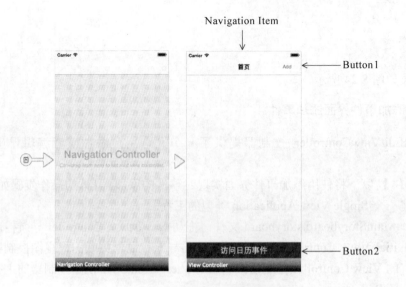

图 8.25 主视图的效果

需要添加的视图以及设置，如表 8-12 所示。

表 8-12 设置主视图

视 图	设 置
Navigation Item	Title：首页
Button1	Name：abtn Title：Add 位置和大小：(0, 0, 46, 30)
Button2	Name：btn Title：访问日历事件 Font：System Bold 21 pt Text Color：白色 Background：深灰色 位置和大小：(0, 525, 320, 43)

（3）打开 8-14ViewController.cs 文件，编写代码，实现对日历事件的添加。代码如下：

```
using System;
using System.Drawing;
using MonoTouch.Foundation;
using MonoTouch.UIKit;
using MonoTouch.EventKitUI;
using MonoTouch.EventKit;
namespace Application
{
    public partial class __14ViewController : UIViewController
    {
        EKEventStore aa ;
        EKEvent newEvent;
        ……                          //这里省略了视图控制器的构造方法和析构方法
        #region View lifecycle
        public override void ViewDidLoad ()
        {
            base.ViewDidLoad ();
            aa = new EKEventStore ();
            //触摸按钮，打开添加事件界面
            abtn.TouchUpInside += (sender, e) => {
                EKEventEditViewController ed=new EKEventEditView-
                Controller();
                ed.EventStore=aa;
                ed.Completed+=this.new_Completed;
                this.PresentViewControllerAsync(ed,true);
            };
            //触摸按钮，实现对日历的访问
            btn.TouchUpInside += async (sender, e) => {
                EKAuthorizationStatus status = EKEventStore.
                GetAuthorizationStatus(EKEntityType.Event);
```

```
            if (status == EKAuthorizationStatus.NotDetermined)
            {
                //请求访问日历事件
                if (await aa.RequestAccessAsync(EKEntityType.Event))
                {
                    this.DisplayEvents(aa);
                } else
                {
                    Console.WriteLine("User denied access to the
                    calendar!");
                }
            } else if (status == EKAuthorizationStatus.Authorized)
            {
                this.DisplayEvents(aa);
            } else
            {
                Console.WriteLine("App does not have access to the
                calendar!");
            }
        };
}
//触摸 Done 按钮后,关闭添加事件界面,并保存事件
public void new_Completed(object sender,EventArgs e){
    this.DismissViewControllerAsync (true);
}
//输出日历事件
private void DisplayEvents (EKEventStore evStore)
{

    NSPredicate evPredicate =
        aa.PredicateForEvents(DateTime.Now, DateTime.Now.AddDays
        (30), evStore.GetCalendars(EKEntityType.Event));
    //枚举日历事件
    evStore.EnumerateEvents(evPredicate, delegate(EKEvent calEvent,
    ref bool stop) {
        //判断 calEvent 是否为空
        if (null != calEvent) {
            stop = false;
            Console.WriteLine("Event title: {0}\n Event start date:
            {1}", calEvent.Title, calEvent.StartDate);
                                                    //输出日历事件
        }
    } );
}
……                              //这里省略了视图加载和卸载前后的一些方法
#endregion
  }
}
```

运行效果如图 8.26 所示。

第 8 章　内置应用程序

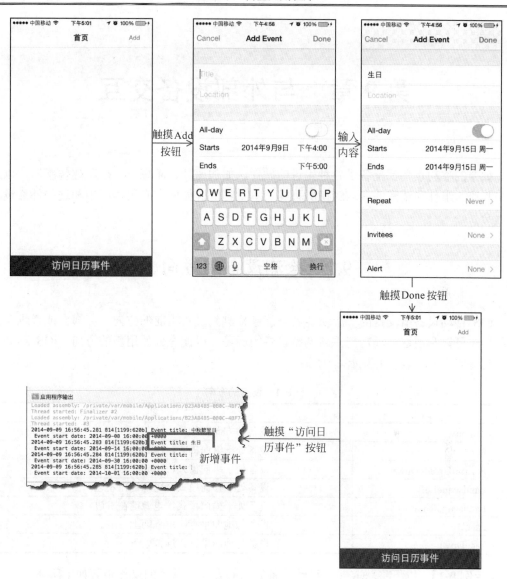

图 8.26　运行效果

第 9 章　与外部设备交互

如今的移动设备都配备了各种先进的硬件，如加速计、陀螺仪、近距离传感器，以及很多其他的组件和精密的多点触摸屏幕等。本章将讲解如何在应用程序中和这些外部设备进行交互。

9.1　检测设备的方向

iOS 移动设备所支持的一个功能就是能够检测到设备当前的方向——横向或者纵向。应用程序可以利用这一项功能重新调整设备的屏幕，以便充分使用新的方向。iOS 移动设备所支持的方向如表 9-1 所示。

表 9-1　设备的方向

方　　向	功　　能
Unknown	设备的方向未知
Portrait	设备处于纵向模式，Home 键在下方
PortraitUpsideDown	设备处于纵向模式，Home 键在上方
LandscapeLeft	设备处于横向模式，电源键在左边
LandscapeRight	设备处于横向模式，电源键在右边
FaceUp	设备与地面平行，屏幕朝上
FaceDown	设备与地面平行，屏幕朝下

【示例 9-1】以下将获取 iOS 移动设备的当前方向。具体的操作步骤如下所述。

（1）创建一个 Single View Application 类型的工程。

（2）打开 MainStoryboard.storyboard 文件，从工具栏中拖动 Label 对象到主视图中，将此对象的 Name 设置为 label，将 Font 设置为 System Bold 26 pt，将 Alignment 设置为居中，将位置和大小设置为（57, 194, 205, 60）。

（3）打开 9-1ViewController.cs 文件，编写代码，实现设备方向的检测。代码如下：

```
using System;
using System.Drawing;
using MonoTouch.Foundation;
using MonoTouch.UIKit;
namespace Application
{
    public partial class __1ViewController : UIViewController
    {
        private NSObject orientationObserver;
        ……                          //这个省略了视图控制器的构造方法和析构方法
        #region View lifecycle
```

```
......                                    //这里省略了视图加载后的方法
public override void ViewWillAppear (bool animated)
{
    base.ViewWillAppear (animated);
    UIDevice.CurrentDevice.BeginGeneratingDeviceOrientation-
    Notifications();
    //使用监听事件监听设备的当前方向
    this.orientationObserver = UIDevice.Notifications.
    ObserveOrientationDidChange((s, e) => {
        label.Text = UIDevice.CurrentDevice.Orientation.ToString();
                                                //显示设备的方向
    } );
}
......                                    //这里省略了视图显示后的方法
public override void ViewWillDisappear (bool animated)
{
    base.ViewWillDisappear (animated);
    //移除监听事件
    NSNotificationCenter.DefaultCenter.RemoveObserver(this.
    orientationObserver);
    UIDevice.CurrentDevice.EndGeneratingDeviceOrientation-
    Notifications();
}
......                                    //这里省略了视图消失后的方法
#endregion
}
```

运行效果如图 9.1 所示。

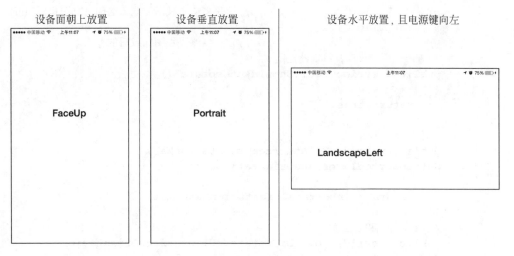

图 9.1　运行效果

注意：对于设备方向的判断是根据电源键而言的。

9.2　调整 UI 的方向

随着设备方向的改变，UI，即视图控制器也将会进行选择并调整到的新的屏幕方向，但是需要注意，设备的方向和 UI 的方向可以是不同的。UI 的方向如表 9-2 所示。

表 9-2 UI的方向

方　　向	功　　能
Portrait	在纵向模式下显示屏幕
PortraitUpsideDown	在纵向模式下显示屏幕，Home 按钮位于屏幕上方
LandscapeLeft	在横向模式下显示屏幕，Home 按钮位于左侧
LandscapeRight	在横向模式下显示屏幕，Home 按钮位于右侧

【示例 9-2】 以下将实现 UI 方向的检测。具体的操作步骤如下所述。

（1）创建一个 Single View Application 类型的工程。

（2）打开 MainStoryboard.storyboard 文件，从工具栏中拖动 Label 对象到主视图中，将此对象的 Name 设置为 label，将 Text 设置为空，将 Font 设置为 System Bold 26 pt，将 Alignment 设置为居中，将位置和大小设置为（57，194，205，60）。

（3）打开 9-2ViewController.cs 文件，编写代码，实现 UI 方向的检测。代码如下：

```
using System;
using System.Drawing;
using MonoTouch.Foundation;
using MonoTouch.UIKit;
namespace Application
{
    public partial class __2ViewController : UIViewController
    {
        ……                              //这里省略了视图控制器的构造方法和析构方法
        #region View lifecycle
        ……                              //这里省略了视图加载后的方法
        //判断是否应该选择特定的控制器
        public override bool ShouldAutorotate ()
        {
            return true;
        }
        //获取支持的方向
        public override UIInterfaceOrientationMask 
        GetSupportedInterfaceOrientations ()
        {
            return UIInterfaceOrientationMask.All;
        }
        //UI 选择结束后触发
        public override void DidRotate (UIInterfaceOrientation 
        fromInterfaceOrientation)
        {
            base.DidRotate (fromInterfaceOrientation);
            label.Text = this.InterfaceOrientation.ToString();
        }
        ……                              //这里省略了视图加载和卸载前后的一些方法
        #endregion
    }
}
```

运行效果如图 9.2 所示。

第 9 章 与外部设备交互

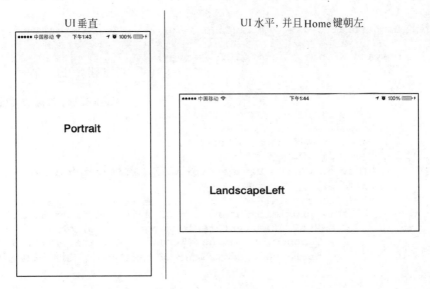

图 9.2 运行效果

⚠️注意：对于 UI 方向的判断是根据 Home 键而言的。UI 方向的调整除了可以使用代码外，还可以在 Info.plist 文件的 Supported Device Orientations 中进行设置，如图 9.3 所示。

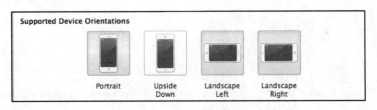

图 9.3 设置 UI 方向

9.3 近距离传感器

近距离传感器是通过红外线进行测距。当手机用户接听电话或者装进口袋时，传感器可以判断出手机贴近了人的脸部或者衣服而关闭屏幕的触控功能，这样就可以防止误操作。同样地，这个功能可以用在所有应用了触摸屏的应用程序上。

【示例 9-3】 以下将实现使用近距离传感器来禁用设备的屏幕。具体的操作步骤如下所述。

（1）创建一个 Single View Application 类型的工程。

（2）打开 9-3ViewController.cs 文件，编写代码，实现近距离传感器禁用设备屏幕的功能。代码如下：

```
using System;
using System.Drawing;
using MonoTouch.Foundation;
using MonoTouch.UIKit;
namespace Application
```

```csharp
{
    public partial class __3ViewController : UIViewController
    {
        private NSObject proximityObserver;
        ……                                    //这里省略了视图控制器的构造方法和析构方法
        #region View lifecycle
        ……                                    //这里省略了视图加载后的方法
        public override void ViewWillAppear (bool animated)
        {
            base.ViewWillAppear (animated);
            //开启近距离传感器
            UIDevice.CurrentDevice.ProximityMonitoringEnabled = true;
            //判断当前设备的近距离传感器是否可用
            if (UIDevice.CurrentDevice.ProximityMonitoringEnabled) {
                this.proximityObserver = UIDevice.Notifications.
                ObserveProximityStateDidChange ((s, e) => {
                    Console.WriteLine ("Proximity state: {0}", UIDevice.
                    CurrentDevice.ProximityState); //输出近距离传感器的状态
                });
            } else {
                UIAlertView alert = new UIAlertView ();
                alert.Title="对不起，你的近距离传感器不可用";
                alert.AddButton ("Cancel");
                alert.Show ();
            }
        }
        ……                                    //这里省略了视图显示后的方法
        public override void ViewWillDisappear (bool animated)
        {
            base.ViewWillDisappear (animated);
            //判断当前设备的近距离传感器是否可用
            if (UIDevice.CurrentDevice.ProximityMonitoringEnabled)
            {
                NSNotificationCenter.DefaultCenter.RemoveObserver(this.
                proximityObserver);
            } else {
                UIAlertView alert = new UIAlertView ();
                alert.Title="对不起，你的近距离传感器不可用";
                alert.AddButton ("Cancel");
                alert.Show ();
            }
        }
        ……                                    //这里省略了视图消失后的方法
        #endregion
    }
}
```

单击运行按钮，运行此应用程序，将一个手指向听筒的方向移动，当到达一定的距离后，手机的屏幕将会关闭，并且在应用程序输出窗口输出如图 9.4 所示的内容。

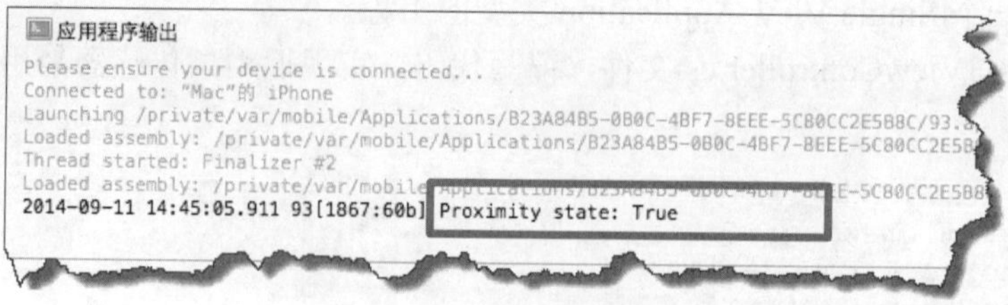

图 9.4　应用程序输出窗口

当手指慢慢移动远离听筒，手机屏幕将会打开，并在应用程序输出窗口输出如图 9.5 所示的内容。

图 9.5　应用程序输出窗口

△注意：此程序必须使用真机进行测试。否则，将会弹出"对不起，你的近距离传感器不可用"的警告视图。

9.4　获取电池信息

在很多时候，我们都需要获取电池的信息。例如，获取电池的电量可以在电池快没有电量时提醒用户进行充电等。

【示例 9-4】以下将实现获取设备中电池的电量级别和电池的状态。具体的操作步骤如下所述。

（1）创建一个 Single View Application 类型的工程。

（2）打开 MainStoryboard.storyboard 文件，对主视图进行设置，效果如图 9.6 所示。

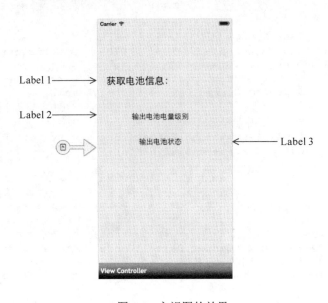

图 9.6　主视图的效果

需要添加的视图以及设置，如表 9-3 所示。

表 9-3 设置主视图

视　　图	设　　置
Label1	Text：获取电池信息： Font：System 23 pt 位置和大小：(20, 129, 185, 29)
Label2	Name：lblOutput Text：输出电池电量级别 Alignment：居中 位置和大小：(40, 209, 217, 37)
Label3	Name：label3 Text：输出电池状态 Alignment：居中 位置和大小：(40, 268, 217, 37)

（3）打开 9-4ViewController.cs 文件，编写代码，实现电池信息的获取。代码如下：

```
using System;
using System.Drawing;
using MonoTouch.Foundation;
using MonoTouch.UIKit;
namespace Application
{
    public partial class __4ViewController : UIViewController
    {
        private NSObject batteryStateChangeObserver;
        ……                                         //这里省略了视图控制器的构造方法和析构方法
        #region View lifecycle
        ……                                         //这里省略了视图加载后的方法
        public override void ViewWillAppear (bool animated)
        {
            base.ViewWillAppear (animated);
            UIDevice.CurrentDevice.BatteryMonitoringEnabled = true;
                                                                //控制电池监控
            //接受电池电量的变化
            batteryStateChangeObserver = UIDevice.Notifications.
            ObserveBatteryStateDidChange((s, e) => {
                lblOutput.Text = string.Format("Battery level: {0}",
                UIDevice.CurrentDevice.BatteryLevel);
                                                     //获取并显示电池电量的级别
                //获取并显示电池的状态
                label.Text=string.Format("Battery state: {0}",UIDevice.
                CurrentDevice.BatteryState);
            } );
        }
        ……                                         //这里省略了视图显示后的方法
        public override void ViewWillDisappear (bool animated)
        {
            base.ViewWillDisappear (animated);
            NSNotificationCenter.DefaultCenter.RemoveObserver
            (batteryStateChangeObserver);
            UIDevice.CurrentDevice.BatteryMonitoringEnabled = true;
        }
        ……                                         //这里省略了视图消失后的方法
        #endregion
    }
}
```

运行效果如图 9.7 所示。

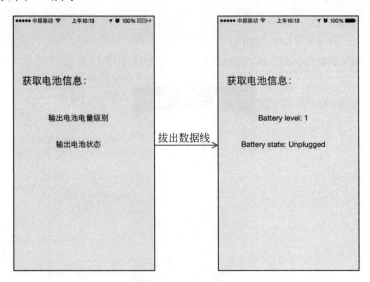

图 9.7　运行效果

在此程序中，需要注意以下两个问题。
（1）电池电量的级别
电池的电量级别可以通过 BatteryLevel 属性进行获取，其中 1 表示电量为 100%。
（2）电池的状态
电池的状态可以通过 BatteryState 属性进行获取。电池的状态如表 9-4 所示。

表 9-4　电池的状态

状　态	功　能
Unknown	该电池状态不能确定或电池监控被禁用
Unplugged	该设备上的电池电源运行
Charging	该设备的电池得到充电，且 USB 线连接
Full	该设备的电池充满，且 USB 线连接

9.5　处理运动事件

运动事件可能在很多的应用程序中使用到，例如音乐播放器和视频播放器。处理运动事件的方法如表 9-5 所示。

表 9-5　运动事件的方法

方　法	功　能
MotionBegan	在运动开始时调用
MotionEnded	在运动结束时调用
MotionCancelled	在运动取消时调用

【示例 9-5】 以下当用户轻轻摇晃手机时，实现换歌曲的功能。具体的操作步骤如下所述。

（1）创建一个 Single View Application 类型的工程。

（2）打开 MainStoryboard.storyboard 文件，对主视图进行设置，效果如图 9.8 所示。

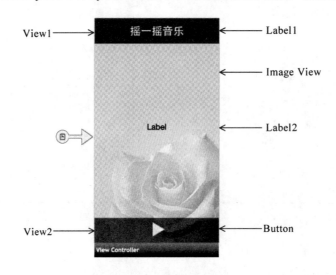

图 9.8 主视图的效果

需要添加的视图以及设置，如表 9-6 所示。

表 9-6 设置主视图

视 图	设 置
View1	Background：深灰色 位置和大小：(122, 250, 75, 75)
Label1	Text：摇一摇音乐 Font：System Bold 28 pt Color：白色 Alignment：居中 位置和大小：(82, 14, 157, 36)
Image View	Image：1.gif 位置和大小：(0, 0, 320, 568)
Label2	Name：Label Font：System 21 pt Alignment：居中 位置和大小：(20, 245, 280, 60)
View2	Background：黑色 Alpha：0.5 位置和大小：(0, 505, 320, 63)
Button	Name：btn Title：（空） Background：2.png 位置和大小：(130, 2, 70, 52)

（3）打开 9-5ViewController.cs 文件，编写代码，实现音乐播放器的功能，并且可以通过摇晃手机更换播放的歌曲。代码如下：

```csharp
using System;
using System.Drawing;
using MonoTouch.Foundation;
using MonoTouch.UIKit;
using MonoTouch.AVFoundation;
namespace Application
{
    public partial class __5ViewController : UIViewController
    {
        AVAudioPlayer audioPlayer;
        int songIndex;
        string[] pNames;
        string[] hNames;
        bool isPlay;
        ……
        #region View lifecycle
        public override void ViewDidLoad ()
        {
            base.ViewDidLoad ();
            pNames = new string[]{"孤单北半球", "天使的翅膀", "狐狸雨","一样爱着你"};
            hNames = new string[]{"孤单北半球—梁静茹","天使的翅膀—安琥","狐狸雨—李善姬", "一样爱着你—By2"};
            songIndex=0;
            isPlay = true;
            this.loadMusic (pNames[songIndex], "mp3");
            label.Text = string.Format("{0}",hNames[songIndex]);
            //触摸按钮实现播放盒暂停
            btn.TouchUpInside += (sender, e) => {
                if(isPlay)
                {
                    audioPlayer.Play();
                    btn.SetBackgroundImage(UIImage.FromFile("3.png"),
                    UIControlState.Normal);
                    isPlay=false;
                    
                } else
                {
                    audioPlayer.Pause();
                    isPlay=true;
                    btn.SetBackgroundImage(UIImage.FromFile("2.png"),
                    UIControlState.Normal);
                }
            };
        }
        //加载歌曲
        public void loadMusic(string name,string type )
        {
            string path = NSBundle.MainBundle.PathForResource (name, type);
            NSUrl url = NSUrl.FromFilename (path);
            audioPlayer = new AVAudioPlayer (url, null);
            audioPlayer.FinishedPlaying += Finished_play;
            audioPlayer.PrepareToPlay ();
        }
        //播放歌曲结束，自动更新为下一首
        public void Finished_play(object sender,EventArgs a)
```

```
    {
        songIndex++;
        //判断当前歌曲的缩影是否等于 pNames 字符串数组的长度
        if (songIndex == pNames.Length)
        {
            songIndex = 0;
        }
        this.loadMusic (pNames[songIndex], "mp3");
        label.Text = string.Format("{0}",hNames[songIndex]);
        audioPlayer.Play ();
    }
    //运行结束
    public override void MotionEnded (UIEventSubtype motion, UIEvent evt)
    {
        base.MotionEnded (motion, evt);
        //判断是否为摇晃事件,如果是,就播放下一首歌曲
        if (motion == UIEventSubtype.MotionShake)
        {
            if (songIndex == hNames.Length - 1)
            {
                songIndex = -1;
            }
            songIndex++;
            audioPlayer.Stop ();
            this.updatePlayerSetting();
        }
    }
    //播放歌曲并显示在标签中
    public void updatePlayerSetting()
    {
        this.loadMusic (pNames[songIndex], "mp3");
        label.Text = string.Format("{0}",hNames[songIndex]);
        audioPlayer.Play ();
    }
    ……
    #endregion
}
```

运行效果如图 9.9 所示。

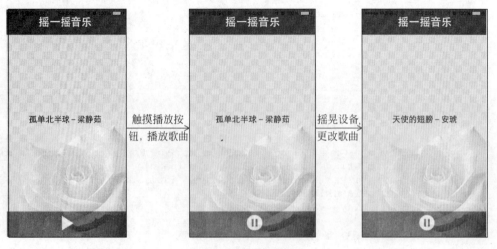

图 9.9 运行效果

9.6 处理触摸事件

在每一个应用程序中,触摸事件是必须有的。触摸(Cocoa Touch)就是用户的手指放在屏幕上一直到手指离开。触摸是在 UIView 上进行的。当用户触摸到屏幕时,触摸事件就会发生。处理触摸使用的方法如表 9-7 所示。

表 9-7 触摸事件的方法

方 法	功 能
TouchesBegan	手指刚触摸到屏幕时调用
TouchesMoved	手指在屏幕上移动时调用
TouchesEnded	手指离开屏幕时调用
TouchesCancelled	触摸取消时调用

【示例 9-6】 以下将在自创的应用程序中实现触摸。具体的操作步骤如下所述。

(1)创建一个 Single View Application 类型的工程。

(2)打开 MainStoryboard.storyboard 文件,从工具栏中拖动 Label 对象到主视图中,将此对象的 Name 设置为 label,将 Alignment 设置为居中,将位置和大小设置为(9,145,302,62)。

(3)打开 9-6ViewController.cs 文件,编写代码,实现触摸屏幕改变主视图的颜色。代码如下:

```
using System;
using System.Drawing;
using MonoTouch.Foundation;
using MonoTouch.UIKit;
namespace Application
{
    public partial class __6ViewController : UIViewController
    {
        ……                                      //这里省略了视图控制器的构造方法和析构方法
        #region View lifecycle
        ……                                      //这里省略了视图加载后的方法
        //触摸开始
        public override void TouchesBegan (NSSet touches, UIEvent evt)
        {
            base.TouchesBegan (touches, evt);
            label.Text = "Touch began!";
        }
        //移动
        public override void TouchesMoved (NSSet touches, UIEvent evt)
        {
            base.TouchesMoved (touches, evt);
            UITouch touch = touches.AnyObject as UITouch;       //创建触摸对象
            UIColor currentColor = this.View.BackgroundColor;
            float red, green, blue, alpha;
            currentColor.GetRGBA(out red, out green, out blue, out alpha);
            //获取先前触摸的位置
            PointF previousLocation = touch.PreviousLocationInView(this.View);
            //获取触摸在 View 这个视图上的位置
            PointF touchLocation = touch.LocationInView(this.View);
```

```
            //判断这两个位置是否相等
            if (previousLocation.X != touchLocation.X)
            {
                label.Text = "Changing background color...";
                float colorValue = touchLocation.X / this.View.Bounds.Width;
                this.View.BackgroundColor = UIColor.FromRGB(colorValue,
                    colorValue, colorValue);
            }
        }
        //触摸结束
        public override void TouchesEnded (NSSet touches, UIEvent evt)
        {
            base.TouchesEnded (touches, evt);
            label.Text = "Touch ended!";
            this.View.BackgroundColor = UIColor.White;
        }
        //触摸取消
        public override void TouchesCancelled (NSSet touches, UIEvent evt)
        {
            base.TouchesCancelled (touches, evt);
            label.Text = "Touch cancelled!";
            this.View.BackgroundColor = UIColor.White;
        }
        ……                      //这里省略了视图加载和卸载前后的一些方法
        #endregion
    }
}
```

运行效果如图 9.10 所示。

图 9.10 运行效果

9.7 手势识别器

在 Xamarin.iOS 中,将常用到的手势封装到一个 UIGestureRecognizer 类中。这个类被称为手势识别器,一个手势对应一个手势识别器。本节将主要讲解常用手势,以及它对应的手势识别器。

9.7.1 轻拍

轻拍手势一般使用 UITapGestureRecognizer 手势识别器进行识别。

【示例 9-7】 以下将实现轻拍手势,实现在轻拍的地方显示小星星的功能。具体的操作步骤如下所述。

(1)创建一个 Single View Application 类型的工程。
(2)添加图像 1.png 到创建工程的 Resources 文件夹中。
(3)打开 MainStoryboard.storyboard 文件,主视图的 Background 属性设置为黑色。
(4)打开 9-7ViewController.cs 文件,编写代码,实现轻拍手势,并且在轻拍过的地方出现小星星。代码如下:

```csharp
using System;
using System.Drawing;
using MonoTouch.Foundation;
using MonoTouch.UIKit;
namespace Application
{
    public partial class __7ViewController : UIViewController
    {
        ……                              //这里省略了视图控制器的构造方法和析构方法
        #region View lifecycle
        public override void ViewDidLoad ()
        {
            base.ViewDidLoad ();
            UITapGestureRecognizer tapGesture = new UITapGestureRecognizer
            (OnTapGesture);
            this.View.AddGestureRecognizer (tapGesture);
        }
        //轻拍手势实现的功能
        private void OnTapGesture(UITapGestureRecognizer tap)
        {
            PointF aa = tap.LocationInView (this.View);
            UIImageView iv = new UIImageView ();
            iv.Frame = new RectangleF (aa.X - 10, aa.Y - 10, 60, 60);
            iv.Image = UIImage.FromFile ("1.png");
            this.View.AddSubview (iv);
        }
        ……                              //这里省略了视图加载和卸载前后的一些方法
        #endregion
    }
}
```

运行效果如图 9.11 所示。

图 9.11　运行效果

9.7.2　捏

捏就是使用两个手指实现向里向外张合从而实现图片的放大和缩小。UIPinchGestureRecognizer 手势识别器可以识别捏的手势。

【示例 9-8】 以下将使用捏的手势实现图像的放大和缩小。具体的操作步骤如下所述。

（1）创建一个 Single View Application 类型的工程。

（2）添加图像 1.jpg 到创建工程的 Resources 文件夹中。

（3）打开 MainStoryboard.storyboard 文件，从工具栏中拖动 Image View 图像视图对象到主视图中。将此对象的 Name 设置为 iv，将 Image 设置为 1.jpg，将位置和大小设置为（0,0, 320, 568）。

（4）打开 9-8ViewController.cs 文件，编写代码，使用捏的手势实现图像的放大和缩小功能。代码如下：

```
using System;
using System.Drawing;
using MonoTouch.Foundation;
using MonoTouch.UIKit;
using MonoTouch.CoreGraphics;
namespace Application
{
    public partial class __8ViewController : UIViewController
    {
        float last;
        ……                              //这里省略了视图控制器的构造方法和析构方法
        #region View lifecycle
        public override void ViewDidLoad ()
        {
            base.ViewDidLoad ();
            UIPinchGestureRecognizer pinchGesture = new UIPinchGestureRecognizer
            (this.OnPinchGesture);
            View.AddGestureRecognizer(pinchGesture);
        }
        //捏手势实现的功能
        private void OnPinchGesture(UIPinchGestureRecognizer pinch)
```

```
        {
            iv.BringSubviewToFront (pinch.View);
            if (pinch.State == UIGestureRecognizerState.Ended)
            {
                last = 1.0f;
                return;
            }
            float scale = 1.0f - (last - pinch.Scale);
            CGAffineTransform current = pinch.View.Transform;
            CGAffineTransform newaa = CGAffineTransform.Scale (current,
            scale, scale);                                    //图像的缩放
            pinch.View.Transform = newaa;
            last = pinch.Scale;
        }
        ……                     //这里省略了视图加载和卸载前后的一些方法
    #endregion
    }
}
```

运行效果如图 9.12 所示。

图 9.12　运行效果

> **注意**：如果开发者在模拟器上运行，可以使用 Command+Alt 键代替两个手指，滑动鼠标实现放大和缩小的功能。

9.7.3 滑动

滑动的手势可以使用 UISwipeGestureRecognizer 手势识别器识别。滑动手势不像其他手势，在使用此手势时，可以使用 Direction 属性设置滑动手势滑动的方向。其中，滑动手势滑动的方向如表 9-8 所示。

表 9-8 滑动手势滑动的方向

方 向	功 能
Left	向左滑动
Right	向右滑动
Up	向上滑动
Down	向下滑动

【示例 9-9】以下将实现滑动手势，并且显示滑动手势滑动的方向。具体的操作步骤如下所述。

（1）创建一个 Single View Application 类型的工程。

（2）打开 MainStoryboard.storyboard 文件，从工具栏中拖动 Label 标签对象到主视图中。将此对象的 Name 设置为 label，将 Text 设置为空，将 Font 设置为 System 23 pt，将 Alignment 设置为居中，将位置和大小设置为（60, 190, 199, 59）。

（3）打开 9-9ViewController.cs 文件，编写代码，实现滑动手势。代码如下：

```
using System;
using System.Drawing;
using MonoTouch.Foundation;
using MonoTouch.UIKit;
namespace Application
{
    public partial class __9ViewController : UIViewController
    {
        ……                              //这里省略了视图控制器的构造方法和析构方法
        #region View lifecycle
        public override void ViewDidLoad ()
        {
            base.ViewDidLoad ();
            UISwipeGestureRecognizer left = new UISwipeGestureRecognizer
            (leftMethods);
            left.Direction = UISwipeGestureRecognizerDirection.Left;
                                                                    //设置方向
            this.View.AddGestureRecognizer (left);
            UISwipeGestureRecognizer right= new UISwipeGestureRecognizer
            (rightMethods);
            right.Direction = UISwipeGestureRecognizerDirection.Right;
            this.View.AddGestureRecognizer (right);
            UISwipeGestureRecognizer up= new UISwipeGestureRecognizer
            (upMethods);
            up.Direction = UISwipeGestureRecognizerDirection.Up;
            this.View.AddGestureRecognizer (up);
```

```
        UISwipeGestureRecognizer down= new UISwipeGestureRecognizer
        (downMethods);
        down.Direction = UISwipeGestureRecognizerDirection.Down;
        this.View.AddGestureRecognizer (down);
    }
    //向左滑动实现的功能
    private void leftMethods(UISwipeGestureRecognizer swipe)
    {
        label.Text = "向左滑动";
    }
    //向右滑动实现的功能
    private void rightMethods(UISwipeGestureRecognizer swipe)
    {
        label.Text = "向右滑动";
    }
    //向上滑动实现的功能
    private void upMethods(UISwipeGestureRecognizer swipe)
    {
        label.Text = "向上滑动";
    }
    //向下滑动实现的功能
    private void downMethods(UISwipeGestureRecognizer swipe)
    {
        label.Text = "向下滑动";
    }
    ……                              //这里省略了视图加载和卸载前后的一些方法
    #endregion
}
```

运行效果如图 9.13 所示。

图 9.13　运行效果

9.7.4 旋转

旋转手势可以使用 UIRotationGestureRecognizer 手势识别器进行识别。

【示例 9-10】 以下将实现使用旋转手势实现图像的旋转。具体的操作步骤如下所述。

（1）创建一个 Single View Application 类型的工程。

（2）添加图像 1.png 到创建工程的 Resources 文件夹中。

（3）打开 MainStoryboard.storyboard 文件，对主视图进行设置，效果如图 9.14 所示。

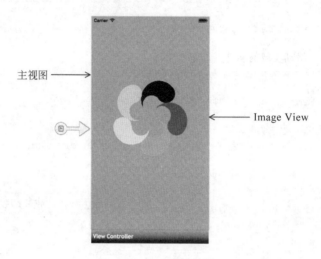

图 9.14　主视图的效果

需要添加的视图以及设置，如表 9-9 所示。

表 9-9　设置主视图

视　　图	设　　置
主视图	Background：浅灰色
Image View	Name：iv Image：1.png 位置和大小：(60, 171, 200, 200)

（4）打开 9-10ViewController.cs 文件，编写代码，实现使用旋转手势旋转图像。代码如下：

```
using System;
using System.Drawing;
using MonoTouch.Foundation;
using MonoTouch.UIKit;
using MonoTouch.CoreGraphics;
namespace Application
{
    public partial class __10ViewController : UIViewController
    {
        ……                        //这里省略了视图控制器的构造方法和析构方法
        #region View lifecycle
        public override void ViewDidLoad ()
        {
```

```
        base.ViewDidLoad ();
        UIRotationGestureRecognizer rotationGesture = new
        UIRotationGestureRecognizer (rotationMethods);
        this.View.AddGestureRecognizer (rotationGesture);
    }
    //旋转手势实现的功能
    private void rotationMethods (UIRotationGestureRecognizer rotation)
    {
        CGAffineTransform transform = CGAffineTransform.MakeRotation
        (rotation.Rotation);
        iv.Transform = transform;
    }
    ……                          //这里省略了视图加载和卸载前后的一些方法
    #endregion
}
```

运行效果如图 9.15 所示。

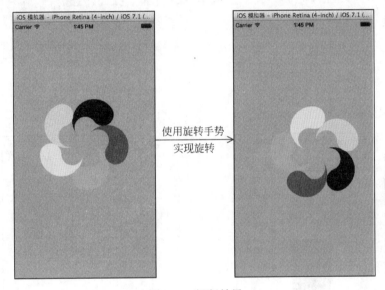

图 9.15　运行效果

9.7.5　移动

移动手势可以使用 UIPanGestureRecognizer 手势识别器进行识别。

【示例 9-11】 以下将实现使用移动手势实现图像的移动功能。具体的操作步骤如下所述。

（1）创建一个 Single View Application 类型的工程。

（2）添加图像 1.png 到创建工程的 Resources 文件夹中。

（3）打开 MainStoryboard.storyboard 文件，从工具栏中拖动 Image View 图像视图对象到主视图中，将此对象的 Name 设置为 iv，将 Image 设置为 1.png，将位置和大小设置为（66，151，187，240）。

（4）打开 9-11ViewController.cs 文件，编写代码，实现移动手势。代码如下：

```
using System;
using System.Drawing;
```

```
using MonoTouch.Foundation;
using MonoTouch.UIKit;
namespace Application
{
    public partial class __11ViewController : UIViewController
    {
        ……                          //这里省略了视图控制器的构造方法和析构方法
        #region View lifecycle
        public override void ViewDidLoad ()
        {
            base.ViewDidLoad ();
            UIPanGestureRecognizer panGesture = new UIPanGestureRecognizer
            (panMethods);
            this.View.AddGestureRecognizer (panGesture);
        }
        //移动手势实现的功能
        private void panMethods (UIPanGestureRecognizer pan)
        {
            PointF aa = pan.LocationInView (this.View);
            iv.Center = aa;
        }
        ……                          //这里省略了视图加载和卸载前后的一些方法
        #endregion
    }
}
```

运行效果如图 9.16 所示。

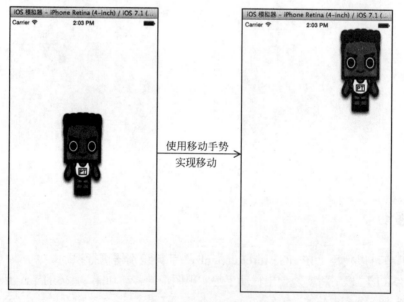

图 9.16　运行效果

9.7.6　长按

长按手势可以通过 **UILongPressGestureRecognizer** 手势识别器进行识别。

【示例 9-12】 以下将使用长按手势实现弹出警告视图的功能。代码如下：
（1）创建一个 Single View Application 类型的工程。

(2) 添加图像 1.jpg、2.png、3.jpg、4.jpg 和 5.jpg 到创建工程的 Resources 文件夹中。

(3) 打开 MainStoryboard.storyboard 文件，从工具栏中拖动 Image View 图像视图对象到主视图中，将此对象的 Name 设置为 iv，将位置和大小设置为 (0, 0, 320, 568)。

(4) 打开 9-12ViewController.cs 文件，编写代码，实现长按后弹出警告视图。代码如下：

```
using System;
using System.Drawing;
using MonoTouch.Foundation;
using MonoTouch.UIKit;
namespace Application
{
    public partial class __12ViewController : UIViewController
    {
        ……                              //这里省略了视图控制器的构造方法和析构方法
        #region View lifecycle
        public override void ViewDidLoad ()
        {
            base.ViewDidLoad ();
            UILongPressGestureRecognizer longPressGesture = new
            UILongPressGestureRecognizer (longPressMethods);
            longPressGesture.MinimumPressDuration = 1.0;
                                                    //设置长按手势需要按下的时间
            this.View.AddGestureRecognizer (longPressGesture);
        }
        //长按手势实现的功能
        private void longPressMethods (UILongPressGestureRecognizer longPress)
        {
            UIAlertView alert = new UIAlertView ();
            alert.Title = "请选择背景";
            alert.AddButton ("第1种背景");
            alert.AddButton ("第2种背景");
            alert.AddButton ("第3种背景");
            alert.AddButton ("第4种背景");
            alert.AddButton ("第5种背景");
            alert.AddButton ("Cancel");
            alert.Show ();
            //警告视图中触摸按钮实现的功能
            alert.Dismissed += (sender, e) => {
                if (e.ButtonIndex == 0) {
                    iv.Image = UIImage.FromFile ("1.jpg");
                } else if (e.ButtonIndex == 1) {
                    iv.Image = UIImage.FromFile ("2.png");
                } else if (e.ButtonIndex == 2) {
                    iv.Image = UIImage.FromFile ("3.jpg");
                } else if (e.ButtonIndex == 3) {
                    iv.Image = UIImage.FromFile ("4.jpg");
                } else if (e.ButtonIndex == 4) {
                    iv.Image = UIImage.FromFile ("5.jpg");
                }
            };
        }
        ……                              //这里省略了视图加载和卸载前后的一些方法
        #endregion
    }
}
```

运行效果如图 9.17 所示。

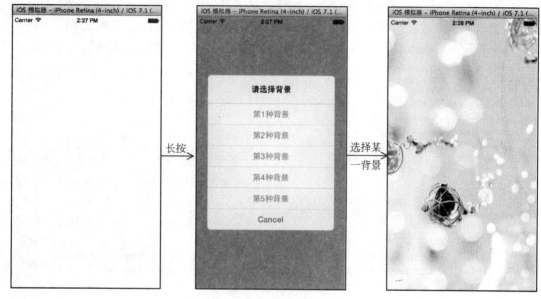

图 9.17　运行效果

9.8　自定义手势

除了可以在 iOS 模拟器上使用常用的手势外，开发者还可以自己定义手势。这些手势在 UIGestureRecognizer 类中是没有的，必须使用触摸方法来实现。

【示例 9-13】　以下程序实现的是一个擦除手势。操作步骤及程序代码介绍如下。

（1）创建一个 Single View Application 类型的工程。

（2）添加图像 1.png 到创建工程的 Resources 文件夹中。

（3）打开 9-13ViewController.cs 文件，编写代码，实现擦除手势。代码如下：

```
using System;
using System.Drawing;
using MonoTouch.Foundation;
using MonoTouch.UIKit;
using MonoTouch.ObjCRuntime;
namespace Application
{
    public partial class __13ViewController : UIViewController
    {
        UIImageView iv;
        ……
        #region View lifecycle
        public override void ViewDidLoad ()
        {
            base.ViewDidLoad ();
            UITapGestureRecognizer tapGesture = new UITapGestureRecognizer
            (OnTapGesture);
            this.View.AddGestureRecognizer (tapGesture);
        }
```

```
private void OnTapGesture(UITapGestureRecognizer tap)
{
    PointF aa = tap.LocationInView (this.View);
    iv = new UIImageView ();
    iv.Frame = new RectangleF (aa.X - 10, aa.Y - 10, 60, 60);
    iv.Image = UIImage.FromFile ("1.png");
    iv.UserInteractionEnabled = true;
    this.View.AddSubview (iv);
    DeleteGesture deleteGesture = new DeleteGesture ();
    deleteGesture.AddTarget (deleteMethods);
    this.View.AddGestureRecognizer (deleteGesture);
}
//擦除手势实现的功能
private void deleteMethods(NSObject gesture)
{
    DeleteGesture de = (DeleteGesture)gesture;
    if (de.State == UIGestureRecognizerState.Recognized)
    {
        de.dv.RemoveFromSuperview ();
    }
    de.reset ();
}
private class DeleteGesture : UIGestureRecognizer
{
    bool b;
    int i;
    public UIView dv;
    //重置
    public void reset()
    {
        base.Reset ();
        b = true;
        i = 0;
        dv = null;
    }
    //开始触摸
    public override void TouchesBegan (NSSet touches, UIEvent evt)
    {
        base.TouchesBegan (touches, evt);
        if (touches.Count != 1)
        {
            this.State = UIGestureRecognizerState.Failed;
            return;
        }
    }
    //移动
    public override void TouchesMoved (NSSet touches, UIEvent evt)
    {
        base.TouchesMoved (touches, evt);
        if (this.State == UIGestureRecognizerState.Failed) return;
        UITouch touch = (UITouch)touches.AnyObject;
        //获取在View视图上的位置
        PointF nowPoint = touch.LocationInView(this.View);
        PointF prevPoint = touch.PreviousLocationInView (this.View);
        if (b == true)
        {
            if (nowPoint.Y < prevPoint.Y)
            {
                b = false;
                i++;
```

```
                } else if (nowPoint.Y > prevPoint.Y)
                {
                    b = true;
                    i++;
                }
                if (dv == null)
                {

                    UIView vie = this.View.HitTest (nowPoint, null);
                    if (vie != null&&vie!=this.View)
                    {
                        dv = vie;
                    }

                }
            }
            //结束触摸
            public override void TouchesEnded (NSSet touches, UIEvent evt)
            {
                base.TouchesEnded (touches, evt);
                if (this.State == UIGestureRecognizerState.Possible) {
                    if (i >= 3)
                    {
                        this.State = UIGestureRecognizerState.Recognized;
                    } else
                    {
                        this.State = UIGestureRecognizerState.Failed;
                    }
                }

            }
            ……
            #endregion
    }
}
```

运行效果如图 9.18 所示。

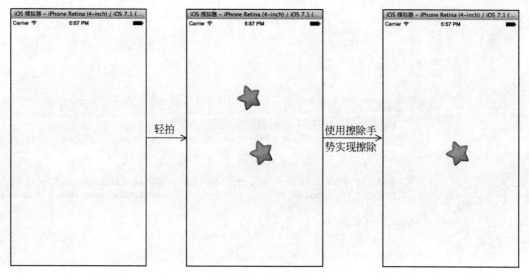

图 9.18 运行效果

9.9 使用加速计

iOS 设备的加速计可用来检测手机收到的加速度的大小和方向。但是由于手机静止的时候是只受到重力加速度,所以很多人把加速计功能又叫做重力感应功能。

【示例 9-14】 以下将实现访问加速计的数据,并显示在标签中。具体的操作步骤如下所述。

(1) 创建一个 Single View Application 类型的工程。

(2) 打开 MainStoryboard.storyboard 文件,对主视图进行设置,效果如图 9.19 所示。

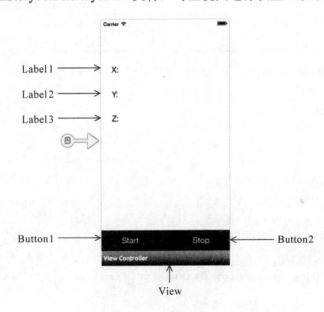

图 9.19 主视图的效果

需要添加的视图以及设置,如表 9-10 所示。

表 9-10 设置主视图

视 图	设 置
Label1	Name:label1 Text:X: Font:System 20 pt 位置和大小:(26, 100, 268, 46)
Label2	Name:label2 Text:Y: Font:System 20 pt 位置和大小:(26, 160, 268, 46)
Label3	Name:label3 Text:Z: Font:System 20 pt 位置和大小:(26, 220, 268, 46)

续表

视　图	设　置
Button1	Name：startbtn Title：Start Font：System 20 pt Text Color：白色 位置和大小：(48, 10, 46, 30)
Button2	Name：stopbtn Title：Stop Font：System 20 pt Text Color：白色 位置和大小：(226, 10, 46, 30)
View	Background：黑色 位置和大小：(0, 518, 320, 50)

（3）打开 9-14ViewController.cs 文件，编写代码，实现对加速计的访问。代码如下：

```
using System;
using System.Drawing;
using MonoTouch.Foundation;
using MonoTouch.UIKit;
namespace Application
{
    public partial class __14ViewController : UIViewController
    {
        ……
        #region View lifecycle
        public override void ViewDidLoad ()
        {
            base.ViewDidLoad ();
            UIAccelerometer.SharedAccelerometer.UpdateInterval = 1 / 10;
                                        //设置加速计的更新频率
            //开始更新加速计的数据
            startbtn.TouchUpInside += delegate {
                UIAccelerometer.SharedAccelerometer.Acceleration += this.
                Acceleration_Received;
            } ;
            //停止更新加速计的数据
            stopbtn.TouchUpInside += delegate {
                UIAccelerometer.SharedAccelerometer.Acceleration -= this.
                Acceleration_Received;
            } ;
        }
        //将加速计的数据显示的标签中
        private void Acceleration_Received (object sender,
        UIAccelerometerEventArgs e)
        {
            label1.Text = string.Format ("X: {0}", e.Acceleration.X);
            label2.Text = string.Format ("Y: {0}", e.Acceleration.Y);
            label3.Text = string.Format ("Z: {0}", e.Acceleration.Z);
        }
        ……
        #endregion
    }
}
```

运行效果如图 9.20 所示。

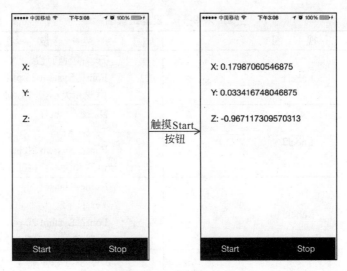

图 9.20 运行效果

9.10 使用陀螺仪

陀螺仪又叫角速度传感器。它不同于加速计，它的测量物理量是偏转、倾斜时的转动角速度。在手机上，仅用加速度计没办法测量或重构出完整的 3D 动作，测不到转动的动作，加速计只能检测轴向的线性动作。但陀螺仪则可以对转动、偏转的动作做很好的测量，这样就可以精确分析判断出使用者的实际动作。而后根据动作，可以对手机做相应的操作。

【示例 9-15】 以下将实现对 X、Y、Z 轴旋转速度的显示。具体的操作步骤如下所述。

（1）创建一个 Single View Application 类型的工程。
（2）打开 MainStoryboard.storyboard 文件，对主视图进行设置，效果如图 9.21 所示。

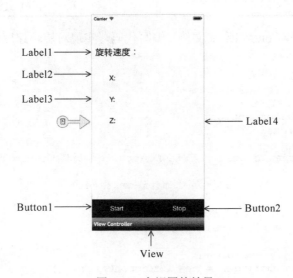

图 9.21 主视图的效果

需要添加的视图以及设置，如表 9-11 所示。

表 9-11 设置主视图

视 图	设 置
Label1	Text：旋转速度： Font：System 25 pt 位置和大小：(10, 84, 130, 41)
Label2	Name：label1 Text：X: Font：System 20 pt 位置和大小：(50, 160, 235, 34)
Label3	Name：label2 Text：Y: Font：System 20 pt 位置和大小：(50, 220, 235, 34)
Label4	Name：label3 Text：Z: Font：System 20 pt 位置和大小：(50, 280, 235, 34)
Button1	Name：startbtn Title：Start Font：System 18 pt Text Color：白色 位置和大小：(48, 10, 46, 30)
Button2	Name：stopbtn Title：Stop Font：System 28 pt Text Color：白色 位置和大小：(226, 10, 46, 30)
View	Background：黑色 位置和大小：(0, 518, 320, 50)

（3）打开 9-15ViewController.cs 文件，编写代码，实现对旋转速度的显示。代码如下：

```
using System;
using System.Drawing;
using MonoTouch.Foundation;
using MonoTouch.UIKit;
using MonoTouch.CoreMotion;
namespace Application
{
    public partial class __15ViewController : UIViewController
    {
        private CMMotionManager motionManager;
        ……
        #region View lifecycle
        public override void ViewDidLoad ()
        {
            base.ViewDidLoad ();
            motionManager = new CMMotionManager();
            motionManager.GyroUpdateInterval = 1 / 10;
```

```
            //触摸按钮,开始更新陀螺仪的值
            startbtn.TouchUpInside += delegate {
                motionManager.StartGyroUpdates(NSOperationQueue.MainQueue,
                this.GyroData_Received);
            } ;
            //触摸按钮,停止更新陀螺仪的值
            stopbtn.TouchUpInside += delegate {
                motionManager.StopGyroUpdates();
            } ;
        }
        //显示X、Y、Z轴的旋转速度
        private void GyroData_Received(CMGyroData gyroData, NSError error)
        {
            label1.Text=string.Format("X: {0}", gyroData.RotationRate.x);
            label2.Text=string.Format("Y: {0}", gyroData.RotationRate.y);
            label3.Text=string.Format("Z: {0}", gyroData.RotationRate.z);
        }
        ......
        #endregion
    }
}
```

运行效果如图 9.22 所示。

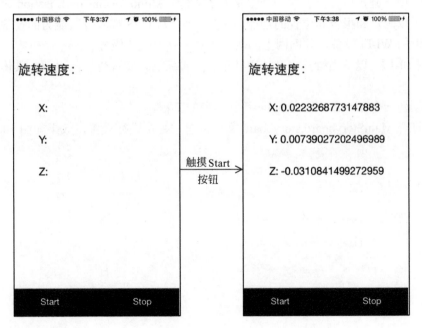

图 9.22 运行效果

注意：在此程序中引入了一个新的命名空间 MonoTouch.CoreMotion。此命名空间用来访问加速度器和陀螺仪的相关数据。它不仅仅提供了获得实时的加速度值和旋转速度值，更重要的是，在其中集成了很多算法，可以直接把重力加速度分量剥离的加速度输出，省去了手动操作。

第 10 章 位置服务和地图

当今，智能手机或者移动设备都配备了高精度的全局定位系统（GPS）的硬件。GPS 硬件从卫星上接收了位置信息。除了卫星外，iOS 设备的优势在于可以获取蜂窝和 WiFi 网络中提供的位置信息给用户。本章将讲解位置服务以及位置服务的典型应用场合——地图。

10.1 确定位置

在 iOS 中，确定所在位置是一个很重要的功能。MonoTouch.CoreLocation 命名空间为设备提供了以确定和报告其位置的功能。即使一个没有 GPS 或者蜂窝能力的设备也可以通过扫描附近的 Wi-Fi 设备，并同网上的数据比对来获取位置服务。

【示例 10-1】 以下当触摸"开始更新位置"按钮后，实现位置的更新。具体的操作步骤如下所述。

（1）创建一个 Single View Application 类型的工程。

（2）打开 MainStoryboard.storyboard 文件，对主视图进行设置，效果如图 10.1 所示。

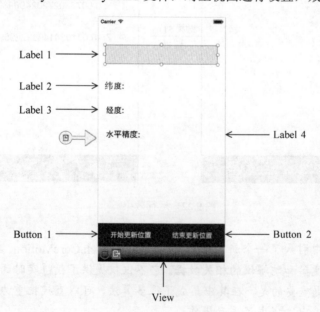

图 10.1 主视图的效果

需要添加的视图以及设置，如表 10-1 所示。

表 10-1 设置主视图

视 图	设 置
Label1	Name：label Text：（空） Font：System 25 pt Alignment：居中 位置和大小：(20, 70, 280, 45)
Label2	Name：label1 Text：纬度： Font：System 20 pt 位置和大小：(20, 149, 280, 41)
Label3	Name：label2 Text：经度： Font：System 20 pt 位置和大小：(20, 209, 280, 41)
Label4	Name：label3 Text：水平精度： Font：System 20 pt 位置和大小：(20, 269, 280, 41)
Button1	Name：btnstart Title：开始更新位置 Font：System 17 pt Text Color：白色 位置和大小：(20, 14, 125, 30)
Button2	Name：btnstop Title：结束更新位置 Font：System 17 pt Text Color：白色 位置和大小：(175, 14, 125, 30)
View	Background：深灰色 位置和大小：(0, 510, 320, 58)

(3) 打开 10-1ViewController.cs 文件，编写代码，实现位置的更新和显示。代码如下：

```
using System;
using System.Drawing;
using MonoTouch.Foundation;
using MonoTouch.UIKit;
using MonoTouch.CoreLocation;
namespace Application
{
    public partial class _0_1ViewController : UIViewController
    {
        private CLLocationManager locationManager;
        ……                              //这里省略了视图控制器的构造方法和析构方法
        #region View lifecycle
        public override void ViewDidLoad ()
        {
            base.ViewDidLoad ();
            locationManager= new CLLocationManager();
```

```csharp
            locationManager.LocationsUpdated += LocationManager_LocationsUpdated;    //更新时调用
            locationManager.Failed += LocationManager_Failed;
            //更新失败调用
            //触摸"开始更新位置"按钮,实现位置的更新
            btnstart.TouchUpInside += delegate {
                label.Text = "Determining location...";
                locationManager.StartUpdatingLocation();    //开启位置服务
            };
            //触摸"结束更新位置"按钮,实现位置的更新
            btnstop.TouchUpInside += delegate {
                locationManager.StopUpdatingLocation();    //结束位置服务
                label.Text = "Location update stopped...";
                label1.Text = "纬度:";
                label2.Text = "经度:";
                label3.Text = "水平精度:";
            };
        }
        //获取位置信息并显示
        private void LocationManager_LocationsUpdated (object sender, CLLocationsUpdatedEventArgs e)
        {
            CLLocation location = e.Locations[0];
            double latitude = Math.Round(location.Coordinate.Latitude, 4);
            //纬度
            double longitude = Math.Round(location.Coordinate.Longitude, 4);    //经度
            double accuracy = Math.Round(location.HorizontalAccuracy, 0);
            //水平精度
            label1.Text = string.Format("纬度: {0}", latitude);
            label2.Text = string.Format("经度: {0}", longitude);
            label3.Text = string.Format("水平精度: {0}", accuracy);
        }
        //输出错误信息
        private void LocationManager_Failed (object sender, NSErrorEventArgs e)
        {
            label.Text = string.Format("Location update failed! Error message: {0}", e.Error.LocalizedDescription);
        }
        ……                            //这里省略了视图加载和卸载前后的一些方法
        #endregion
    }
}
```

运行效果如图 10.2 所示。

> 注意:在此程序中使用到两个类,CLLocation 类和 CLLocationManager 类。其中,CLLocation 类代表了一个位置信息,其中包含速度、水平精度、经度和纬度等。CLLocationManager 类是用来管理和提供位置服务的。

第 10 章 位置服务和地图

图 10.2 运行效果

10.2 确定方向

在 iOS 移动设备中都会有一个罗盘来判断设备的方向。本节中将使用罗盘来获取位置方向。

【示例 10-2】 以下将自制一个简单的指南针。在触摸"开始更新方向"按钮后，将会显示方向，并会出现一个指南针。具体的操作步骤如下所述。

（1）创建一个 Single View Application 类型的工程。
（2）添加图像 1.png 到创建工程的 Resources 文件夹中。
（3）打开 MainStoryboard.storyboard 文件，对主视图进行设置，效果如图 10.3 所示。

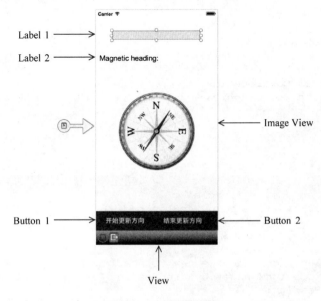

图 10.3 主视图的效果

需要添加的视图以及设置，如表 10-2 所示。

表 10-2　设置主视图

视　　图	设　　置
Label1	Name：label1 Text：（空） Alignment：居中 位置和大小：(45, 56, 230, 21)
Label2	Name：label1 Text：Magnetic heading Font：System 20 pt 位置和大小：(9, 105, 302, 39)
Image View	Name：iv Image：1.png 位置和大小：(60, 313, 200, 200)
Button1	Name：btnStart Title：开始更新方向 Font：System 17 pt Text Color：白色 位置和大小：(14, 10, 123, 30)
Button2	Name：btnStop Title：结束更新方向 Font：System 17 pt Text Color：白色 位置和大小：(166, 10, 123, 30)
View	Background：深灰色 位置和大小：(0, 518, 320, 50)

（4）打开 10-2ViewController.cs 文件，编写代码，实现方向的更新以及指南针的效果。代码如下：

```
using System;
using System.Drawing;
using MonoTouch.Foundation;
using MonoTouch.UIKit;
using MonoTouch.CoreLocation;
using MonoTouch.CoreGraphics;
namespace Application
{
    public partial class _0_2ViewController : UIViewController
    {
        private CLLocationManager locationManager;
        ……                    //这里省略了视图控制器的构造方法和析构方法
        #region View lifecycle
        public override void ViewDidLoad ()
        {
            base.ViewDidLoad ();
            iv.Hidden = true;
            locationManager = new CLLocationManager();
            locationManager.UpdatedHeading += LocationManager_ UpdatedHeading;
            locationManager.Failed += (sender, e) => Console.WriteLine
            ("Failed! { 0}", e.Error.LocalizedDescription);
```

```csharp
//触摸"开始更新方向"按钮,实现方向的更新
btnStart.TouchUpInside += delegate {
    //判断设备是否支持heading
    if(CLLocationManager.HeadingAvailable)
    {
        iv.Hidden=false;
        label1.Text = "Starting updating heading...";
        locationManager.StartUpdatingHeading();
        //开始更新方向
    } else
    {
        UIAlertView alert=new UIAlertView();
        alert.Title="对不起,指南针不可用";
        alert.AddButton("Cancel");
        alert.Show();
    }
} ;
btnStop.TouchUpInside += delegate {
    iv.Hidden=true;
    locationManager.StopUpdatingHeading();        //结束方向更新
    label1.Text = "Stopped updating heading...";
    label2.Text = "Magnetic heading: ";
} ;
}
//实现方向的获取,并旋转图像
private void LocationManager_UpdatedHeading (object sender,
CLHeading UpdatedEventArgs e)
{
    label2.Text = string.Format("Magnetic heading: {0}", Math.
Round(e.NewHeading.MagneticHeading, 1));
    double head = 1.0 * 3.1415926 * e.NewHeading.MagneticHeading /
180.0f;
    float heading = (float)head;
    iv.Transform = CGAffineTransform.MakeRotation (heading);
}
……                              //这里省略了视图加载和卸载前后的一些方法
#endregion
}
}
```

运行效果如图 10.4 所示。

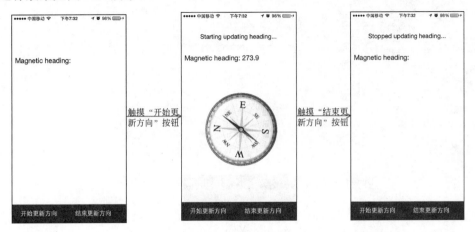

图 10.4　运行效果

> 注意：在标签中显示的数据对应的方向，如表 10-3 所示。

表 10-3　数据对应的方向

度　数	方　向
0 或者 360 度	该设备向北
90 度	该设备向东
180 度	该设备向南
270 度	该设备向西

10.3　使用区域监测

如果希望 iOS 设备进入某个区域发出通知，就要使用到区域监测。这种区域监测的功能也被称为临近警告。所谓临近警告的示意图如图 10.5 所示。

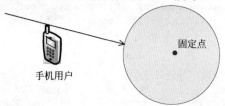

图 10.5　临近警告

用户设备渐渐地接近指定固定点，当与该固定点的距离小于指定范围时，系统可以触发相应的处理。用户设备离开指定固定点，当与该固定点的距离大于指定范围时，系统也可以触发相应的处理。iOS 的区域监测需要使用 CLLocationManager 来实现，监听设备是否进入或离开某个区域。

【示例 10-3】以下将实现区域监测的功能。具体的操作步骤如下所述。

（1）创建一个 Single View Application 类型的工程。

（2）打开 MainStoryboard.storyboard 文件，对主视图进行设置，效果如图 10.6 所示。

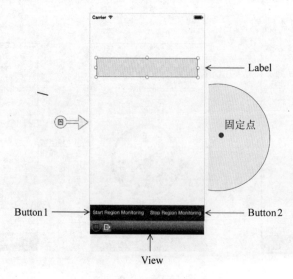

图 10.6　主视图的效果

需要添加的视图以及设置，如表 10-4 所示。

表 10-4　设置主视图

视　　图	设　　置
Label1	Name：label Text：（空） Font：System 21 pt Alignment：居中 位置和大小：(20, 124, 280, 49)
Button1	Name：startbtn Title：Start Region Monitoring Font：System 13 pt Text Color：白色 位置和大小：(8, 6, 144, 30)
Button2	Name：stopbtn Title：Stop Region Monitoring Font：System 13 pt Text Color：白色 位置和大小：(168, 6, 144, 30)
View	Background：深灰色 位置和大小：(0, 525, 320, 43)

（3）打开 10-3ViewController.cs 文件，编写代码，实现区域监测。代码如下：

```
using System;
using System.Drawing;
using MonoTouch.Foundation;
using MonoTouch.UIKit;
using MonoTouch.CoreLocation;
namespace Application
{
    public partial class _0_3ViewController : UIViewController
    {
        private CLLocationManager locationManager;
        private CLCircularRegion region;
        ……                          //这里省略了视图控制器的构造方法和析构方法
        #region View lifecycle
        public override void ViewDidLoad ()
        {
            base.ViewDidLoad ();
            this.locationManager = new CLLocationManager();
            locationManager.RegionEntered+=this.LocationManager_Region Entered;
            locationManager.RegionLeft+=this.LocationManager_RegionLeft;
            locationManager.LocationsUpdated+=LocationManager_Locations
            Updated;
            startbtn.TouchUpInside += (sender, e) => {
                locationManager.StartUpdatingLocation();
            };
            stopbtn.TouchUpInside += (sender, e) => {
                locationManager.StopMonitoring(this.region);
            };
        }
        private void LocationManager_LocationsUpdated (object sender, CLLocationsUpdatedEventArgs e)
        {
```

```
            CLLocation location = e.Locations[0];
            label.Text=string.Format("Got location! {0}", location.
            HorizontalAccuracy);
            if (location.HorizontalAccuracy < 100)
            {
                //使用 CLCircularRegion 创建一个圆形区域,半径为 100 米
                this.region = new CLCircularRegion(location.Coordinate,
                100, "Home");
                this.locationManager.StartMonitoring(this.region);
                //开始监听 region 区域
                this.locationManager.StopUpdatingLocation();
                //结束监听 region 区域
            }
        }
        //进入指定区域以后将弹出提示框提示用户
        private void LocationManager_RegionEntered (object sender, CLRegion
        EventArgs e)
        {
            UIAlertView alert = new UIAlertView ();
            alert.Title = "区域检测提示";
            alert.Message = "您已经【进入】视频监视区域!";
            alert.AddButton ("Cancel");
            alert.Show ();
        }
        //离开指定区域以后将弹出提示框提示用户
        private void LocationManager_RegionLeft (object sender, CLRegion
        EventArgs e)
        {
            UIAlertView alert = new UIAlertView ();
            alert.Title = "区域检测提示";
            alert.Message = "您已经【离开】视频监视区域!";
            alert.AddButton ("Cancel");
            alert.Show ();
        }
        ……                              //这里省略了视图加载和卸载前后的一些方法
        #endregion
    }
}
```

运行效果如图 10.7 所示。

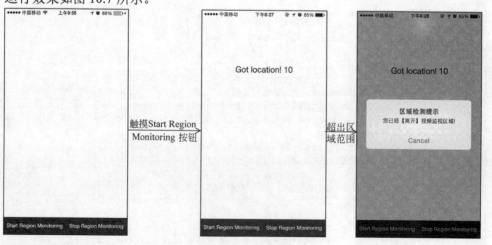

图 10.7 运行效果

10.4 使用 significant-change 位置服务

significant-change 位置服务提供了低耗电的方法来获取当前位置,当前位置改变时会发出通知。此服务的开启需要使用 StartMonitoringSignificantLocationChanges() 方法。

【示例 10-4】 以下将使用 significant-change 位置服务检测当前的位置。具体的操作步骤如下所述。

(1) 创建一个 Single View Application 类型的工程。
(2) 打开 MainStoryboard.storyboard 文件,对主视图进行设置,效果如图 10.8 所示。

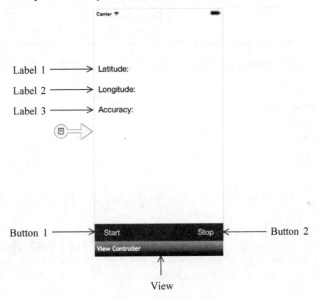

图 10.8 主视图的效果

需要添加的视图以及设置,如表 10-5 所示。

表 10-5 设置主视图

视 图	设 置
Label1	Name:label1 Text:Latitude: Font:System 20 pt 位置和大小:(10, 130, 300, 30)
Label2	Name:label2 Text:Longitude: Font:System 20 pt 位置和大小:(10, 180, 300, 30)
Label3	Name:label3 Text:Accuracy: Font:System 20 pt 位置和大小:(10, 230, 300, 30)

视图	设置
Button1	Name：startbtn Title：Start Font：System 19 pt Text Color：白色 位置和大小：(20, 6, 46, 30)
Button2	Name：stopbtn Title：Stop Font：System 19 pt Text Color：白色 位置和大小：(254, 6, 46, 30)
View	Background：深灰色 位置和大小：(0, 525, 320, 43)

（3）打开 10-4ViewController.cs 文件，编写代码，实现位置的更新和显示。代码如下：

```
using System;
using System.Drawing;
using MonoTouch.Foundation;
using MonoTouch.UIKit;
using MonoTouch.CoreLocation;
namespace Application
{
    public partial class _0_4ViewController : UIViewController
    {
        private CLLocationManager locationManager;
        ……                            //这里省略了视图控制器的构造方法和析构方法
        #region View lifecycle
        public override void ViewDidLoad ()
        {
            base.ViewDidLoad ();
            this.locationManager = new CLLocationManager();
            this.locationManager.LocationsUpdated += LocationManager_
            LocationsUpdated;
            //触摸 Start 按钮，实现开启 significant-change 位置服务
            startbtn.TouchUpInside += (s, e) => {
                //开启 significant-change 位置服务
                this.locationManager.StartMonitoringSignificantLocation
                Changes();
            } ;
            //触摸 Stop 按钮，实现关闭 significant-change 位置服务
            stopbtn.TouchUpInside += (s, e) => {
                //关闭 significant-change 位置服务
                this.locationManager.StopMonitoringSignificantLocation
                Changes();
                label1.Text = "Latitude:";
                label2.Text = "Longitude:";
                label3.Text = "Accuracy:";
            } ;
        }
        //更新位置，并显示
```

第 10 章 位置服务和地图

```
private void LocationManager_LocationsUpdated (object sender,
CLLocationsUpdatedEventArgs e)
{
    CLLocation location = e.Locations[0];
    double latitude = Math.Round(location.Coordinate.Latitude, 4);
    double longitude = Math.Round(location.Coordinate.Longitude,
    4);
    double accuracy = Math.Round(location.HorizontalAccuracy, 0);
    label1.Text = string.Format("Latitude: {0}", latitude);
    label2.Text = string.Format("Longitude: {0}", longitude);
    label3.Text = string.Format("Accuracy: {0}", accuracy);
}
    ……                    //这里省略了视图加载和卸载前后的一些方法
#endregion
}
```

运行效果如图 10.9 所示。

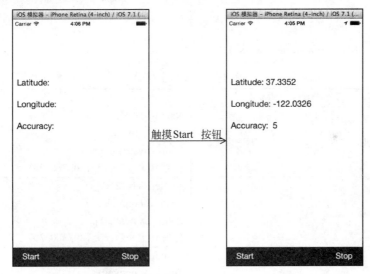

图 10.9 运行效果

10.5 在后台运行位置服务

significant-change 位置服务启动后，即使应用没有被启动也可以接收通知，并且进入到后台状态来处理事件。但是不能让程序的后台任务超过 10 分钟。如果想要让程序的后台任务超过 10 分钟，就需要在后台运行应用程序。

【示例 10-5】 以下将实现当应用程序进入后台后，使用位置服务的功能。具体的操作步骤如下所述。

（1）创建一个 Single View Application 类型的工程。
（2）打开 MainStoryboard.storyboard 文件，对主视图进行设置，效果如图 10.10 所示。

第 2 篇　资源使用篇

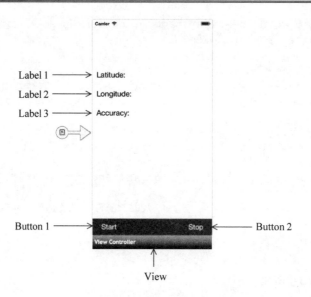

图 10.10　主视图的效果

需要添加的视图以及设置，如表 10-6 所示。

表 10-6　设置主视图

视　　图	设　　置
Label1	Name：label1 Text：Latitude: Font：System 20 pt 位置和大小：(10, 130, 300, 30)
Label2	Name：label2 Text：Longitude: Font：System 20 pt 位置和大小：(10, 180, 300, 30)
Label3	Name：label3 Text：Accuracy: Font：System 20 pt 位置和大小：(10, 230, 300, 30)
Button1	Name：startbtn Title：Start Font：System 19 pt Text Color：白色 位置和大小：(20, 6, 46, 30)
Button2	Name：stopbtn Title：Stop Font：System 19 pt Text Color：白色 位置和大小：(254, 6, 46, 30)
View	Background：深灰色 位置和大小：(0, 525, 320, 43)

（3）打开 Info.plist 文件，在 Source 选项卡中添加 Required background modes 属性，并

在此属性中添加 App registers for location updates，如图 10.11 所示。

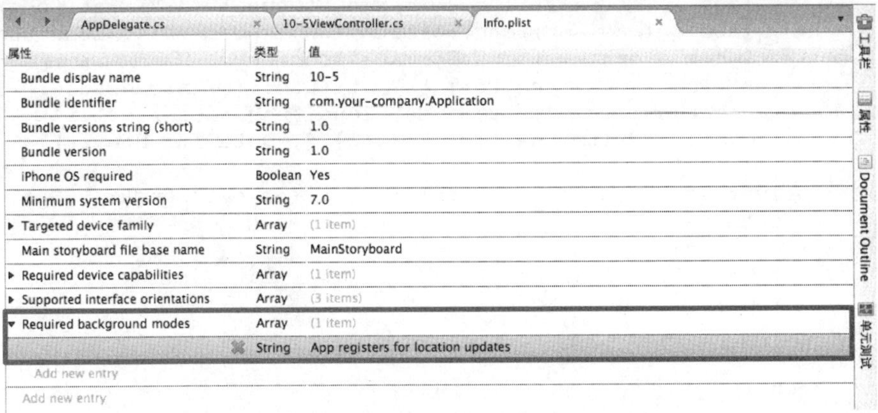

图 10.11　添加属性

（4）打开 10-5ViewController.cs 文件，编写代码，实现位置的更新和显示。代码如下：

```csharp
using System;
using System.Drawing;
using MonoTouch.Foundation;
using MonoTouch.UIKit;
using MonoTouch.CoreLocation;
namespace Application
{
    public partial class _0_5ViewController : UIViewController
    {
        private CLLocationManager locationManager;
        ……
        #region View lifecycle
        public override void ViewDidLoad ()
        {
            base.ViewDidLoad ();
            locationManager = new CLLocationManager();
            locationManager.LocationsUpdated += LocationManager_LocationsUpdated;
            locationManager.Failed += this.LocationManager_Failed;
            //触摸 Start 按钮，实现开启位置服务
            startbtn.TouchUpInside += delegate {
                this.locationManager.StartUpdatingLocation();
            } ;
            //触摸 Stop 按钮，实现关闭位置服务
            stopbtn.TouchUpInside += delegate {
                this.locationManager.StopUpdatingLocation();
                label1.Text = "Latitude:";
                label2.Text = "Longitude:";
                label3.Text = "Accuracy:";
            } ;
        }
        //更新位置，并显示
        private void LocationManager_LocationsUpdated (object sender, CLLocationsUpdatedEventArgs e)
        {
            CLLocation location = e.Locations[0];
            double latitude = Math.Round(location.Coordinate.Latitude, 4);
            double longitude = Math.Round(location.Coordinate.Longitude,
```

```
        4);
        double accuracy = Math.Round(location.HorizontalAccuracy, 0);
        label1.Text = string.Format("Latitude: {0}", latitude);
        label2.Text = string.Format("Longitude: {0}", longitude);
        label3.Text = string.Format("Accuracy: {0}", accuracy);
        string s = string.Format("Latitude: {0}\nLongitude: {1},\n
        Accuracy: {2}m", latitude, longitude, accuracy);
        Console.WriteLine("{0}:\n\t{1} ", DateTime.Now, s);
    }
    private void LocationManager_Failed (object sender, NSError
EventArgs e)
    {
        Console.WriteLine("Location update failed! Error message: {0}",
        e.Error.LocalizedDescription);
    }
    ……
    #endregion
}
```

运行效果如图 10.12 所示。

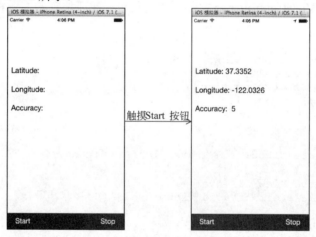

图 10.12 运行效果

如果退出此应用程序,开发者还可以在应用程序输出窗口看到位置的更新,如图 10.13 所示。

图 10.13 运行效果

10.6 使 用 地 图

当今,人们的出行已经离不开对地图的需要了。地图是智能手机中的一个重要功能。在很多的手机中都内嵌了地图服务。在程序中使用地图服务,不仅可以实现常规的导航功能,还能增强社交类应用程序的用户黏性。本节将讲解有关地图的一些操作。

10.6.1 显示地图

如果想要在一个应用程序中显示地图,只需要打开此应用程序的工程,在 MainStoryboard.storyboard 文件的主视图中添加一个 Map Kit View 对象即可。单击运行按钮,就可以看到如图 10.14 所示的运行效果。

图 10.14 运行效果

10.6.2 改变地图的类型

在图 10.14 中所看到的地图类型是 Xamarin.iOS 中默认的地图类型。在 Xamarin.iOS 中,提供了 3 种地图类型,如表 10-7 所示。

表 10-7 地图的类型

地 图 类 型	特　　　点
Standard	显示所有的街道和一些路名的位置
Satellite	显示卫星图像
Hybrid	在显示公路和道路名称信息的基础上使用卫星图像显示

【示例 10-6】 以下将显示 3 种地图类型。具体的操作步骤如下所述。

(1) 创建一个 Single View Application 类型的工程。

(2) 打开 MainStoryboard.storyboard 文件,对主视图进行设置,效果如图 10.15 所示。

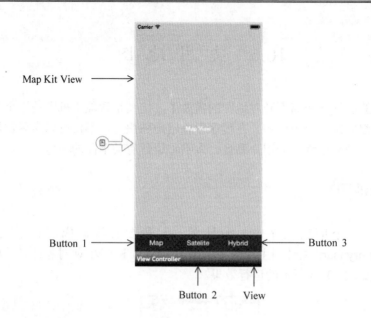

图 10.15 主视图的效果

需要添加的视图以及设置，如表 10-8 所示。

表 10-8 设置主视图

视　　图	设　　置
Map Kit View	Name：map 位置和大小：(0, 0, 320, 525)
Button1	Name：mbtn Title：Map Font：System 17 pt Text Color：白色 位置和大小：(28, 6, 52, 30)
Button2	Name：sbtn Title：Satelite Font：System 17 pt Text Color：白色 位置和大小：(130, 6, 60, 30)
Button3	Name：hbtn Title：Hybrid Font：System 17 pt Text Color：白色 位置和大小：(234, 6, 52, 30)
View	Background：深灰色 位置和大小：(0, 525, 320, 43)

（3）打开 10-6ViewController.cs 文件，编写代码，实现地图类型的显示。代码如下：

```
using System;
using System.Drawing;
using MonoTouch.Foundation;
using MonoTouch.UIKit;
```

```
namespace Application
{
    public partial class _0_6ViewController : UIViewController
    {
        ……
        #region View lifecycle
        public override void ViewDidLoad ()
        {
            base.ViewDidLoad ();
            //触摸 Map 按钮，显示 Standard 类型的地图
            mbtn.TouchUpInside += (sender, e) => {
                map.MapType=MonoTouch.MapKit.MKMapType.Standard;
            };
            //触摸 Satellite 按钮，显示 Satellite 类型的地图
            sbtn.TouchUpInside += (sender, e) => {
                map.MapType=MonoTouch.MapKit.MKMapType.Satellite;
            };
            //触摸 Hybrid 按钮，显示 Hybrid 类型的地图
            hbtn.TouchUpInside += (sender, e) => {
                map.MapType=MonoTouch.MapKit.MKMapType.Hybrid;
            };
        }
        ……
        #endregion
    }
}
```

运行效果如图 10.16 所示。

图 10.16　运行效果

10.6.3　在地图上显示当前位置

当用户不知道自己所处的地理位置时，就可以在地图上进行当前位置的获取。要获取

当前位置,可以对 showsUserLocation 属性进行设置。

【示例 10-7】以下当用户触摸"显示当前位置"按钮后,在地图上会标出用户所处的当前位置。具体的操作步骤如下所述。

(1)创建一个 Single View Application 类型的工程。

(2)打开 MainStoryboard.storyboard 文件,对主视图进行设置,效果如图 10.17 所示。

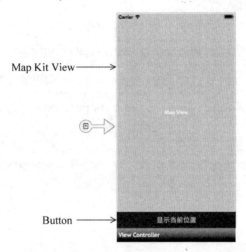

图 10.17 主视图的效果

需要添加的视图以及设置,如表 10-9 所示。

表 10-9 设置主视图

视 图	设 置
Map Kit View	Name:map 位置和大小:(0, 0, 320, 525)
Button	Name:btn Title:显示当前位置 Font:System 17 pt Text Color:白色 Background:深灰色 位置和大小:(0, 525, 320, 43)

(3)打开 10-7ViewController.cs 文件,编写代码,实现触摸按钮,显示当前位置。代码如下:

```
using System;
using System.Drawing;
using MonoTouch.Foundation;
using MonoTouch.UIKit;
namespace Application
{
    public partial class _0_7ViewController : UIViewController
    {
```

```
……
#region View lifecycle
public override void ViewDidLoad ()
{
    base.ViewDidLoad ();
    //触摸按钮，显示当前位置
    btn.TouchUpInside += (sender, e) => {
        map.ShowsUserLocation=true;
    };
}
……
#endregion
}
```

运行效果如图 10.18 所示。

图 10.18　运行效果

注意：在图中出现的蓝色小圆圈就是用户当前的位置。

10.6.4　指定位置

在地图中不仅可以实现获取当前的位置，还可以根据某一个经度和纬度进行定位，将此位置在地图上进行对应的显示。要实现指定位置这一功能，就要使用 CLLocationCoordinate2D 数据类型。

【示例 10-8】以下通过使用 CLLocationCoordinate2D 数据类型来指定首都的经纬度，并在触摸按钮后将界面移动到指定的位置上。具体的操作步骤如下所述

（1）创建一个 Single View Application 类型的工程。

（2）打开 MainStoryboard.storyboard 文件，对主视图进行设置，效果如图 10.19 所示。

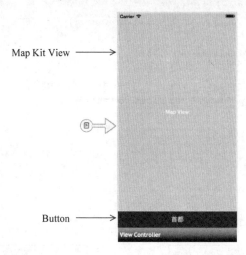

图 10.19　主视图的效果

需要添加的视图以及设置如表 10-10 所示。

表 10-10　设置主视图

视　　图	设　　置
Map Kit View	Name：map 位置和大小：(0, 0, 320, 525)
Button	Name：btn Title：首都 Font：System 17 pt Text Color：白色 Background：深灰色 位置和大小：(0, 525, 320, 43)

（3）打开 10-8ViewController.cs 文件，编写代码，实现触摸按钮到达指定位置。代码如下：

```
using System;
using System.Drawing;
using MonoTouch.Foundation;
using MonoTouch.UIKit;
using MonoTouch.CoreLocation;
namespace Application
{
    public partial class _0_8ViewController : UIViewController
    {
        ……
        #region View lifecycle
        public override void ViewDidLoad ()
        {
            base.ViewDidLoad ();
            btn.TouchUpInside += (sender, e) => {
                CLLocationCoordinate2D mapCoordinate = new CLLocation
                Coordinate2D(39.908605,116.398019);        //指定经纬度
                map.SetCenterCoordinate(mapCoordinate,true);
```

第 10 章 位置服务和地图

```
            //设置地图的中心坐标
        };
    }
    ……
    #endregion
    }
}
```

运行效果如图 10.20 所示。

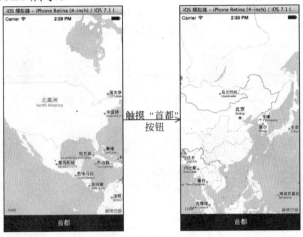

图 10.20 运行效果

10.6.5 添加标记

在图 10.20 所示的运行结果中，虽然指定了一个位置，但是跳转到这个位置后，还是不知道自己指定的到底是哪个地方，这时就需要在指定的位置上添加上一个标记，要实现标记的添加，就要使用 MKPlacemark 类。

【示例 10-9】以下将在指定的位置处添加标记。具体的操作步骤如下所述。

（1）创建一个 Single View Application 类型的工程。

（2）打开 MainStoryboard.storyboard 文件，对主视图进行设置，效果如图 10.21 所示。

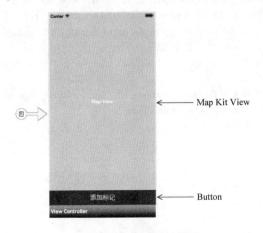

图 10.21 主视图的效果

需要添加的视图以及设置，如表 10-11 所示。

表 10-11　设置主视图

视　　图	设　　置
Map Kit View	Name：map 位置和大小：(0, 0, 320, 525)
Button	Name：btn Title：添加标记 Font：System 19 pt Text Color：白色 位置和大小：(0, 525, 320, 43)

（3）打开 10-12ViewController.cs 文件，编写代码，实现为指定的位置添加标记。代码如下：

```
using System;
using System.Drawing;
using MonoTouch.Foundation;
using MonoTouch.UIKit;
using MonoTouch.CoreLocation;
using MonoTouch.MapKit;
namespace Application
{
    public partial class _0_12ViewController : UIViewController
    {
        ……
        #region View lifecycle
        public override void ViewDidLoad ()
        {
            base.ViewDidLoad ();
            //触摸按钮，为指定的位置添加标记
            btn.TouchUpInside += (sender, e) => {
                NSString str1=new NSString("中国");
                NSString str2=new NSString("Country");
                CLLocationCoordinate2D mapCoordinate = new CLLocation
                Coordinate2D(39.908605,116.398019);         //指定经纬度
                map.SetCenterCoordinate(mapCoordinate,true);
                //设置地图的中心坐标
                NSDictionary dic=NSDictionary.FromObjectAnd Key(str1,
                str2);
                MKPlacemark lun=new MKPlacemark(mapCoordinate,dic);
                map.AddPlacemark(lun);                      //添加标记
            };
        }
        ……
        #endregion
    }
}
```

运行效果如图 10.22 所示。

第 10 章 位置服务和地图

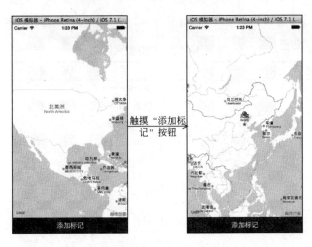

图 10.22 运行效果

注意：在此程序中添加了一个新的命名空间 MonoTouch.MapKit，此命名空间主要提供了以下 4 个功能。
- 显示地图。
- CLLocation 和地址之间的转换。
- 支持在地图上做标记。
- 把一个位置解析成地址。

10.6.6 添加标注

在地图上除了可以进行添加标记外，还可以添加标注。这时，就要使用到 MKPointAnnotation 类。标注是与地图上的位置关联的标记，所以也是标记的一种。

【示例 10-10】 以下为指定的位置添加标注。具体的操作步骤如下所述。
（1）创建一个 Single View Application 类型的工程。
（2）打开 MainStoryboard.storyboard 文件，对主视图进行设置，效果如图 10.23 所示。

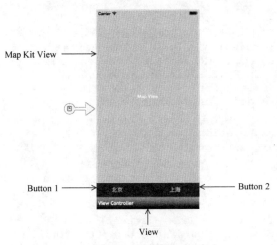

图 10.23 主视图的效果

• 369 •

需要添加的视图以及设置，如表 10-12 所示。

表 10-12 设置主视图

视　　图	设　　置
Map Kit View	Name：map 位置和大小：(0, 0, 320, 525)
Button1	Name：bbtn Title：北京 Font：System 17 pt Text Color：白色 位置和大小：(41, 6, 46, 30)
Button2	Name：sbtn Title：上海 Font：System 17 pt Text Color：白色 位置和大小：(221, 6, 46, 30)
View	Background：深灰色 位置和大小：(0, 525, 320, 43)

（3）打开 10-9ViewController.cs 文件，编写代码，实现为地图添加标注的功能。代码如下：

```
using System;
using System.Drawing;
using MonoTouch.Foundation;
using MonoTouch.UIKit;
using MonoTouch.MapKit;
using MonoTouch.CoreLocation;
namespace Application
{
    public partial class _0_9ViewController : UIViewController
    {
        ……
        #region View lifecycle
        public override void ViewDidLoad ()
        {
            base.ViewDidLoad ();
            //触摸"北京"按钮，为指定的位置添加标注
            bbtn.TouchUpInside += (sender, e) => {
                CLLocationCoordinate2D mapCoordinate = new CLLocation Coor
                dinate2D(39.908605,116.398019);
                MKPointAnnotation myAnnotation = new MKPointAnnotation();
                myAnnotation.Coordinate = mapCoordinate;
                //设置标注的坐标位置
                myAnnotation.Title = "北京";                   //设置标注的标题
                myAnnotation.Subtitle = "它是中国的首都";//设置标注的副标题
                map.AddAnnotation(myAnnotation);              //添加标注
                map.SetCenterCoordinate(mapCoordinate,true);
            };
            //触摸"上海"按钮，为指定的位置添加标注
            sbtn.TouchUpInside += (sender, e) => {
                CLLocationCoordinate2D mapCoordinate = new CLLocation Coor
                dinate2D (31.240948,121.485958);
                MKPointAnnotation myAnnotation = new MKPointAnnotation();
```

```
                    myAnnotation.Coordinate = mapCoordinate;
                    myAnnotation.Title = "上海";
                    myAnnotation.Subtitle = "它有"东方巴黎"的美称,是中华人民共和
            国直辖市之一";
                    map.AddAnnotation(myAnnotation);
                    map.SetCenterCoordinate(mapCoordinate,true);
                };
            }
            ......
            #endregion
        }
}
```

运行效果如图 10.24 所示。

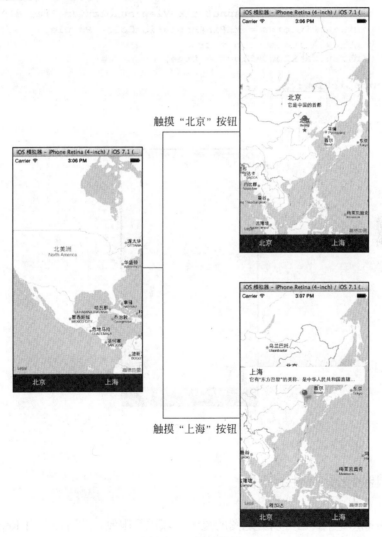

图 10.24 运行效果

注意:用户除了可以使用系统的标注外,还可以自己去定义一个标注,如以下的代码,就将【示例 10-10】中的标注改为紫色,并且在添加此标注时,伴随有一个下落效果的动画。代码如下:

第 2 篇 资源使用篇

```
public override void ViewDidLoad ()
{
    base.ViewDidLoad ();
    map.WeakDelegate = this;
    ......
}
[Export ("mapView:viewForAnnotation:")]
public MKAnnotationView GetViewForAnnotation (MKMapView mapView, NSObject
annotation)
{
    string reuseID = "myAnnotation";
    MKPinAnnotationView pinView = mapView.DequeueReusableAnnotation(reuseID)
    as MKPinAnnotationView;
    if (null == pinView)
    {
        pinView = new MKPinAnnotationView(annotation, reuseID);
        pinView.PinColor = MKPinAnnotationColor.Purple;      //标注的颜色
        pinView.AnimatesDrop = true;                          //动画效果
        pinView.CanShowCallout = true;
    }
    return pinView;
}
```

运行效果如图 10.25 所示。

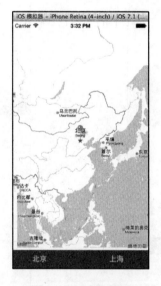

图 10.25 运行效果

10.6.7 限制地图的显示范围

在 Xamarin.iOS 中，也可以对地图的显示区域进行限制，此时需要对 Region 属性进行设置。

【示例 10-11】以下程序通过对 Region 属性进行设置，实现在一定的范围内显示指定的位置。具体的操作步骤如下所述。

（1）创建一个 Single View Application 类型的工程。

（2）打开 MainStoryboard.storyboard 文件，对主视图进行设置，效果如图 10.26 所示。

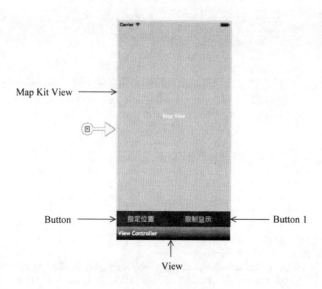

图 10.26 运行效果

需要添加的视图以及设置,如表 10-13 所示。

表 10-13 设置主视图

视 图	设 置
Map Kit View	Name:map 位置和大小:(0, 0, 320, 525)
Button1	Name:zbth Title:指定位置 Font:System 17 pt Text Color:白色 位置和大小:(41, 6, 46, 30)
Button2	Name:sbtn Title:限制显示 Font:System 17 pt Text Color:白色 位置和大小:(221, 6, 46, 30)
View	Background:深灰色 位置和大小:(0, 525, 320, 43)

(3)打开 10-13ViewController.cs 文件,编写代码,实现标记的添加,以及限定显示范围。代码如下:

```
using System;
using System.Drawing;
using MonoTouch.Foundation;
using MonoTouch.UIKit;
using MonoTouch.CoreLocation;
using MonoTouch.MapKit;
namespace Application
{
    public partial class _0_13ViewController : UIViewController
    {
        ……
        #region View lifecycle
```

```
public override void ViewDidLoad ()
{
    base.ViewDidLoad ();
    CLLocationCoordinate2D mapCoordinate = new CLLocation Coor
    dinate2D(39.908605,116.398019);
    //触摸"指定位置"按钮,实现跳转到指定的位置,并添加标记
    zbtn.TouchUpInside += (sender, e) => {
        NSString str1=new NSString("中国");
        NSString str2=new NSString("Country");
        map.SetCenterCoordinate(mapCoordinate,true);
        //设置地图的中心坐标
        NSDictionary dic=NSDictionary.FromObject AndKey(str1,
        str2);
        MKPlacemark lun=new MKPlacemark(mapCoordinate,dic);
        map.AddPlacemark(lun);
    };
    //触摸"限制显示"按钮,实现在一定范围内显示指定位置
    sbtn.TouchUpInside += (sender, e) => {
        MKCoordinateRegionreg=MKCoordinateRegion.FromDistance
        (map Coordinate,1000.0f,1000.0f);
        map.Region=reg;
    };
    ......
    #endregion
}
```

运行效果如图 10.27 所示。

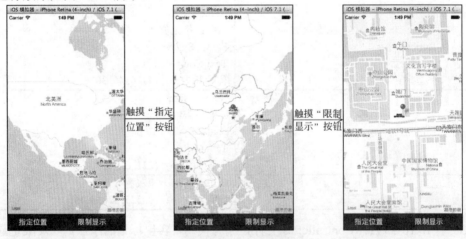

图 10.27 运行效果

10.6.8 添加覆盖图

覆盖图是 iOS 4 开始引入的。覆盖图与标注的区别是,当缩放时,标注的尺寸保存一样,覆盖图的大小与地图视图的缩放关联。

【示例 10-12】 以下将实现在地图的当前位置上添加一个覆盖图。此覆盖图是一个红色的圆。具体的操作步骤如下所述。

(1)创建一个 Single View Application 类型的工程。

（2）打开 MainStoryboard.storyboard 文件，对主视图进行设置，效果如图 10.28 所示。

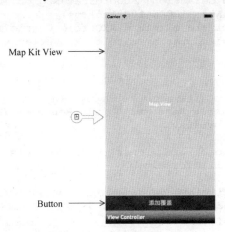

图 10.28　主视图的效果

需要添加的视图以及设置，如表 10-14 所示。

表 10-14　设置主视图

视　　图	设　　置
Map Kit View	Name：map 位置和大小：(0, 0, 320, 525)
Button	Name：abtn Title：添加覆盖 Font：System 17 pt Text Color：白色 Background：深灰色 位置和大小：(0, 525, 320, 43)

（3）打开 10-10ViewController.cs 文件，编写代码，实现触摸按钮，添加覆盖图。代码如下：

```
using System;
using System.Drawing;
using MonoTouch.Foundation;
using MonoTouch.UIKit;
using MonoTouch.MapKit;
using MonoTouch.CoreLocation;
namespace Application
{
    public partial class _0_10ViewController : UIViewController
    {
        ……
        #region View lifecycle
        public override void ViewDidLoad ()
        {
            base.ViewDidLoad ();
            map.ShowsUserLocation = true;
            map.WeakDelegate = this;
            //触摸按钮，添加覆盖图
            abtn.TouchUpInside += (sender, e) => {
                CLLocationCoordinate2D mapCoordinate = map.UserLocation.
```

```
            Coordinate;
            map.SetRegion(MKCoordinateRegion.FromDistance(mapCoor
            dinate, 1000, 1000), true);
            MKCircle circle = MKCircle.Circle(mapCoordinate, 250);
            map.AddOverlay(circle, MKOverlayLevel.AboveRoads);
        } ;
    }
    //绘制覆盖层控件
    [Export ("mapView:rendererForOverlay:")]
    public MKOverlayRenderer OverlayRenderer (MKMapView mapView,
    IMKOverlay overlay)
    {
        MKCircle circle = overlay as MKCircle;
        if (null != circle)
        {
            MKCircleRenderer renderer = new MKCircleRenderer(circle);
            renderer.FillColor = UIColor.FromRGBA(1.0f, 0.5f, 0.5f,
            0.5f);          //填充颜色
            renderer.StrokeColor = UIColor.Red;              //边框颜色
            renderer.LineWidth = 2f;
            return renderer;
        } else
        {
            return null;
        }
    }
    ......
    #endregion
}
```

运行效果如图 10.29 所示。

图 10.29　运行效果

10.7　地 理 编 码

地理编码（Geocoding）是基于空间定位技术的一种编码方法，它提供了一种把描述成地址的地理位置信息转换成可以被用于 GIS（地理信息系统）的地理坐标的方式，即根据经纬度获取实际的地址信息。

【示例 10-13】 以下将实现使用地理编码将当前的地理位置信息转换为可以被用于 GIS（地理信息系统）的地理坐标。代码如下：

（1）创建一个 Single View Application 类型的工程。

（2）打开 MainStoryboard.storyboard 文件，对主视图进行设置，效果如图 10.30 所示。

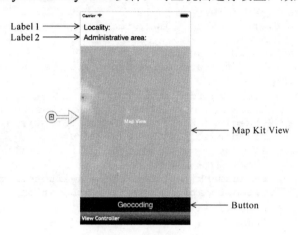

图 10.30　主视图的效果

需要添加的视图以及设置，如表 10-15 所示。

表 10-15　设置主视图

视　　图	设　　置
Label1	Name：label1 Text：Locality: Font：System 21 pt 位置和大小：(10, 33, 290, 21)
Label2	Name：label2 Text：Administrative area: Font：System 21 pt 位置和大小：(10, 62, 290, 21)
Button	Name：btn Title：Geocoding Font：System 21 pt Text Color：白色 Background：深灰色 位置和大小：(0, 527, 320, 41)
Map Kit View	Name：map 位置和大小：(0, 96, 320, 432)

（3）打开 10-11ViewController.cs 文件，编写代码，实现位置转换为可以被用于 GIS（地理信息系统）的地理坐标。代码如下：

```
using System;
using System.Drawing;
using MonoTouch.Foundation;
using MonoTouch.UIKit;
using MonoTouch.CoreLocation;
```

```
using MonoTouch.MapKit;
namespace Application
{
    public partial class _0_11ViewController : UIViewController
    {
        private CLGeocoder geocoder;
        ……
        #region View lifecycle
        public override void ViewDidLoad ()
        {
            base.ViewDidLoad ();
            map.ShowsUserLocation = true;
            //触摸按钮，实现转换
            btn.TouchUpInside += async (sender, e) => {
                CLLocation currentLocation=map.UserLocation.Location;
                map.SetRegion(MKCoordinateRegion.FromDistance(currentLoca
                tion.Coordinate, 1000, 1000), true);
                geocoder = new CLGeocoder ();
                //检索地理编码的数据
                CLPlacemark[] placemarks = await this.geocoder.Reverse
                GeocodeLocationAsync(currentLocation);
                //判断检索的数据是否不为空
                if (null != placemarks)
                {
                    CLPlacemark placemark = placemarks[0];
                    label1.Text = string.Format("Locality: {0}", placemark.
                    Locality);           //显示城市
                    label2.Text = string.Format(" Administrative area: {0}",
                    placemark.AdministrativeArea);                //显示省
                }
            };
        }
        ……
        #endregion
    }
}
```

运行效果如图 10.31 所示。

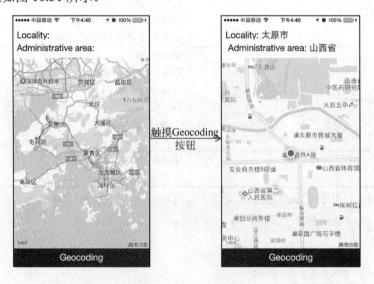

图 10.31　运行效果

第 3 篇　高级应用篇

- 第 11 章　图形和动画
- 第 12 章　多任务处理
- 第 13 章　本地化
- 第 14 章　发布应用程序
- 第 15 章　高级功能

第 11 章　图形和动画

在 iOS SDK 中包含两个非常有用的框架：Core Graphics 和 Core Animation。这两个框架简化了动画的 UI 元素和绘制 2D 图形的过程。使用这两个框架可以制作出令人惊叹的应用程序。在 Xamarin.iOS 中这两个框架对应的命名空间为 MonoTouch.CoreGraphics 和 MonoTouch.CoreAnimation。本节将讲解使用这两个命名空间实现图形的绘制以及动画效果。

11.1　视图动画

在 UIView 类中可以实现一些既有趣又好看的动画效果。本节将主要讲解如何使用 UIView 来实现动画。

11.1.1　动画块

UIView 动画是以块的形式出现的。如果想实现 UIView 动画的播放，首先要创建动画块。一个动画块是由开始动画 BeginAnimations ()方法和结束动画 CommitAnimations ()方法构成。

【示例 11-1】　以下将使用 UIView 来实现一个简单的动画，当用户触摸屏幕后，会以动画的形式出现一个日期选择器对象。当用户选择日期后，此选择器会以动画的形式隐藏。具体的操作步骤如下所述。

（1）创建一个 Single View Application 类型的工程。

（2）打开 MainStoryboard.storyboard 文件，对主视图进行设置，效果如图 11.1 所示。

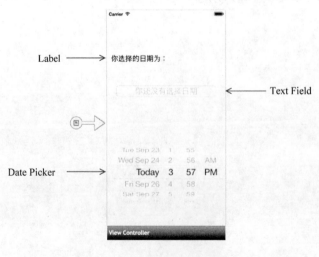

图 11.1　主视图的效果

需要添加的视图以及设置，如表 11-1 所示。

表 11-1 设置主视图

视 图	设 置
Label	Text：你选择的日期为： Font：System 19 pt 位置和大小：(12, 113, 203, 31)
Text Field	Name：tf Text：（空） Font：System 21 pt Alignment：居中 Placeholder：你还没有选择日期 位置和大小：(25, 199, 270, 30)
Date Picker	Name：datePicker 位置和大小：(0, 288, 320, 216)

注意：在图 11.1 中的 Date Picker 在运行时，需要将其位置和大小改为(0,568,320,216)。

（3）打开 11-0ViewController.cs 文件，编写代码，实现日期的选择和日期选择器的动画效果。代码如下：

```
using System;
using System.Drawing;
using MonoTouch.Foundation;
using MonoTouch.UIKit;
namespace Application
{
    public partial class _1_0ViewController : UIViewController
    {
        ……                              //这里省略了视图控制器的构造方法和析构方法
        #region View lifecycle
        public override void ViewDidLoad ()
        {
            base.ViewDidLoad ();
            datePicker.Frame = new RectangleF (0, 568, 320, 216);
            datePicker.ValueChanged += (sender, e) => {
                NSDateFormatter formatter=new NSDateFormatter();
                formatter.DateFormat="YYYY/MM/dd HH:mm";
                //设置日期的格式
                string a=formatter.ToString(datePicker.Date);
                            //将日期转换为字符串
                tf.Text=a;
                //日期选择器隐藏的动画效果
                UIView.BeginAnimations ("");
                datePicker.Frame = new RectangleF (0, 568, 320, 216);
                UIView.CommitAnimations ();
            };
        }
        //触摸屏幕
        public override void TouchesBegan (NSSet touches, UIEvent evt)
        {
            base.TouchesBegan (touches, evt);
            //日期选择器出现的动画效果
            UIView.BeginAnimations ("");
```

```
            datePicker.Frame = new RectangleF (0, 288, 320, 216);
            UIView.CommitAnimations ();
        }
        public override void ViewWillAppear (bool animated)
        ……                       //这里省略了视图加载和卸载前后的一些方法
        #endregion
    }
}
```

运行效果如图 11.2 所示。

图 11.2　运行效果

11.1.2　修改动画块

在动画块中，开发者可以修改动画的各种属性，如动画的持续时间和动画重复的次数等。以下将讲解几个动画的常用属性设置。

1．持续时间

持续时间就是动画从开头到结束所需要的时间。SetAnimationDuration()方法可以实现对动画持续时间的设置。其语法形式如下：

```
UIView.SetAnimationDuration(时间);
```

其中，时间的数据类型是一个双精度类型。

2．动画重复播放的次数

SetAnimationDuration()方法可以用来对动画重复播放的次数进行设置。其语法形式如下：

```
UIView.SetAnimationDuration(重复播放次数);
```

3．速度的改变

在一个动画中，速度的改变也是很重要的。要实现对动画相对速度的修改，需使用SetAnimationCurve()方法，其语法形式如下：

```
UIView.SetAnimationCurve(速度的改变);
```

其中,速度的改变可以有 4 种选择,如表 11-2 所示。

表 11-2 速度的改变

相 对 速 度	功　　能
EaseInOut	动画开始缓慢,中间加速,最后再变为缓慢
EaseIn	动画开始缓慢,然后慢慢加快
EaseOut	动画开始快速,然后再慢下来
Linear	动画从开始到结束一直保持匀速

11.1.3 动画属性

UIKit 动画支持一组特定的 UIView 属性,这些属性被称为动画属性。这些动画属性如表 11-3 所示。

表 11-3 动画属性

动 画 属 性	功　　能
Frame	可以用来制作位置和大小改变的动画效果
Bounds	
Center	
Alpha	可以用来指定透明度发生改变的动画效果
BackgroundColor	可以用来制作视图颜色发生改变的动画效果
Transform	可以用来制作视图的转换动画

注意:表 11-3 所示的这些动画属性是开发者最常使用到的。

11.1.4 基于块的视图动画

除了在【示例 11-1】中看到的实现视图动画的方式外,还有一种可以实现视图动画的方式,那就是基于块的视图动画,它可以使用 Animate()方法来实现。其语法形式如下:

```
UIView.Animate(
    double duration,
    double dely,
    UiviewAnimationOption options,
    NSAction animation,
    NSAction completion
);
```

其中,参数说明如下。
- duration:设置动画的持续时间
- dely:设置动画的延迟时间。在默认情况下,没有延迟时间。
- options:是一种位掩码,用来描述额外的选项。
- animation:用来设置动画块。
- completion:用来设置动画完成后所执行的效果。

【示例 11-2】 以下将使用 Animate()方法来实现一个动画效果。具体的操作步骤如下所述。

（1）创建一个 Single View Application 类型的工程。

（2）打开 MainStoryboard.storyboard 文件，对主视图进行设置，效果如图 11.3 所示。

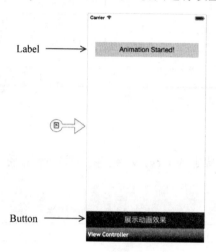

图 11.3　主视图的效果

需要添加的视图以及设置，如表 11-4 所示。

表 11-4　设置主视图

视　图	设　置
Label	Name：label Text：Animation Started! Font：System 18 pt Alignment：居中 Background：青色 位置和大小：(20, 77, 280, 38)
Button	Name：btn Title：展示动画效果 Font：System 19 pt Text Color：白色 Background：深灰色 位置和大小：(0, 525, 320, 43)

（3）打开 11-1ViewController.cs 文件，编写代码，实现动画效果。代码如下：

```
using System;
using System.Drawing;
using MonoTouch.Foundation;
using MonoTouch.UIKit;
namespace Application
{
    public partial class _1_1ViewController : UIViewController
    {
        ……                        //这里省略了视图控制器的构造方法和析构方法
        #region View lifecycle
```

```
public override void ViewDidLoad ()
{
    base.ViewDidLoad ();
    btn.TouchUpInside += (sender, e) => {
        RectangleF labelFrame = label.Frame;
        labelFrame.Y = 456f;
        //基于块的视图动画
        UIView.Animate(1d,0d,UIViewAnimation Options. Curve EaseIn
        Out,
            () => label.Frame = labelFrame,
            () => {
                label.Text = "Animation ended!";
                label.BackgroundColor = UIColor.Orange;
            } );
    } ;
}
……                    //这里省略了视图加载和卸载前后的一些方法
#endregion
}
```

运行效果如图 11.4 所示。

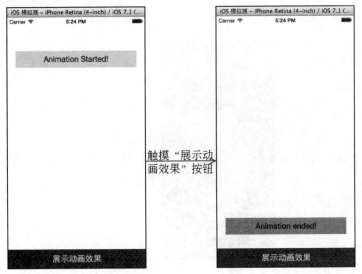

图 11.4　运行效果

11.2　视图的过渡动画

过渡是使用动画形式重画视图。一般使用过渡动画的原因是，通过视图动画来突出视图上的改变。视图的过渡动画需要使用 SetAnimationTransition()方法进行设置。其语法形式如下：

`UIView.SetAnimationTransition` (过渡动画 过渡的视图，布尔值)；

其中，过渡动画有 5 种，如表 11-5 所示。

第 3 篇 高级应用篇

表 11-5 过渡动画

过 渡 动 画	功　　能
None	无过渡动画
FlipFromLeft	从左向右旋转
FlipFromRight	从右向左旋转
CurlUp	卷曲翻页，从下往上
CurlDown	卷曲翻页，从上往下

以下将使用这些过渡动画来实现一些功能。

11.2.1 旋转动画

旋转的动画效果一般都会使用在图片的浏览上。

【示例 11-3】以下将实现在浏览图像时出现旋转动画的功能。具体的操作步骤如下所述。

（1）创建一个 Single View Application 类型的工程。

（2）添加图像 1.jpg、2.jpg、3.jpg、4.jpg 和 5.jpg 到创建工程的 Resources 文件夹中。

（3）打开 MainStoryboard.storyboard 文件，对主视图进行设置，效果如图 11.5 所示。

图 11.5 主视图的效果

需要添加的视图以及设置，如表 11-6 所示。

表 11-6 设置主视图

视　　图	设　　置
主视图	Background：黑色
Image View	Name：iv Image：1.jpg 位置和大小：(8, 32, 304, 485)
Page Control	Name：pageControl # of Pages 5 位置和大小：(89, 525, 142, 36)

（4）打开 11-15ViewController.cs 文件，编写代码，实现在图像视图与图像视图切换时，出现旋转动画的效果。代码如下：

```csharp
using System;
using System.Drawing;
using MonoTouch.Foundation;
using MonoTouch.UIKit;
namespace Application
{
    public partial class _1_15ViewController : UIViewController
    {
        ……
        #region View lifecycle
        public override void ViewDidLoad ()
        {
            base.ViewDidLoad ();
            //创建并设置手势识别器对象
            UISwipeGestureRecognizer left = new UISwipeGestureRecognizer
            (leftMethods);
            left.Direction = UISwipeGestureRecognizerDirection.Left;
            //设置方向
            this.View.AddGestureRecognizer (left);
            UISwipeGestureRecognizer right = new UISwipeGestureRecognizer
            (rightMethods);
            right.Direction = UISwipeGestureRecognizerDirection.Right;
            //设置方向
            this.View.AddGestureRecognizer (right);
        }
        //实现向左滑动时，出现从右向左旋转的旋转效果
        private void leftMethods(UISwipeGestureRecognizer swipe)
        {
            int j = pageControl.CurrentPage;
            j--;
            if (j > -1)
            {
                string file = string.Format ("{0}.jpg", j+1);
                iv.Image = UIImage.FromFile (file);
                UIView.BeginAnimations ("aaa");
                UIView.SetAnimationDuration (5);
                UIView.SetAnimationTransition (UIViewAnimationTransition.
                FlipFromRight, iv, true);
                UIView.CommitAnimations ();
                pageControl.CurrentPage = j;
            }
        }
        //实现向右滑动时，出现从左向右旋转的旋转效果
        private void rightMethods(UISwipeGestureRecognizer swipe)
        {
            int c = pageControl.CurrentPage;
            c++;
            if (c < 5)
            {
                string file = string.Format ("{0}.jpg", c+1);
                iv.Image = UIImage.FromFile (file);
                UIView.BeginAnimations ("aaa");
                UIView.SetAnimationDuration (5);
                UIView.SetAnimationTransition (UIViewAnimationTransition.
                FlipFromLeft, iv, true);
                UIView.CommitAnimations ();
```

```
                    pageControl.CurrentPage = c;
            }
    }
    ......
    #endregion
}
```

运行效果如图 11.6 所示。

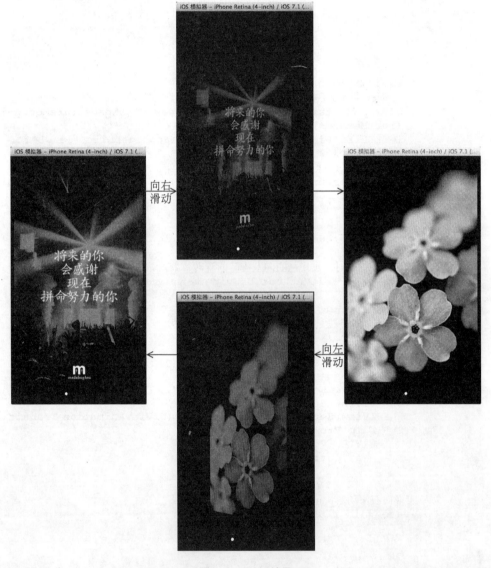

图 11.6 运行效果

11.2.2 卷页动画

在阅读器中，为了在切换界面时使读者有更深刻的体会，所以就添加了卷页效果。

【示例 11-4】 以下将实现卷页的效果。当用户触摸按钮后，就出现卷页的动画效果。

具体的操作步骤如下所述。

（1）创建一个 Single View Application 类型的工程。

（2）添加图像 1.jpg、2.jpg、3.jpg、4.jpg、5.jpg、6.png 和 7.png 到创建工程的 Resources 文件夹中。

（3）打开 MainStoryboard.storyboard 文件，对主视图进行设置，效果如图 11.7 所示。

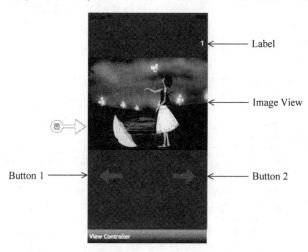

图 11.7　主视图的效果

需要添加的视图以及设置，如表 11-7 所示。

表 11-7　设置主视图

视　　图	设　　置
主视图	Background：深灰色
Label	Name：label Text：1 Color：白色 Font：System Bold 20 pt Alignment：居右 位置和大小：(206, 69, 104, 21)
Image View	Name：iv Image：1.jpg 位置和大小：(0, 117, 320, 242)
Button1	Name：Ubtn Title：（空） Background：6.png 位置和大小：(30, 406, 63, 44)
Button2	Name：Nbtn Title：（空） Background：7.png 位置和大小：(227, 406, 63, 44)

（4）打开 11-16ViewController.cs 文件，编写代码，实现在图像视图与图像视图切换时的过渡动画。代码如下：

```
using System;
using System.Drawing;
using MonoTouch.Foundation;
using MonoTouch.UIKit;
namespace Application
{
    public partial class _1_16ViewController : UIViewController
    {
        ……                                    //这里省略了视图控制器的构造方法和析构方法
        #region View lifecycle
        public override void ViewDidLoad ()
        {
            base.ViewDidLoad ();
            //触摸"上一页"按钮,实现卷曲翻页,从上往下
            Ubtn.TouchUpInside += (sender, e) => {
                int j=int.Parse(label.Text);
                j--;
                if(j>0)
                {
                    string file = string.Format ("{0}.jpg", j);
                    iv.Image = UIImage.FromFile (file);
                    UIView.BeginAnimations ("aaa");
                    UIView.SetAnimationDuration (0.5);
                    UIView.SetAnimationTransition (UIViewAnimationTransi
                    tion.CurlDown, iv, true);
                    UIView.CommitAnimations ();
                    label.Text=string.Format("{0}",j);
                } else
                {
                    UIAlertView alert=new UIAlertView();
                    alert.Title=("对不起,你已到达首页");
                    alert.AddButton("Cancel");
                    alert.Show();
                }
            };
            //触摸"下一页"按钮,实现卷曲翻页,从下往上
            Nbtn.TouchUpInside += (sender, e) => {
                int i=int.Parse(label.Text);
                i++;
                if (i < 6)
                {
                    string file = string.Format ("{0}.jpg", i);
                    iv.Image = UIImage.FromFile (file);
                    UIView.BeginAnimations ("aaa");
                    UIView.SetAnimationDuration (0.5);
                    UIView.SetAnimationTransition (UIViewAnimationTransi
                    tion.CurlUp, iv, true);
                    UIView.CommitAnimations ();
                    label.Text=string.Format("{0}",i);
                } else
                {
                    UIAlertView alert=new UIAlertView();
                    alert.Title=("对不起,你已到达最后一页");
                    alert.AddButton("Cancel");
                    alert.Show();
                }
            };
        }
        ……                                    //这里省略了视图加载和卸载前后的一些方法
        #endregion
```

```
    }
}
```

运行效果如图 11.8 所示。

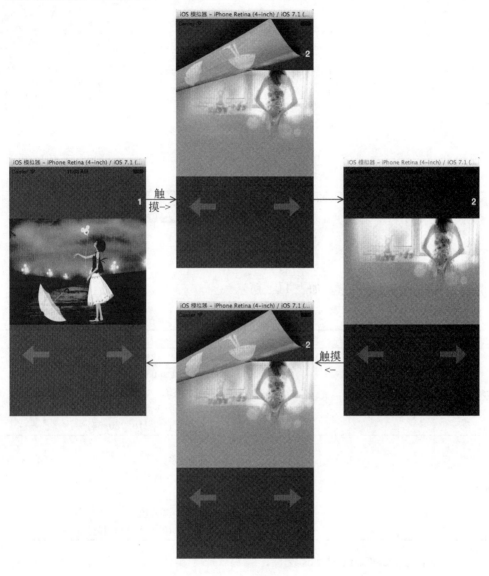

图 11.8　运行效果

11.3　转　换　视　图

在 11.1.3 节的表 11-3 中提到了 Transform 动画属性,此属性用来制作一些视图的转换动画,如旋转、缩放和平移等。

【示例 11-5】　以下将实现标签对象的旋转动画。具体的操作步骤如下所述。

(1) 创建一个 Single View Application 类型的工程。

（2）打开 MainStoryboard.storyboard 文件，对主视图进行设置，效果如图 11.9 所示。

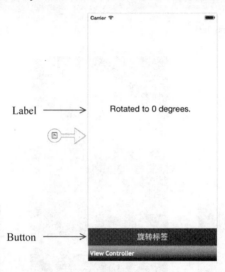

图 11.9　主视图的效果

需要添加的视图以及设置，如表 11-8 所示。

表 11-8　设置主视图

视　　图	设　　置
Label	Name：label Text：Rotated to 0 degrees. Font：System 21 pt Alignment：居中 位置和大小：(15, 216, 280, 34)
Button	Name：btn Title：旋转标签 Font：System 19 pt Text Color：白色 Background：深灰色 位置和大小：(0, 525, 320, 43)

（3）打开 11-2ViewController.cs 文件，编写代码，实现动画效果。代码如下：

```
using System;
using System.Drawing;
using MonoTouch.Foundation;
using MonoTouch.UIKit;
using MonoTouch.CoreGraphics;
namespace Application
{
    public partial class _1_2ViewController : UIViewController
    {
        private double rotationAngle;
        ……                      //这里省略了视图控制器的构造方法和析构方法
        #region View lifecycle
        public override void ViewDidLoad ()
        {
```

```
        base.ViewDidLoad ();
        //触摸按钮,将标签进行 180 度的旋转
        btn.TouchUpInside += async (sender, e) => {
            rotationAngle += 180;
            CGAffineTransform rotation = CGAffineTransform.MakeRot
            ation ((float)DegreesToRadians(this.rotationAngle));
            //实现旋转的动画效果
            await UIView.AnimateAsync(0.5d, () => label.Transform =
            rotation);
            label.Text = string.Format("Rotated to {0} degrees.",
            rotationAngle);
            if (rotationAngle >= 360) {
                rotationAngle = 0;
                label.Transform = CGAffineTransform.MakeIdentity();
            }
        } ;
    }
    //将角度转换为弧度
    public double DegreesToRadians(double degrees)
    {
        return (degrees * Math.PI / 180);
    }
    ……                          //这里省略了视图加载和卸载前后的一些方法
    #endregion
    }
}
```

运行效果如图 11.10 所示。

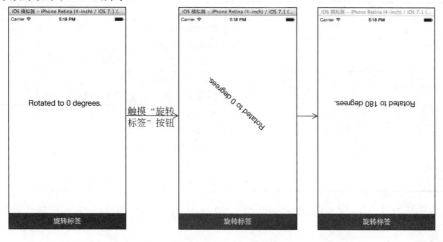

图 11.10　运行效果

11.4　计时器动画

NSTimer 计时器可以每隔一段时间将图像进行更新,这样也可以使图片有一种动态的感觉。

【示例 11-6】 以下将使用 NSTimer 计时器实现图像缩放的动画效果。具体的操作步骤如下所述。

（1）创建一个 Single View Application 类型的工程。

（2）添加图像 1.png 到创建工程的 Resources 文件夹中。

（3）打开 MainStoryboard.storyboard 文件，对主视图进行设置，效果如图 11.11 所示。

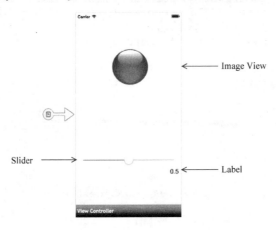

图 11.11 主视图的效果

需要添加的视图以及设置，如表 11-9 所示。

表 11-9 设置主视图

视　　图	设　　置
Image View	Name：iv Image：1.jpg 位置和大小：(103, 104, 114, 114)
Slider	Name：slider 位置和大小：(20, 420, 280, 31)
Label	Name：label Text：0.5 Font：System 19 pt Alignment：居右 位置和大小：(216, 459, 96, 21)

（4）打开 11-17ViewController.cs 文件，编写代码，实现缩放的动画效果。代码如下：

```
using System;
using System.Drawing;
using MonoTouch.Foundation;
using MonoTouch.UIKit;
using MonoTouch.CoreGraphics;
using MonoTouch.ObjCRuntime;
namespace Application
{
    public partial class _1_17ViewController : UIViewController
    {
        PointF position;
        NSTimer timer;
        float angle;
        float ballRadius;
        PointF translation;
        ……                    //这里省略了视图控制器的构造方法和析构方法
```

```
#region View lifecycle
public override void ViewDidLoad ()
{
    base.ViewDidLoad ();
    ballRadius = iv.Frame.Size.Width / 2;
    position = new PointF (12.0f, 0.0f);
    translation = new PointF (0, 0);
    timer = NSTimer.CreateScheduledTimer (slider.Value, this, new
    Selector ("onTimer:"), null, true);              //创建计时器对象
    //滑动滑块改变计时器的时间,从而改变动画的速度
    slider.ValueChanged += (sender, e) => {
        timer.Invalidate();
        timer = NSTimer.CreateScheduledTimer (slider.Value, this,
        new Selector ("onTimer:"), null, true);
        label.Text=string.Format("{0}",slider.Value);
    };
}
[Export("onTimer:")]
//实现缩放的动画效果
public void onTimer(NSTimer timer)
{
    angle += 0.03f;
    if (iv.Center.X>320-ballRadius)
    {
        angle = 0;
    }
    UIView.BeginAnimations ("");
    iv.Transform = CGAffineTransform.MakeScale (angle, angle);
    UIView.CommitAnimations();
}
……                          //这里省略了视图加载和卸载前后的一些方法
#endregion
```

运行效果如图 11.12 所示。

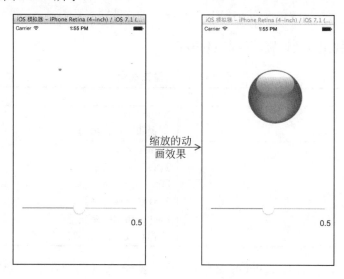

图 11.12　运行效果

🔔注意：拖动滑块控件中的滑块可以改变动画的播放速度。

11.5 图像动画

在 UIImageView 类中，内置了动画。本节将讲解图像视图内置的动画效果。

【示例 11-7】 以下使用 UIImageView 内置的动画，实现图像的切换。具体的操作步骤如下所述。

（1）创建一个 Single View Application 类型的工程。

（2）添加图像 1.jpg、2.jpg、3.jpg、4.jpg 和 5.jpg 到创建工程的 Resources 文件夹中。

（3）打开 MainStoryboard.storyboard 文件，对主视图进行设置，效果如图 11.13 所示。

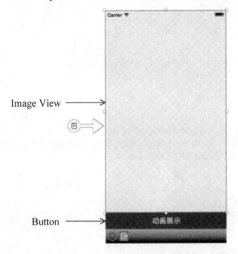

图 11.13 主视图的效果

需要添加的视图以及设置，如表 11-10 所示。

表 11-10 设置主视图

视　　图	设　　置
Image View	Name：iv 位置和大小：(0, 0, 320, 525)
Button	Name：btn Title：动画展示 Font：System 19 pt Text Color：白色 Background：深灰色 位置和大小：(0, 525, 320, 43)

（4）打开 11-3ViewController.cs 文件，编写代码，实现动画效果。代码如下：

```
using System;
using System.Drawing;
using MonoTouch.Foundation;
```

```
using MonoTouch.UIKit;
namespace Application
{
    public partial class _1_3ViewController : UIViewController
    {
        ……                              //这里省略了视图控制器的构造方法和析构方法
        #region View lifecycle
        public override void ViewDidLoad ()
        {
            base.ViewDidLoad ();
            //加载要实现动画的图像
            iv.AnimationImages = new UIImage[] {
                UIImage.FromFile ("1.jpg"),
                UIImage.FromFile ("2.jpg"),
                UIImage.FromFile ("3.jpg"),
                UIImage.FromFile ("4.jpg"),
                UIImage.FromFile ("5.jpg")
            };
            iv.AnimationDuration = 3;
            iv.AnimationRepeatCount = 10;
            //触摸按钮，实现动画的播放或暂停
            btn.TouchUpInside += (sender, e) => {
                if (!iv.IsAnimating)
                {
                    iv.StartAnimating();                          //开始动画
                    btn.SetTitle("结束展示",UIControlState.Normal);
                }else{
                    iv.StopAnimating();                           //结束动画
                    btn.SetTitle("动画展示",UIControlState.Normal);
                }
            } ;
        }
        ……                              //这里省略了视图加载和卸载前后的一些方法
        #endregion
    }
}
```

运行效果如图 11.14 所示。

图 11.14　运行效果

11.6 图层动画

视图可以用属于自己的动画,图层也不例外。图层是一个非常强大和有用的对象。它可以用于绘图和动画。本节将讲解图层动画的实现。

【示例 11-8】 以下将实现复制标签内容的动画效果。具体的操作步骤如下所述。

(1)创建一个 Single View Application 类型的工程。

(2)打开 MainStoryboard.storyboard 文件,对主视图进行设置,效果如图 11.15 所示。

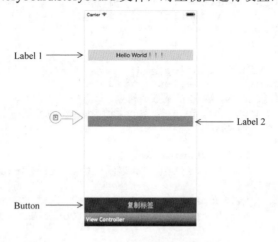

图 11.15 主视图的效果

需要添加的视图以及设置,如表 11-11 所示。

表 11-11 设置主视图

视　　图	设　　置
Label1	Name:lblSource Text:Hello World!!! Alignment:居中 Background:青色 位置和大小:(10, 109, 300, 30)
Label2	Name:lblTarget Text:(空) Alignment:居中 Background:橘黄色 位置和大小:(10, 295, 300, 30)
Button	Name:btn Title:复制标签 Font:System 19 pt Text Color:白色 Background:深灰色 位置和大小:(0, 525, 320, 43)

(3)打开 11-4ViewController.cs 文件,编写代码,实现动画效果。代码如下:

```csharp
using System;
using System.Drawing;
using MonoTouch.Foundation;
using MonoTouch.UIKit;
using MonoTouch.CoreAnimation;
namespace Application
{
    public partial class _1_4ViewController : UIViewController
    {
        private CALayer copyLayer;
        ......
        #region View lifecycle
        public override void ViewDidLoad ()
        {
            base.ViewDidLoad ();
            //触摸按钮,实现动画效果
            btn.TouchUpInside += (s, e) => {
                lblTarget.Text = string.Empty;
                lblTarget.BackgroundColor = UIColor.Orange;
                copyLayer = new CALayer();
                copyLayer.Frame = lblSource.Frame;
                copyLayer.Contents = lblSource.Layer.Contents;
                View.Layer.AddSublayer(this.copyLayer);
                //构建动画
                CABasicAnimation positionAnimation = CABasicAnimation.FromKeyPath("position");
                positionAnimation.To = NSValue.FromPointF (lblTarget.Center);      //设置动画的目标值
                positionAnimation.Duration = 1;
                //设置动画的持续时间
                positionAnimation.RemovedOnCompletion = true;
                //动画完成后移除动画
                positionAnimation.TimingFunction = CAMediaTimingFunction.FromName(CAMediaTimingFunction.EaseInEaseOut);
                //设定动画的速度变化
                //动画停止后执行复制标签的功能
                positionAnimation.AnimationStopped += delegate {
                    lblTarget.BackgroundColor = this.lblSource.BackgroundColor;
                    lblTarget.Text = lblSource.Text;
                    lblTarget.TextColor = lblSource.TextColor;
                    copyLayer.RemoveFromSuperLayer();
                } ;
                CABasicAnimation sizeAnimation = CABasicAnimation.FromKeyPath("bounds");
                sizeAnimation.To = NSValue.FromRectangleF(new RectangleF(0f, 0f, lblSource.Bounds.Width * 2f, lblSource.Bounds.Height * 2));
                sizeAnimation.Duration = positionAnimation.Duration / 2;
                sizeAnimation.RemovedOnCompletion = true;
                sizeAnimation.AutoReverses = true;
                //结束后执行逆动画
                //添加动画效果
                copyLayer.AddAnimation(positionAnimation, "Position Animation");
                copyLayer.AddAnimation(sizeAnimation, "SizeAnimation");
            } ;
```

```
            }
            ……
            #endregion
        }
}
```

运行效果如图 11.16 所示。

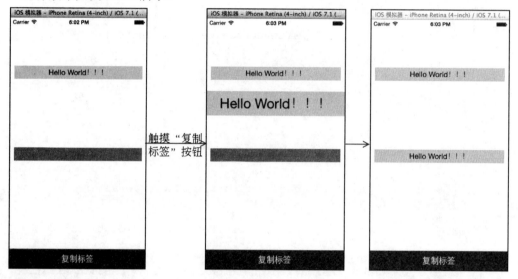

图 11.16　运行效果

11.7　图层的过渡动画

在 iOS 中，除了可以使用 UIView 实现过渡动画以外，还可以使用 CATransition 类实现过渡动画效果。它的过渡动画效果分为两大类：一类是公开的动画效果，另一类是非公开动画效果。本节将讲解这两类动画效果。

11.7.1　公开的过渡动画

公开动画是可以在帮助文档中查到的动画效果，这些动画效果有 4 种，如表 11-12 所示。

表 11-12　公开的动画效果

公开动画效果	功　　能
TransitionFade	渐渐消失
TransitionMoveIn	覆盖进入
TransitionPush	推出
TransitionReveal	揭开

🔔 注意：如果想要设置这些动画效果，需要使用到 Type 属性。

【示例 11-9】　以下将实现两个公开的过渡动画效果。具体的操作步骤如下所述。

（1）创建一个 Single View Application 类型的工程。
（2）添加图像 1.jpg 和 2.jpg 到创建工程的 Resources 文件夹中。
（3）打开 MainStoryboard.storyboard 文件，对主视图进行设置，效果如图 11.17 所示。

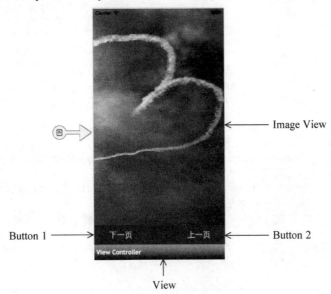

图 11.17　主视图的效果

需要添加的视图以及设置，如表 11-13 所示。

表 11-13　设置主视图

视　　图	设　　置
Image View	Name：iv Image：1.jpg 位置和大小：(0, 0, 320, 524)
Button1	Name：Nbtn Title：下一页 Font：System 19 pt Text Color：白色 位置和大小：(20, 8, 88, 30)
Button2	Name：Ubtn Title：上一页 Font：System 19 pt Text Color：白色 位置和大小：(212, 8, 88, 30)
View	Background：深灰色 位置和大小：(0, 522, 320, 46)

（4）打开 11-18ViewController.cs 文件，编写代码，实现过渡动画效果。代码如下：

```
using System;
using System.Drawing;
using MonoTouch.Foundation;
using MonoTouch.UIKit;
```

```
using MonoTouch.CoreAnimation;
namespace Application
{
    public partial class _1_18ViewController : UIViewController
    {
        ……                                    //这里省略了视图控制器的构造方法和析构方法
        #region View lifecycle
        public override void ViewDidLoad ()
        {
            base.ViewDidLoad ();
            //触摸按钮，实现推出的动画效果
            Nbtn.TouchUpInside += (sender, e) => {
                CATransition transition = CATransition.CreateAnimation ();
                transition.Duration = 10.0f;
                transition.Type = CATransition.TransitionPush;
                //设置动画效果
                transition.Subtype = CATransition.TransitionFromLeft;
                //设置动画方向
                iv.Image=UIImage.FromFile("2.jpg");
                this.iv.Layer.AddAnimation (transition, null);
            };
            //触摸按钮，实现覆盖进入的动画效果
            Ubtn.TouchUpInside += (sender, e) => {
                CATransition transition = CATransition.CreateAnimation ();
                transition.Duration = 10.0f;
                transition.Type = CATransition.TransitionMoveIn;
                transition.Subtype = CATransition.TransitionFromTop;
                iv.Image=UIImage.FromFile("1.jpg");
                this.iv.Layer.AddAnimation (transition, null);
            };
        }
        ……                                    //这里省略了视图加载和卸载前后的一些方法
        #endregion
    }
}
```

运行效果如图 11.18 所示。

注意：Subtype 属性可以实现对动画播放方向的设置。这些方向如表 11-14 所示。

表 11-14 动画方向

动 画 方 向	功　　能
TransitionFromRight	从过渡层右侧开始实现动画
TransitionFromLeft	从过渡层左侧开始实现动画
TransitionFromTop	从过渡层顶部开始实现动画
TransitionFromBottom	从过渡层底部开始实现动画

11.7.2 非公开的过渡动画

在帮助文档找不到的动画那就是非公开的动画效果，这些动画效果有 7 种，如表 11-15 所示。

第 11 章 图形和动画

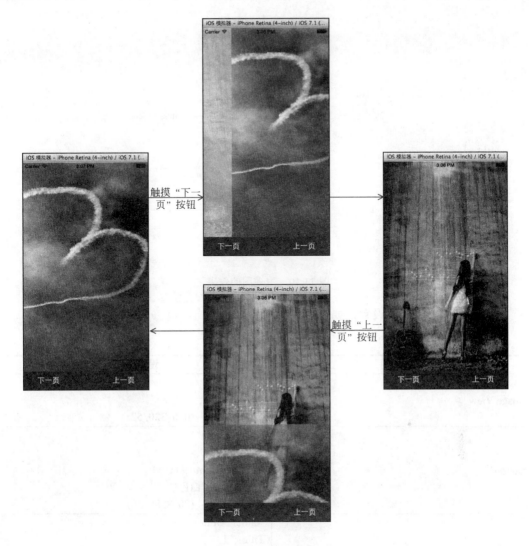

图 11.18 运行效果

表 11-15 非公开动画效果

非公开动画效果	功　能
"cube"	立方体
"suckEffect"	吸收
"oglFlip"	翻转
"rippleEffect"	波纹
"pageCurl"	卷页
"cameraIrisHollowOpen"	镜头开
"cameraIrisHollowClose"	镜头关

【示例 11-10】以下将实现两个非公开的过渡动画效果。具体的操作步骤如下所述。

（1）创建一个 Single View Application 类型的工程。

（2）添加图像 1.jpg 和 2.jpg 到创建工程的 Resources 文件夹中。

（3）打开 MainStoryboard.storyboard 文件，对主视图进行设置，效果如图 11.19 所示。

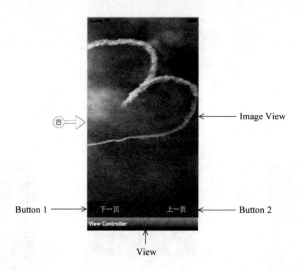

图 11.19 主视图的效果

需要添加的视图以及设置，如表 11-16 所示。

表 11-16 设置主视图

视 图	设 置
Image View	Name：iv Image：1.jpg 位置和大小：(0, 0, 320, 525)
Button1	Name：Nbtn Title：下一页 Font：System 19 pt Text Color：白色 位置和大小：(20, 8, 88, 30)
Button2	Name：Ubtn Title：上一页 Font：System 19 pt Text Color：白色 位置和大小：(212, 8, 88, 30)
View	Background：深灰色 位置和大小：(0, 525, 320, 43)

（4）打开 11-19ViewController.cs 文件，编写代码，实现过渡动画效果。代码如下：

```
using System;
using System.Drawing;
using MonoTouch.Foundation;
using MonoTouch.UIKit;
using MonoTouch.CoreAnimation;
namespace Application
{
    public partial class _1_19ViewController : UIViewController
    {
        ……                              //这里省略了视图控制器的构造方法和析构方法
        #region View lifecycle
        public override void ViewDidLoad ()
```

```
{
    base.ViewDidLoad ();
    //触摸按钮，实现立方体的动画效果
    Nbtn.TouchUpInside += (sender, e) => {
        CATransition transition = CATransition.CreateAnimation ();
        transition.Duration = 10.0f;
        transition.Type = "cube";
        transition.Subtype = CATransition.TransitionFromLeft;
        iv.Image=UIImage.FromFile("2.jpg");
        this.iv.Layer.AddAnimation (transition, null);
    };
    //触摸按钮，实现吸收的动画效果
    Ubtn.TouchUpInside += (sender, e) => {
        CATransition transition = CATransition.CreateAnimation ();
        transition.Duration = 10.0f;
        transition.Type = "suckEffect";
        transition.Subtype = CATransition.TransitionFromTop;
        iv.Image=UIImage.FromFile("1.jpg");
        this.iv.Layer.AddAnimation (transition, null);
    };
}
……                          //这里省略了视图加载和卸载前后的一些方法
#endregion
}
```

运行效果如图 11.20 所示。

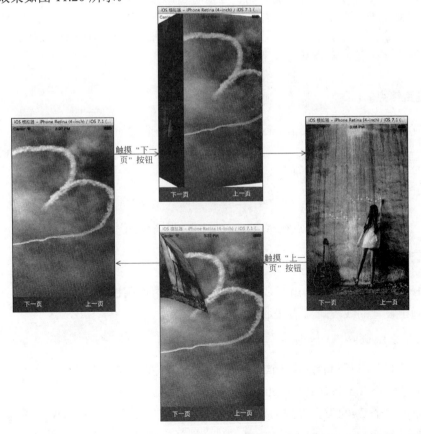

图 11.20　运行效果

11.8 绘制路径

路径是使用贝赛尔曲线所构成的一段闭合或者开放的曲线段。本节将实现对路径的绘制。

11.8.1 绘制线段

在绘制路径中，线段是最为简单的一种。绘制线段的步骤如下所述。

1．设置起点

要设置起点，就要使用 MoveTo()方法，其语法形式如下：

```
图形上下文.MoveTo (
    float x,
    float y
);
```

其中，x 用来指定线段起点在 X 轴的位置；y 用来指定线段起点在 Y 轴的位置。

> **注意**：绘图是在图形上下文中进行的，可以将图形上下文理解为一个画布。一个图形上下文表示一个绘制目标。它包含绘制系统用于完成绘制指令的绘制参数和设备相关信息。图形上下文定义基本图形属性，如颜色、剪切区域、线的宽度、样式信息、字体信息和合成选项等。

2．设置终点

要设置终点，就要使用 AddLineTo()方法，其语法形式如下：

```
图形上下文.AddLineToPoint (
    float x,
    float y
);
```

其中，x 用来指定线段起点在 X 轴的位置； y 用来指定线段起点在 Y 轴的位置。

3．填充

设置好起点和终点之后，就可以进行绘制了，所谓绘制也就是填充，它的功能就像一支画笔，画出设置的线段。其语法形式如下：

```
图形上下文.StrokePath ();
```

【**示例 11-11**】 以下将绘制一个线段。具体的操作步骤如下所述。

（1）创建一个 Single View Application 类型的工程。

（2）添加一个 C#的类文件到创建的工程中，并命名为 DrawingView。

（3）打开 MainStoryboard.storyboard 文件，选择主视图后，将 Class 设置为创建的类文

件名 DrawingView。

（4）打开 DrawingView.cs 文件，编写代码，实现对线段的绘制。代码如下：

```
using System;
using MonoTouch.Foundation;
using MonoTouch.UIKit;
using System.CodeDom.Compiler;
using MonoTouch.CoreGraphics;
using System.Drawing;
namespace Application
{
    partial class DrawingView : UIView
    {
        ……                              //这里省略了视图的构造器方法
        public override void Draw (RectangleF rect)
        {
            base.Draw (rect);
            CGContext context = UIGraphics.GetCurrentContext();
                                                //获取当前图形上下文
            context.MoveTo (10, 160);           //设置线段的起点
            context.AddLineToPoint (310, 160);  //设置线段的终点
            context.StrokePath ();              //填充
        }
    }
}
```

运行效果如图 11.21 所示。

为了让绘制的线段可以更加丰富，开发者可以对线段的填充颜色以及线宽进行设置。其中设置线段的填充颜色需要使用到 SetStrokeColorWithColor()方法，其语法形式如下：

图形上下文.SetStrokeColorWithColor(颜色);

设置线的宽度需要使用到 SetLineWidth()方法，其语法形式如下：

图形上下文. SetLineWidth(宽度值);

【示例 11-12】 以下以【示例 11-11】为基础，为绘制的线段设置线宽为 10，填充颜色为绿色。代码如下：

```
public override void Draw (RectangleF rect)
{
    base.Draw (rect);
    CGContext context = UIGraphics.GetCurrentContext();
    context.SetLineWidth(10f);
                                            //设置线宽
    ontext.SetStrokeColorWithColor(UIColor.Green.CGColor);
                                            //设置线的填充颜色
    context.MoveTo (10, 160);
    context.AddLineToPoint (310, 160);
    context.StrokePath ();
}
```

运行效果如图 11.22 所示。

图 11.21　运行效果

图 11.22　运行效果

11.8.2　绘制水平线

StrokeLineSegments()方法可以将相邻的两个点绘制成直线。其语法形式如下：

图形上下文.StrokeLineSegments (数组);

其中，在数组中存放了多个点。

【示例 11-13】 以下将使用 StrokeLineSegments()方法来实现对水平线的绘制。具体的操作步骤如下所述。

（1）创建一个 Single View Application 类型的工程。

（2）添加一个 C#的类文件到创建的工程中，并命名为 DrawingView。

（3）打开 MainStoryboard.storyboard 文件，选择主视图后，将 Class 设置为创建的类文件名 DrawingView。

（4）打开 DrawingView.cs 文件，编写代码，实现对水平线的绘制。代码如下：

```
using System;
using MonoTouch.Foundation;
using MonoTouch.UIKit;
using System.CodeDom.Compiler;
using MonoTouch.CoreGraphics;
using System.Drawing;
namespace Application
{
    partial class DrawingView : UIView
    {
        ……                                       //这里省略了视图的构造器方法
        public override void Draw (RectangleF rect)
        {
            base.Draw (rect);
            Console.WriteLine("DrawingView draw!");
            CGContext context = UIGraphics.GetCurrentContext();
            context.SetLineWidth(5f);
```

```
            context.SetStrokeColorWithColor(UIColor.Green.CGColor);
            //设置多个点,将这多个点放置在数组 addlines 中
            PointF[] addlines = new PointF[] {
                new PointF (10.0f, 350.0f),
                new PointF (70.0f, 50.0f),
                new PointF (130.0f, 350.0f),
                new PointF (190.0f, 50.0f),
                new PointF (250.0f, 350.0f),
                new PointF (310.0f, 50.0f),
            };
            context.StrokeLineSegments (addlines);                          //绘制
        }
}
```

运行效果如图 11.23 所示。

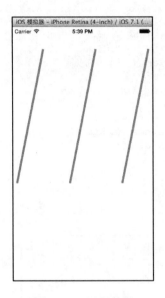

图 11.23 运行效果

11.8.3 绘制折线

AddLines()方法可以用来对多个点进行连接。其语法形式如下:

```
图形上下文.AddLines (数组);
```

其中,在数组中存放了多个点。

【示例 11-14】 以下将绘制一个折线。具体的操作步骤如下所述。

(1) 创建一个 Single View Application 类型的工程。

(2) 添加一个 C#的类文件到创建的工程中,并命名为 DrawingView。

(3) 打开 MainStoryboard.storyboard 文件,选择主视图后,将 Class 设置为创建的类文件名 DrawingView。

(4) 打开 DrawingView.cs 文件,编写代码,实现对折线的绘制。代码如下:

```
using System;
```

```
using MonoTouch.Foundation;
using MonoTouch.UIKit;
using System.CodeDom.Compiler;
using MonoTouch.CoreGraphics;
using System.Drawing;
namespace Application
{
    partial class DrawingView : UIView
    {
        ……                                         //这里省略了视图的构造器方法
        public override void Draw (RectangleF rect)
        {
            base.Draw (rect);
            Console.WriteLine("DrawingView draw!");
            CGContext context = UIGraphics.GetCurrentContext();
            context.SetLineWidth(5f);
            context.SetStrokeColorWithColor(UIColor.Green.CGColor);
            PointF[] addlines = new PointF[] {
                new PointF (10.0f, 260.0f),
                new PointF (70.0f, 100.0f),
                new PointF (130.0f, 260.0f),
                new PointF (190.0f, 100.0f),
                new PointF (250.0f, 260.0f),
                new PointF (310.0f, 100.0f),
            };
            context.AddLines (addlines);                    //连接每一个点
            context.StrokePath ();
        }
    }
}
```

运行效果如图 11.24 所示。

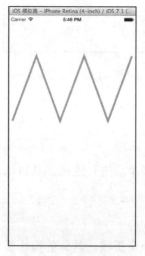

图 11.24　运行效果

11.8.4　绘制曲线

路径不仅有直线段，也有弯曲的线段，被称为曲线或者贝塞尔曲线。确定一条贝塞尔曲线需要 4 个点：开始点、第一个控制点、第二个控制点和结束点，如图 11.25 所示。

第 11 章 图形和动画

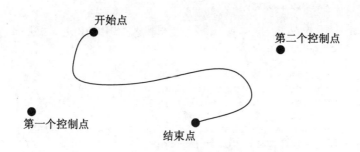

图 11.25 贝塞尔曲线

> **注意**：第一个控制点是用来控制开始点附近的曲线的；第二个控制点是用来控制结束点附近的曲线的。

MoveTo()方法可以对贝塞尔曲线开始点进行设置，AddCurveToPoint()方法可以实现对第一个控制点、第二个控制点以及结束点进行设置。AddCurveToPoint()方法的语法形式如下：

```
图形上下文.AddCurveToPoint (
    float cp1x,
    float cp1y,
    float cp2x,
    float cp2y,
    float x,
    float y
);
```

其中，这些参数的说明如下。
- cp1x：设置第一个控制点的 x 值。
- cp1y：设置第一个控制点的 y 值。
- cp2x：设置第二个控制点的 x 值。
- cp2y：设置第二个控制点的 y 值。
- x：设置结束点的 x 值。
- y：设置结束点的 y 值。

【示例 11-15】 以下将绘制一条贝塞尔曲线。具体的操作步骤如下所述。
（1）创建一个 Single View Application 类型的工程。
（2）添加一个 C#的类文件到创建的工程中，并命名为 DrawingView。
（3）打开 MainStoryboard.storyboard 文件，选择主视图后，将 Class 设置为创建的类文件名 DrawingView。
（4）打开 DrawingView.cs 文件，编写代码，实现对曲线的绘制。代码如下：

```
using System;
using MonoTouch.Foundation;
using MonoTouch.UIKit;
using System.CodeDom.Compiler;
using MonoTouch.CoreGraphics;
using System.Drawing;
namespace Application
```

```
{
    partial class DrawingView : UIView
    {
        ……                          ////这里省略了视图的构造器方法
        public override void Draw (RectangleF rect)
        {
            base.Draw (rect);
            Console.WriteLine("DrawingView draw!");
            CGContext context = UIGraphics.GetCurrentContext();
            context.SetLineWidth(5f);
            context.SetStrokeColorWithColor(UIColor.Green.CGColor);
            context.MoveTo (0, this.Bounds.Height);        //设置开始点
            context.AddCurveToPoint(0f, this.Bounds.Height, 50f, this.
            Bounds.Height / 2f, this.Bounds.Width, 0f);
                //设置第一个控制点、第二个控制点以及结束点
            context.StrokePath();
        }
    }
}
```

运行效果如图 11.26 所示。

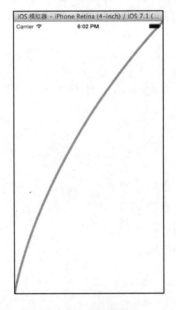

图 11.26　运行效果

11.9　绘制形状

除了可以绘制一些简单的路径外，还可以绘制一些形状，如矩形和圆等。这些形状是路径的一种。本节将讲解如何实现对形状的绘制。

【示例 11-16】 以下将绘制一个矩形和一个圆。具体的操作步骤如下所述。

（1）创建一个 Single View Application 类型的工程。

（2）添加一个 C#的类文件到创建的工程中，并命名为 DrawingView。

（3）打开 MainStoryboard.storyboard 文件，选择主视图后，将 Class 设置为创建的类文件名 DrawingView。

（4）打开 DrawingView.cs 文件，编写代码，实现对圆和矩形的绘制。代码如下：

```
using System;
using MonoTouch.Foundation;
using MonoTouch.UIKit;
using System.CodeDom.Compiler;
using MonoTouch.CoreGraphics;
using System.Drawing;
namespace Application
{
    partial class DrawingView : UIView
    {
        ……                                      //这里省略了视图的构造方法
        public override void Draw (RectangleF rect)
        {
            base.Draw (rect);
            CGContext context = UIGraphics.GetCurrentContext();
            context.SetFillColorWithColor(UIColor.Blue.CGColor);
            context.SetShadow(new SizeF(10f, 10f), 5f);
                                                            //设置阴影
            context.AddEllipseInRect(new RectangleF(100f, 100f,100f,
            100f));                                //设置圆的位置和大小
            context.FillPath();
            context.SetFillColorWithColor(UIColor.Red.CGColor);
            context.AddRect(new RectangleF(150f, 150f, 100f, 100f));
                                                //设置矩形的位置和大小
            context.FillPath();
        }
    }
}
```

运行效果如图 11.27 所示。

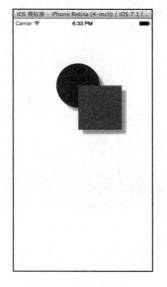

图 11.27　运行效果

11.10 绘制位图

本节将讲解绘制位图的两种方法：一种是绘制单个位图；另一种是绘制多个位图。

11.10.1 绘制单个位图

单个位图的绘制，要使用到 DrawImage()方法，其语法形式如下：

```
图形上下文.DrawImage (
    RectangleF rect,
    CGImage image
);
```

其中，rect 用来指定绘制区域；image 用来指定相应绘制的位图。

【示例 11-17】 以下程序通过使用 DrawImage()方法，实现绘制一个位图。具体的操作步骤如下所述。

（1）创建一个 Single View Application 类型的工程。

（2）添加图像 1.png 到创建工程的 Resources 文件夹中。

（3）添加一个 C#的类文件到创建的工程中，并命名为 DrawingView。

（4）打开 MainStoryboard.storyboard 文件，选择主视图后，将 Class 设置为创建的类文件名 DrawingView。

（5）打开 DrawingView.cs 文件，编写代码，实现对图像的绘制。代码如下：

```
using System;
using MonoTouch.Foundation;
using MonoTouch.UIKit;
using System.CodeDom.Compiler;
using MonoTouch.CoreGraphics;
using System.Drawing;
namespace Application
{
    partial class DrawingView : UIView
    {
        ……                                      //这里省略了视图的构造方法
        public override void Draw (RectangleF rect)
        {
            base.Draw (rect);
            CGContext context = UIGraphics.GetCurrentContext();
            UIImage im = UIImage.FromFile ("1.png");
            RectangleF imageRect = new RectangleF (60, 200, 200, 200);
            context.DrawImage (imageRect, im.CGImage);          //绘制位图
        }
    }
}
```

运行效果如图 11.28 所示。

图 11.28　运行效果

11.10.2　绘制多个位图

多个位图的绘制需要使用 DrawTiledImage()方法，其语法形式如下：

```
图形上下文.DrawTiledImage (
    RectangleF rect,
    CGImage image
);
```

其中，rect 用来指定绘制区域；image 用来指定相应绘制的位图。

【示例 11-18】 以下将使用 DrawTiledImage()方法，实现绘制多个位图。具体的操作步骤如下所述。

（1）创建一个 Single View Application 类型的工程。
（2）添加图像 1.png 到创建工程的 Resources 文件夹中。
（3）添加一个 C#的类文件到创建的工程中，并命名为 DrawingView。
（4）打开 MainStoryboard.storyboard 文件，选择主视图后，将 Class 设置为创建的类文件名 DrawingView。
（5）打开 DrawingView.cs 文件，编写代码，实现对图像的绘制。代码如下：

```
using System;
using MonoTouch.Foundation;
using MonoTouch.UIKit;
using System.CodeDom.Compiler;
using MonoTouch.CoreGraphics;
using System.Drawing;
namespace Application
{
    partial class DrawingView : UIView
    {
        ……
        public override void Draw (RectangleF rect)
```

```
        {
            base.Draw (rect);
            CGContext context = UIGraphics.GetCurrentContext();
            UIImage im = UIImage.FromFile ("1.png");
            RectangleF imageRect = new RectangleF (0, 0, 50, 50);
            //设置绘制区域的大小,并带有裁剪功能
            context.ClipToRect (new RectangleF (0, 100, 350, 250));
            context.DrawTiledImage (imageRect, im.CGImage);
            //平铺的方式绘制位图
        }
    }
}
```

运行效果如图 11.29 所示。

图 11.29 运行效果

11.11 绘 制 文 字

文字在 iOS 应用程序中起着相当重要的作用,可以将想要表达的意思通过文字展示给使用它的用户。本节将讲解文字的绘制。

【示例 11-19】以下将绘制"This text is drawn!"的文字信息。具体的操作步骤如下所述。

(1)创建一个 Single View Application 类型的工程。

(2)添加一个 C#的类文件到创建的工程中,并命名为 DrawingView。

(3)打开 MainStoryboard.storyboard 文件,选择主视图后,将 Class 设置为创建的类文件名 DrawingView。

(4)打开 DrawingView.cs 文件,编写代码,实现对文本的绘制。代码如下:

```
using System;
using MonoTouch.Foundation;
```

```
using MonoTouch.UIKit;
using System.CodeDom.Compiler;
using MonoTouch.CoreGraphics;
using System.Drawing;
namespace Application
{
    partial class DrawingView : UIView
    {
        ……                                       //这里省略了视图的构造方法
        public override void Draw (RectangleF rect)
        {
            base.Draw (rect);
            CGContext context = UIGraphics.GetCurrentContext();
            PointF location = new PointF(10f, 200f);
            UIFont font = UIFont.FromName("Verdana-Bold", 28f);
                                                              //设置字体
            NSString drawText = new NSString("This text is drawn!");
            context.SetTextDrawingMode(CGTextDrawingMode.Stroke);
                                                              //设置绘制模式

            context.SetStrokeColorWithColor(UIColor.Black.CGColor);
                                                              //设置边框颜色
            context.SetLineWidth(4f);
            drawText.DrawString(location, font);
                                                              //绘制文本
            context.SetTextDrawingMode(CGTextDrawingMode.Fill);
            context.SetFillColorWithColor(UIColor.Yellow.CGColor);
                                                              //设置填充颜色
            drawText.DrawString(location, font);
        }
    }
}
```

运行效果如图 11.30 所示。

图 11.30 运行效果

11.12 创建一个简单的绘制应用程序——画板

本节将讲解画板的实现。

【示例 11-20】 以下将使用触摸以及绘制中学到的一些内容，实现一个画板的功能。具体的操作步骤如下所述。

（1）创建一个 Single View Application 类型的工程。

（2）添加一个 C#的类文件到创建的工程中，并命名为 CanvasView。

（3）打开 MainStoryboard.storyboard 文件，选择主视图后，将 Class 设置为创建的类文件名 CanvasView。

（4）打开 CanvasView.cs 文件，编写代码，实现画板的功能。代码如下：

```csharp
using System;
using MonoTouch.Foundation;
using MonoTouch.UIKit;
using System.CodeDom.Compiler;
using System.Drawing;
using MonoTouch.CoreGraphics;
namespace Application
{
    partial class CanvasView : UIView
    {
        public CanvasView (IntPtr handle) : base (handle)
        {
            this.drawPath = new CGPath();
        }
        private PointF touchLocation;
        private PointF previousTouchLocation;
        private CGPath drawPath;
        private bool fingerDraw;
        //开始触摸
        public override void TouchesBegan (NSSet touches, UIEvent evt)
        {
            base.TouchesBegan (touches, evt);
            UITouch touch = touches.AnyObject as UITouch;
            this.fingerDraw = true;
            //绘制触摸的位置
            this.touchLocation = touch.LocationInView(this);
            this.previousTouchLocation = touch.PreviousLocationInView(this);
            this.SetNeedsDisplay();                     //更新图层内容
        }
        //移动触摸
        public override void TouchesMoved (NSSet touches, UIEvent evt)
        {
            base.TouchesMoved (touches, evt);
            UITouch touch = touches.AnyObject as UITouch;
            //获取触摸的位置
            this.touchLocation = touch.LocationInView(this);
            this.previousTouchLocation = touch.Previous LocationInView(this);
            this.SetNeedsDisplay();
        }
```

```
//绘制
public override void Draw (RectangleF rect)
{
    base.Draw (rect);
    if (this.fingerDraw) {
        using (CGContext context = UIGraphics.GetCurrentContext()){
            context.SetStrokeColorWithColor (UIColor.Blue.
            CGColor);
            context.SetLineWidth (5f);                      //设置线宽
            context.SetLineJoin (CGLineJoin.Round);   //设置线的样式
            context.SetLineCap (CGLineCap.Round);
                                              //设置画线的端点的样式设置
            this.drawPath.MoveToPoint (this.previousTouch Loca
            tion);                                //设置开始位置
            this.drawPath.AddLineToPoint (this.touchLocation);
                                                  //设置结束位置
            context.AddPath (this.drawPath);         //添加路径
            context.DrawPath (CGPathDrawingMode.Stroke);   //绘制
        }
    }
}
```

运行效果如图 11.31 所示。

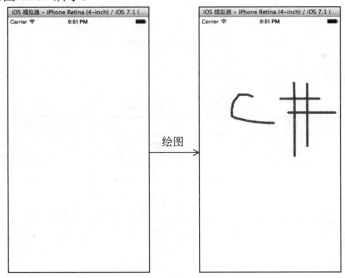

图 11.31　运行效果

11.13　创建位图图形上下文

上一节中实现了画板的功能，但是在画板中绘制的内容是不可以进行保存的，需要将这些内容转换为图形。位图图形上下文就是用来专门处理图像的图像上下文，包括转换。

【示例 11-21】 以下将会扩展画板应用程序的功能，实现对所画内容的保存。具体的操作步骤如下所述。

（1）创建一个 Single View Application 类型的工程。

（2）添加一个 C#的类文件到创建的工程中，并命名为 CanvasView。

（3）打开 MainStoryboard.storyboard 文件，对主视图进行设置，效果如图 11.32 所示。

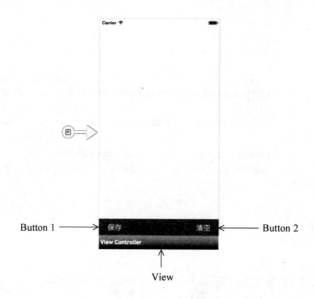

图 11.32　运行效果

需要添加的视图以及设置，如表 11-17 所示。

表 11-17　设置主视图

视　图	设　置
Button1	Name：sbtn Title：保存 Font：System 19 pt Text Color：白色 位置和大小：(20, 533, 46, 30)
Button2	Name：cbtn Title：清空 Font：System 19 pt Text Color：白色 位置和大小：(254, 533, 46, 30)
View	Background：深灰色 位置和大小：(0, 525, 320, 43)

（4）打开 CanvasView.cs 文件，实现画板的功能，以及将绘制的内容转换为图像。代码如下：

```
using System;
using MonoTouch.Foundation;
using MonoTouch.UIKit;
using System.CodeDom.Compiler;
using System.Drawing;
```

```csharp
using MonoTouch.CoreGraphics;
namespace Application
{
    partial class CanvasView : UIView
    {
        private PointF touchLocation;
        private PointF previousTouchLocation;
        private CGPath drawPath;
        private bool fingerDraw;
        public CanvasView (RectangleF frame) : base(frame)
        {
            this.drawPath = new CGPath();
        }
        //开始触摸
        public override void TouchesBegan (NSSet touches, UIEvent evt)
        {
            base.TouchesBegan (touches, evt);
            UITouch touch = touches.AnyObject as UITouch;
            this.fingerDraw = true;
            this.touchLocation = touch.LocationInView(this);
            this.previousTouchLocation = touch.PreviousLocation InView
            (this);
            this.SetNeedsDisplay();
        }
        //移动触摸
        public override void TouchesMoved (NSSet touches, UIEvent evt)
        {
            base.TouchesMoved (touches, evt);
            UITouch touch = touches.AnyObject as UITouch;
            this.touchLocation = touch.LocationInView(this);
            this.previousTouchLocation = touch.PreviousLocationInView
            (this);
            this.SetNeedsDisplay();
        }
        //绘制
        public override void Draw (RectangleF rect)
        {
            base.Draw (rect);
            if (this.fingerDraw)
            {
                using (CGContext context = UIGraphics.GetCurrentContext())
                {
                    context.SetStrokeColorWithColor(UIColor.Red.
                    CGColor);
                    context.SetLineWidth(5f);
                    context.SetLineJoin(CGLineJoin.Round);
                    context.SetLineCap(CGLineCap.Round);
                    this.drawPath.MoveToPoint(this.previousTouch
                    Location);
                    this.drawPath.AddLineToPoint(this.touchLocation);
                    context.AddPath(this.drawPath);
                    context.DrawPath(CGPathDrawingMode.Stroke);
                }
            }
        }
        //获取绘制的图像
        public UIImage GetDrawingImage()
        {
            UIImage toReturn = null;
            UIGraphics.BeginImageContext(this.Bounds.Size);
```

```
                //创建一个基于位图的图形上下文
                using (CGContext context = UIGraphics.GetCurrentContext())
                {
                    context.SetStrokeColorWithColor(UIColor.Red.CGColor);
                    context.SetLineWidth(10f);
                    context.SetLineJoin(CGLineJoin.Round);
                    context.SetLineCap(CGLineCap.Round);
                    context.AddPath(this.drawPath);
                    context.DrawPath(CGPathDrawingMode.Stroke);
                    toReturn = UIGraphics.GetImageFromCurrentImageContext();
                                                        //获取图像
                }
                UIGraphics.EndImageContext();
                return toReturn;
        }
        //清除绘制的内容
        public void ClearDrawing()
        {
            this.fingerDraw = false;
            this.drawPath.Dispose();
            this.drawPath = new CGPath();
            this.SetNeedsDisplay();
        }
    }
}
```

（5）打开 11-14ViewController.cs 文件，编写代码，实现图像的保存以及画板的清空。代码如下：

```
using System;
using System.Drawing;
using MonoTouch.Foundation;
using MonoTouch.UIKit;
namespace Application
{
    public partial class _1_11ViewController : UIViewController
    {
        ……
        #region View lifecycle
        public override void ViewDidLoad ()
        {
            base.ViewDidLoad ();
           CanvasView canvasView = new CanvasView(new RectangleF (0, 0, 320,
            525));
            canvasView.BackgroundColor = UIColor.White;
            this.View.AddSubview(canvasView);
            //触摸按钮，实现保存
            sbtn.TouchUpInside += (sender, e) => {
                UIImage drawingImage = canvasView.GetDrawingImage();
                //获取图像
                //将图像进行保存
                drawingImage.SaveToPhotosAlbum((img, err) => {
                    //判断是否存在错误
                    if (null != err)
                    {
                        Console.WriteLine("Error saving image! { 0}",
                        err.LocalizedDescription);
                    }
                } );
```

```
        };
        //触摸按钮,实现画板的清空
        cbtn.TouchUpInside += (sender, e) => {
            canvasView.ClearDrawing ();
        };
    }
    ……
    #endregion
}
```

运行效果如图 11.33 所示。

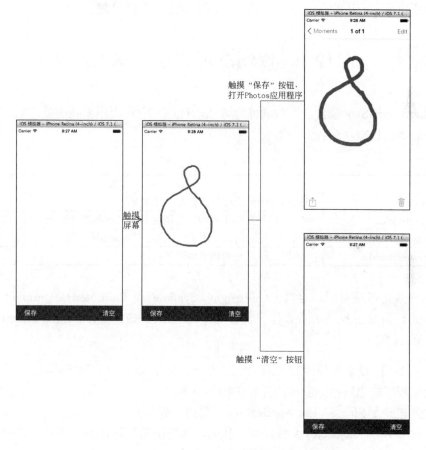

图 11.33　运行效果

第 12 章 多任务处理

多任务处理是 iOS 的一项功能。通过此功能，当应用软件在后台执行某些任务，用户可以同时使用另一应用软件。本章将讲解如何实现多任务的处理。

12.1 检测应用程序的状态

在 Xamarin.iOS 中包含了一些额外的方法，可以在应用程序委托中处理它们，以监视应用程序的当前状态。这些方法如表 12-1 所示。

表 12-1 方法总结

方 法	功 能
OnActivated	当应用程序激活时调用。如解锁屏幕
OnResignActivation	当应用程序变为非活动状态调用。如当屏幕被锁定或当有来电发生
DidEnterBackground	当应用程序进入后台时调用。如按 home 键的时候，此时该应用被暂停
WillEnterForeground	当应用程序即将回归前台时调用

💡**注意**：当应用程序移到后台时，OnResignActivation()和 DidEnterBackground()方法被调用；当应用程序由后台转向前台时，OnActivated()和 WillEnterForeground()方法被调用。

【示例 12-1】以下将使用表 12-1 中的方法对当前设备的应用程序的状态进行检测，并输出到应用程序输出窗口。具体的操作步骤如下所述。

（1）创建一个 Single View Application 类型的工程。

（2）打开 AppDelegate.cs 文件，编写代码，实现应用程序状态的检测。代码如下：

```
using System;
using System.Collections.Generic;
using System.Linq;
using MonoTouch.Foundation;
using MonoTouch.UIKit;
namespace Application
{
    [Register ("AppDelegate")]
    public partial class AppDelegate : UIApplicationDelegate
    {
        ……                          //这里省略了对 Window 属性的设置
        //当应用程序激活时调用
        public override void OnActivated (UIApplication application)
        {
            Console.WriteLine("Activated, application state: {0}",
```

```
        application.ApplicationState);
}
//当应用程序变为非活动状态调用
public override void OnResignActivation (UIApplication application)
{
    Console.WriteLine("Resign activation, application state: {0}",
    application.ApplicationState);
}
//当应用程序进入后台时调用
public override void DidEnterBackground (UIApplication application)
{
    Console.WriteLine("Entered background, application state: {0}",
    application.ApplicationState);
}
//当应用程序即将回归前台时调用
public override void WillEnterForeground (UIApplication
application)
{
    Console.WriteLine("Will enter foreground, application state:
    {0}", application.ApplicationState);
}
……                      //这里省略了WillTerminate()方法
}
```

当单击"运行"按钮后,打开 12-1 应用程序,会在应用程序输出窗口输出如图 12.1 所示的运行效果。

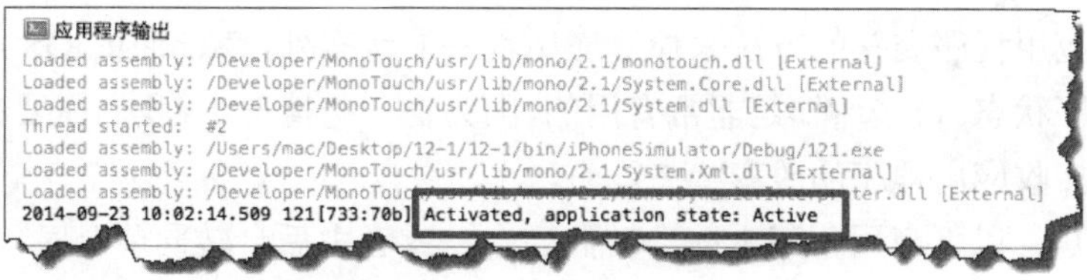

图 12.1 应用程序输出窗口 1

选择菜单栏中的"硬件"|"首页"命令,或者按键盘上的 Shift+ Command+ H 键,会在应用程序输出窗口输出如图 12.2 所示的运行效果。

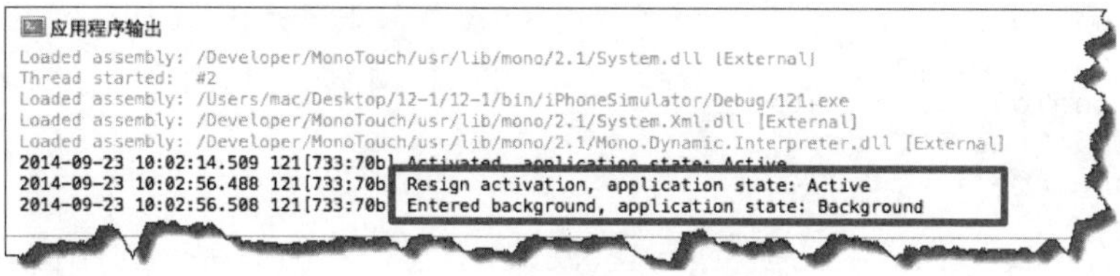

图 12.2 应用程序输出窗口 2

在首页中选择 12-1 应用程序,将其打开,会在应用程序输出窗口输出如图 12.3 所示的运行效果。

> 注意:在此程序中可以看到,所有的这些方法都包含了一个参数,这个参数是 UIApplication 的实例。UIApplication 类包含了 ApplicationState 属性,它可以返回应用程序的状态值,这些值如表 12-2 所示。

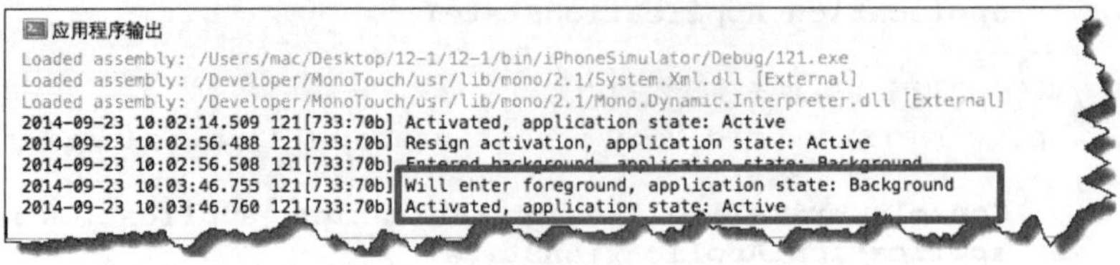

```
应用程序输出
Loaded assembly: /Users/mac/Desktop/12-1/12-1/bin/iPhoneSimulator/Debug/121.exe
Loaded assembly: /Developer/MonoTouch/usr/lib/mono/2.1/System.Xml.dll [External]
Loaded assembly: /Developer/MonoTouch/usr/lib/mono/2.1/Mono.Dynamic.Interpreter.dll [External]
2014-09-23 10:02:14.509 121[733:70b] Activated, application state: Active
2014-09-23 10:02:56.488 121[733:70b] Resign activation, application state: Active
2014-09-23 10:02:56.508 121[733:70b] Entered background, application state: Background
2014-09-23 10:03:46.755 121[733:70b] Will enter foreground, application state: Background
2014-09-23 10:03:46.760 121[733:70b] Activated, application state: Active
```

图 12.3　应用程序输出窗口 3

表 12-2　应用程序的状态值

状　态　值	功　　能
Active	应用程序激活
Inactive	应用程序处于非活动状态
Background	应用程序在后台

在【示例 12-1】的应用程序中省略了一个 WillTerminate()方法，此方法是 iOS 应用程序受到限制时调用的，如发出内容警告，或者应用程序不释放资源时调用，一般是用不到的。

12.2　接收应用程序状态的通知

除了调用应用程序委托的方法来检测应用程序的状态外，Xamarin.iOS 还提供了通知，为接收应用程序状态。这些通知是非常有用的，因为在一些情况下，我们需要在 AppDelegate 类以外的地方接收应用程序的改变状态。

【示例 12-2】　以下应用程序由前台转向后台，或者由后台转向前台时，接收应用程序的通知。具体操作步骤如下所述。

（1）创建一个 Single View Application 类型的工程。

（2）打开 12-2ViewController.cs 文件，编写代码，实现接收应用程序状态的通知，并在应用程序输出窗口输出。代码如下：

```
using System;
using System.Drawing;
using MonoTouch.Foundation;
using MonoTouch.UIKit;
namespace Application
{
    public partial class _2_2ViewController : UIViewController
    {
        private NSObject appDidEnterBackgroundObserver;
        private NSObject appWillEnterForegroundObserver;
        ……                        //这里省略了视图控制器的构造方法和析构方法
        #region View lifecycle
        ……                        //这里省略了视图加载后的方法
        public override void ViewWillAppear (bool animated)
        {
            base.ViewWillAppear (animated);
            this.AddNotificationObservers();
        }
```

```
……                          //这里省略了视图显示后的方法
public override void ViewWillDisappear (bool animated)
{
    base.ViewWillDisappear (animated);
    this.RemoveNotificationObservers();
}
……                          //这里省略了视图消失后的方法
private void AddNotificationObservers()
{
    //添加应用程序在后台的通知
    appDidEnterBackgroundObserver =
        UIApplication.Notifications.ObserveDidEnterBackground((s,
        e) => Console.WriteLine("App did enter background! App state:
        {0}", UIApplication.SharedApplication.ApplicationState));
    //添加应用程序由后台转向前台的通知
    appWillEnterForegroundObserver =
        UIApplication.Notifications.ObserveWillEnterForeground
        ((s, e) => Console.WriteLine("App will enter foreground! App
        state: {0}", UIApplication.SharedApplication.Application
        State));
}
//移除通知
private void RemoveNotificationObservers()
{
    NSNotificationCenter.DefaultCenter.RemoveObservers(new [] {
        this.appDidEnterBackgroundObserver,
        this.appWillEnterForegroundObserver
    } );
}
#endregion
```

当单击"运行"按钮后，打开 12-2 应用程序，选择菜单栏中的"硬件"|"首页"命令，或者按 Shift+ Command+ H 键，会在应用程序输出窗口输出如图 12.4 所示的运行效果。

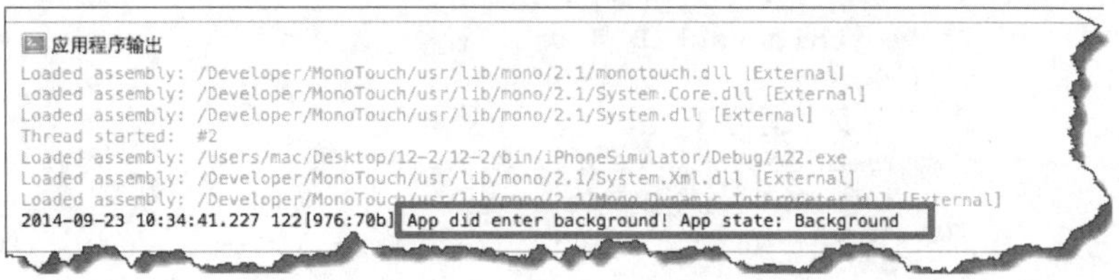

图 12.4　应用程序输出窗口 1

在首页中选择 12-2 应用程序，将其打开，会在应用程序输出窗口输出如图 12.5 所示的运行效果。

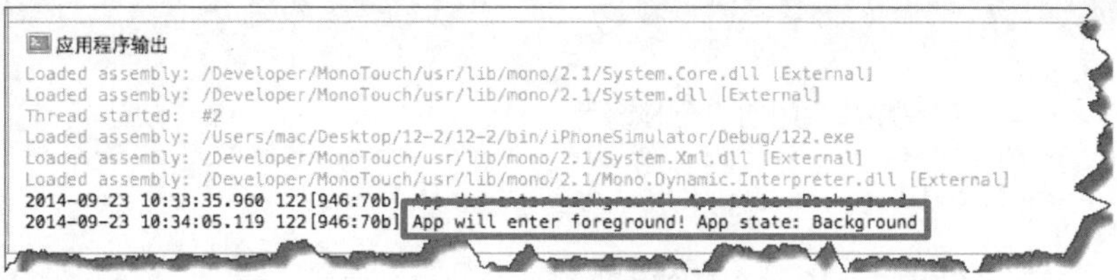

图 12.5　应用程序的输出窗口 2

12.3 在后台运行代码

在 iOS 中支持后台执行代码的功能。本节将讲解如何在后台执行代码。

【示例 12-3】 以下将实现在后台运行代码的功能。具体的操作步骤如下所述。

（1）创建一个 Single View Application 类型的工程。

（2）打开 AppDelegate.cs 文件，编写代码，实现在后台运行代码的功能。代码如下：

```csharp
using System;
using System.Collections.Generic;
using System.Linq;
using MonoTouch.Foundation;
using MonoTouch.UIKit;
using System.Threading;
namespace Application
{
    [Register ("AppDelegate")]
    public partial class AppDelegate : UIApplicationDelegate
    {
        ……                              //这里省略了对 Window 属性的设置
        private int taskID;
        ……                              //这里省略了当应用程序激活时调用的方法
        //当应用程序进入后台时调用
        public override void DidEnterBackground (UIApplication application)
        {
            //判断 taskID 是否为 0
            if (this.taskID == 0)
            {
                //请求的应用程序被允许在后台处理
                this.taskID = application.BeginBackgroundTask(() => {
                    application.EndBackgroundTask(this.taskID);
                    //结束后台任务
                    this.taskID = 0;
                } );
                //队列执行方法
                ThreadPool.QueueUserWorkItem(delegate {
                    //输出 60 次内容
                    for (int i = 0; i < 60; i++)
                    {
                        Console.WriteLine("Task {0} - Current time {1}",
                            this.taskID, DateTime.Now);
                        Thread.Sleep(1000);
                    }
                    application.EndBackgroundTask(this.taskID);
                    this.taskID = 0;
                } );
            }
        }
        //当应用程序即将回归前台时调用
        public override void WillEnterForeground (UIApplication application)
        {
            //判断 taskID 是否为 0
            if (this.taskID != 0)
```

```
            {
                UIAlertView alert = new UIAlertView ();
                alert.Title = "Background task is running!";
                alert.AddButton ("Cancel");
                alert.Show ();
            } else
            {
                UIAlertView alert = new UIAlertView ();
                alert.Title = "Background task completed!";
                alert.AddButton ("Cancel");
                alert.Show ();
            }
        }
        ……                          //这里省略了WillTerminate()方法
    }
}
```

当单击"运行"按钮后,打开 12-3 应用程序,选择菜单栏中的"硬件"|"首页"命令,或者按 Shift+ Command+ H 键,会在应用程序输出窗口输出如图 12.6 所示的运行效果。

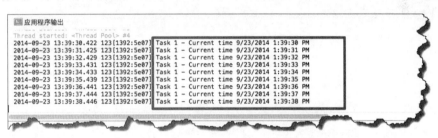

图 12.6 应用程序的输出窗口

在首页中选择 12-3 应用程序,将其打开,会看到如图 12.7 所示的运行效果。

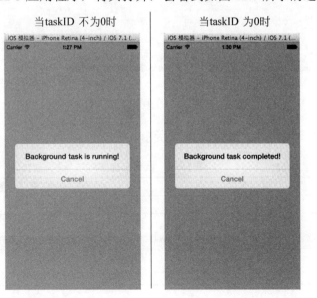

图 12.7 运行效果

注意:当 taskID 为 0 时,表明输出的内容已输出了 60 次。

12.4 在后台播放音频

QQ音乐应用程序可以支持在后台播放歌曲。本节将讲解这一功能是如何实现的。

【示例12-4】 以下将实现在后台播放音频的功能。具体的操作步骤如下所述。

（1）创建一个 Single View Application 类型的工程。

（2）添加图像 1.jpg 到创建工程的 Resources 文件夹中。

（3）添加音频文件"孤单北半球.mp3"、"狐狸雨.mp3"、"天使的翅膀.mp3"和"一样爱着你.mp3"到创建的工程中。

（4）打开 MainStoryboard.storyboard 文件，拖动 Navigation Controller 导航控制器到画布中，将 Is Initial View Controller 复选框选中，将导航控制器关联的根视图控制器设置为画布中原先有的 View Controller 控制器，对 View Controller 视图控制器的主视图进行设置，效果如图 12.8 所示。

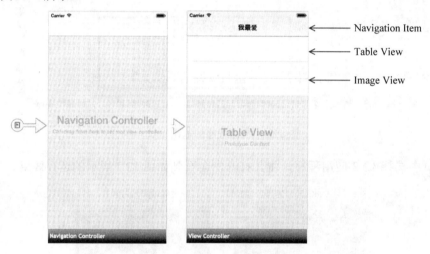

图 12.8 主视图的效果

需要添加的视图以及设置，如表 12-3 所示。

表 12-3 设置主视图

视　　图	设　　置
Navigation Item	Title：我最爱
Table View	Name：tv 位置和大小：(0, 65, 320, 503)
Image View	Image：1.jpg 位置和大小：(0, 64, 320, 456)

（5）打开 12-4ViewController.cs 文件，编写代码，实现音乐的播放。代码如下：

```
using System;
using System.Drawing;
using MonoTouch.Foundation;
```

第12章 多任务处理

```csharp
using MonoTouch.UIKit;
using MonoTouch.AVFoundation;
using System.Collections.Generic;
namespace Application
{
    public partial class _2_4ViewController : UIViewController
    {
        string[] pNames;
        ……
        #region View lifecycle
        public override void ViewDidLoad ()
        {
            base.ViewDidLoad ();
            pNames = new string[]{"孤单北半球", "天使的翅膀", "狐狸雨","一样爱着你"};
            tv.Source = new myViewSource(pNames);
            tv.BackgroundColor = UIColor.Clear;
        }
        private class myViewSource : UITableViewSource {
            private AVAudioPlayer audioPlayer;
            private string[] dupPremTeams;
            public myViewSource(string[] prems) {
                dupPremTeams = prems;
            }
            //设置行数
            public override int RowsInSection(UITableView table,
                int section) {
                return dupPremTeams.Length;
            }
            //设置每一行的内容
            public override UITableViewCell GetCell(UITableView tableView,
            NSIndexPath index) {
                UITableViewCell theCell = new UITableViewCell();
                theCell.BackgroundColor = UIColor.Clear;
                theCell.TextLabel.Text = dupPremTeams[index.Row];
                //设置内容
                return theCell;
            }
            //实现行的选择以及歌曲的播放
            public override void RowSelected (UITableView tableView,
            NSIndexPath indexPath)
            {
                int rowIndex = indexPath.Row;
                NSError error = null;
                //设置类型

                AVAudioSession.SharedInstance().SetCategory(AVAudioSession.
                Category Playback, out error);
                //判断是否存在错误
                if (error != null)
                {
                    Console.WriteLine("Error setting audio session
                    category: {0}", error.LocalizedDescription);
                }
                string name = dupPremTeams [rowIndex];
                string path = NSBundle.MainBundle.PathForResource (name,
                "mp3");
                NSUrl url = NSUrl.FromFilename (path);
                audioPlayer = AVAudioPlayer.FromUrl (url);
                UITableViewCell cellView = tableView.CellAt (indexPath);
```

```
                    //判断单元格的风格是否为 None 风格
                    if (cellView.Accessory == UITableViewCellAccessory.None) {
                        cellView.Accessory = UITableViewCellAccessory.
                        Checkmark;
                        this.audioPlayer.Play();              //播放音频
                    } else {
                        cellView.Accessory = UITableViewCellAccessory.None;
                        tableView.DeselectRow (indexPath, true);
                        this.audioPlayer.Stop();              //结束播放的音频
                    }
                }
            }
            ……
            #endregion
        }
}
```

（6）打开 Info.plist 文件，在 Source 选项卡中，添加 Required background modes 属性，如图 12.9 所示。

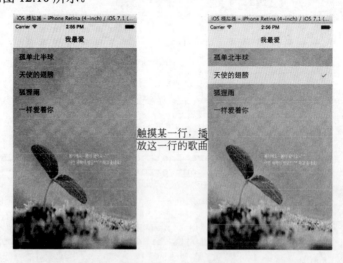

图 12.9　Info.plist 文件

运行效果如图 12.10 所示。

图 12.10　运行效果

> **注意**：当选择菜单栏中的"硬件"|"首页"命令后，或者按 Shift+ Command+ H 键后，播放的歌曲仍会继续播放。

12.5 在后台更新数据

在很多的应用程序中都支持在后台更新数据，如 iOS 内置的天气和地图应用等。本节将讲解如何在后台实现数据的更新。

【示例 12-5】以下将实现在后台更新网站 http://software.tavlikos.com 中的数据的功能。具体的操作步骤如下所述。

（1）创建一个空类型的工程。

（2）在创建的工程中添加一个类型为 iPhone View Controller 的文件，并命名为 Background FetchApp ViewController。

（3）打开 BackgroundFetchAppViewController.xib 文件，对主视图进行设置，效果如图 12.11 所示。

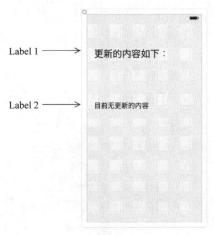

图 12.11　主视图的效果

需要添加的视图以及设置，如表 12-4 所示。

表 12-4　设置主视图

视　　图	设　　置
主视图	Background：浅灰色
Label1	Text：更新的内容如下： Font：System 26.0 位置和大小：(20,82,215, 51)
Label2	Text：目前无更新的内容 Font：System 29.0 Line：5 位置和大小：(20,172,280, 154) 为标签声明插座变量 label

（4）打开 Info.plist 文件，在 Source 选项卡中，添加 Required background modes 属性，并在此属性中添加 fetch，如图 12.12 所示。

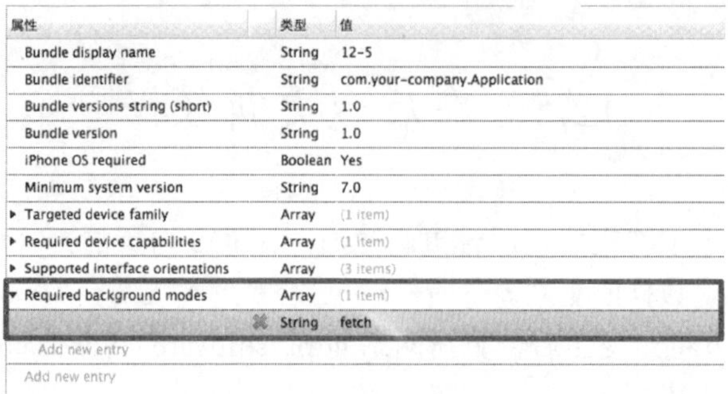

属性	类型	值
Bundle display name	String	12-5
Bundle identifier	String	com.your-company.Application
Bundle versions string (short)	String	1.0
Bundle version	String	1.0
iPhone OS required	Boolean	Yes
Minimum system version	String	7.0
▶ Targeted device family	Array	(1 item)
▶ Required device capabilities	Array	(1 item)
▶ Supported interface orientations	Array	(3 items)
▼ Required background modes	Array	(1 item)
String	fetch	
Add new entry		
Add new entry		

图 12.12　Info.plist 文件的设置

（5）打开 12-5ViewController.cs 文件，编写代码，实现对 LabelStatus 属性的设置。代码如下：

```
using System;
using System.Drawing;
using MonoTouch.Foundation;
using MonoTouch.UIKit;
namespace Application
{
    public partial class BackgroundFetchAppViewController : UIViewController
    {
        ……
        //设置属性，获取标签
        public UILabel LabelStatus
        {
            get
            {
                return label;
            }
        }
    }
}
```

（6）打开 AppDelegate.cs 文件，编写代码，实现在后台更新数据的功能。代码如下：

```
using System;
using System.Collections.Generic;
using System.Linq;
using MonoTouch.Foundation;
using MonoTouch.UIKit;
using System.Net;
using System.IO;
namespace Application
{
    [Register ("AppDelegate")]
    public partial class AppDelegate : UIApplicationDelegate
    {
        UIWindow window;
        BackgroundFetchAppViewController viewController;
        public override bool FinishedLaunching (UIApplication app,
        NSDictionary options)
        {
```

第 12 章 多任务处理

```
UIApplication.SharedApplication.SetMinimumBackgroundFetchInterval(UI
Application.BackgroundFetchIntervalMinimum);
            //启动 Background Fetch 支持，设置时间间隔
    window = new UIWindow (UIScreen.MainScreen.Bounds);
    viewController = new BackgroundFetchAppViewController ();
    window.RootViewController = viewController;//设置根视图控制器
    window.MakeKeyAndVisible ();
    return true;
}
private int updateCount;
//实现在后台更新数据的功能
public override void PerformFetch (UIApplication application,
Action<UIBackgroundFetchResult> completionHandler)
{
    try
    {
        HttpWebRequest request = WebRequest.Create ("http://
        software. tavlikos.com") as HttpWebRequest;
        using (StreamReader sr = new StreamReader (request.
        GetResponse(). GetResponse Stream()))
        {
            Console.WriteLine("Received response: {0}", sr.
            ReadToEnd());
        }
        this.viewController.LabelStatus.Text =
            string.Format("Update count: {0}/n{1}", ++updateCount,
            DateTime.Now);
        completionHandler(UIBackgroundFetchResult.NewData);
                                        //完成处理后调用
    } catch
    {
        this.viewController.LabelStatus.Text =
            string.Format("Update {0} failed on {1}!", ++update
            Count,DateTime.Now);
        completionHandler(UIBackgroundFetchResult.Failed);
    }
}
}
```

运行效果如图 12.13 所示。

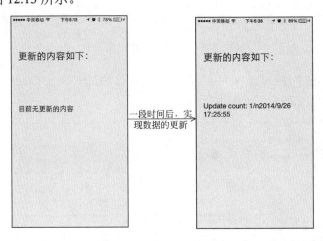

图 12.13　运行效果

12.6 禁用后台模式

虽然 Xamarin Studio 编译的所有应用程序的默认行为都支持后台模式，但是通过在应用程序的 Info.plist 文件中添加一项 Application does not run in background，可以重写这种行为。

【示例 12-6】 以【示例 12-1】为基础，实现对后台模式的禁用。具体的操作如下所述。

（1）打开示例 12-1 的工程。

（2）打开 Info.plist 文件，在 Source 选项卡中，添加 Application does not run in background 属性，如图 12.14 所示。

图 12.14　Info.plist 文件的设置

当单击"运行"按钮后，打开 12-1 应用程序，会在应用程序输出窗口输出如图 12.15 所示的运行效果。

图 12.15　应用程序输出窗口 1

选择菜单栏中的"硬件"|"首页"命令，或者按 Shift+ Command+ H 键，会在应用程序输出窗口输出如图 12.16 所示的运行效果。

图 12.16　应用程序输出窗口 2

在首页中选择 12-1 应用程序，将其打开，会在应用程序输出窗口输出如图 12.17 所示的运行效果。

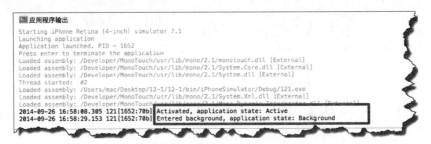

图 12.17　应用程序输出窗口 3

在第三个应用程序输出窗口输出的内容和第二个中的内容一样，没有改变，这和禁用后台模式有关。

第13章 本 地 化

本地化封装了语言、区域,以及技术约定和规范的信息。它用于提供用户所处地域相关的定制化信息和首选项信息的设置。通过获取用户的本地化信息设置,我们可以为用户提供更加友好的、人性化的界面设置,包括更改应用程序的界面的语言、货币类型、数字和日期格式,并提供正确的地理位置显示等。iOS 为应用程序的开发提供了很好的本地化机制。良好的本地化意味着应用程序可以为更多的用户提供服务。其中,NSLocale 类的主要作用便是用来封装本地化相关的各种信息。本章将讲解一些本地化的内容。

13.1 创建一个具有多种语言的应用程序

在 App Store 中很流行在一个应用程序中包含多个语言版本。这种应用程序之所以流行,是因为可以便于多个国家的用户阅读此应用程序中的内容。例如,一个香港人,可以将当前设备的语言设置为繁体中文,此时具有多种语言的应用程序中的内容也会相应的变为繁体字。

【示例 13-1】 以下将实现支持两种不同语言的应用程序。具体的操作步骤如下所述。
(1)创建一个 Single View Application 类型的工程。
(2)添加文件夹 en.lproj 到创建的工程中。添加文件夹的步骤如下所述。
首先,右击"引用"文件夹上方的 13-1,如图 13.1 所示。
选择 Add|New Folder 命令,在导航栏出现一个新的文件夹,文件夹的名称为"新建文件夹",将此名称改为 en.lproj。此时一个名为 en.lproj 的文件夹就创建好了。
(3)在 en.lproj 文件夹中添加一个名为 Localizable.strings 的文件。添加文件的步骤如下所述。
首先在 en.lproj 文件夹上右击鼠标,如图 13.2 所示。

图 13.1 添加文件夹

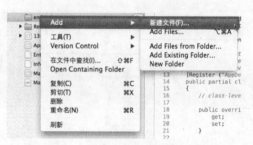

图 13.2 添加文件 1

选择 Add|新建文件(F)…命令，弹出 New Files 对话框，如图 13.3 所示。

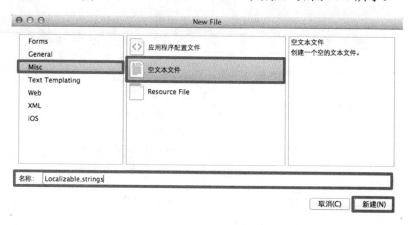

图 13.3　添加文件 2

选择 Misc 中的"空文本文件"后，输入文件名称 Localizable.strings，然后单击"新建"按钮，此时将会在 en.lproj 文件夹中添加一个 Localizable.strings 的文件。

（4）添加文件夹 es.lproj 到创建的工程中。

（5）在 es.lproj 文件夹中添加一个名为 Localizable.strings 的文件。

（6）打开 en.lproj 文件夹中的 Localizable.strings 文件，在此文件中输入以下的内容：

```
//Localized output on MultipleLanguageAppViewController
"Have a nice day!" = "Have a nice day!";
```

（7）打开 es.lproj 文件夹中的 Localizable.strings 文件，在此文件中输入以下的内容：

```
//Localized output on MultipleLanguageAppViewController
"Have a nice day!" = "¡uéenga un buen dí!";
```

（8）打开 MainStoryboard.storyboard 文件，对主视图进行设置，效果如图 13.4 所示。

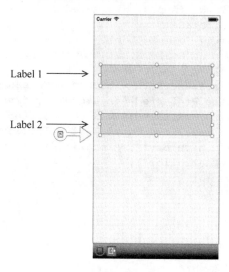

图 13.4　主视图的效果

需要添加的视图以及设置，如表 13-1 所示。

表 13-1　设置主视图

视　　图	设　　置
主视图	Background：浅灰色
Label1	Name：lblLocale Text：（空） Font：System 19 pt Alignment：居中 位置和大小：(20,126, 280, 49)
Label2	Name：lblLocalizedOutput Text：（空） Font：System 19 pt Alignment：居中 位置和大小：(20,248, 280, 49)

（9）打开 13-1ViewController.cs 文件，编写代码，实现一个包含两种语言的应用程序。代码如下：

```
using System;
using System.Drawing;
using MonoTouch.Foundation;
using MonoTouch.UIKit;
namespace Application
{
    public partial class _3_1ViewController : UIViewController
    {
        ……
        #region View lifecycle
        ……
        public override void ViewWillAppear (bool animated)
        {
            base.ViewWillAppear (animated);
            this.lblLocale.Text =
                string.Format("Locale: {0} - Language: {1}", NSLocale.
                CurrentLocale.LocaleIdentifier, NSLocale.Preferred Langua
                ges[0]);                              //获取当前使用语言
            string resourcePath = NSBundle.MainBundle.PathFor Resource
            (NSLocale.Preferred Languages[0], "lproj");
            NSBundle localeBundle = NSBundle.FromPath(resourcePath);
            this.lblLocalizedOutput.Text = localeBundle.LocalizedString
            ("Have a nice day!", "Localized output on MultipleLanguage
            AppViewController");                     //检索本地化字符串
        }
        ……
        #endregion
    }
}
```

运行效果如图 13.5 所示。

图 13.5 运行效果

需要注意的是，对于 iOS Simulator 语言的切换步骤如下所述。

（1）选择"硬件"|"首页"命令，回到 iOS Simulator 首页，如图 13.6 所示。

（2）单击 Settings 应用程序图标，将应用程序打开，如图 13.7 所示。

图 13.6 操作步骤 1

图 13.7 操作步骤 2

（3）进入 Settings 窗口中，选择 General 后，打开 General 窗口，如图 13.8 所示。

（4）选择 International 选项，进入 International 窗口，如图 13.9 所示。

（5）选择 Language 选项，进入 Language 窗口，如图 13.10 所示。

（6）选择 Español 这一选项，触摸 Done 按钮，对 iOS Simulator 的语言进行设置，如图 13.11 所示。

图 13.8　操作步骤 3　　　　图 13.9　操作步骤 4

图 13.10　操作步骤 5　　　　图 13.11　操作步骤 6

（7）设置完成后，在首页的应用程序标题就变为了西班牙语。

13.2　本地化资源

本地化资源其实就是一些类似于图像和声音文件的内容，这些内容都是对于特定于语言环境而言的。

【示例 13-2】以下将根据用户本地化的偏好设置实现加载和显示资源的功能。具体的操作步骤如下所述。

（1）创建一个 Single View Application 类型的工程。
（2）添加文件夹 en.lproj 和 es.lproj 到创建的工程中。
（3）在 en.lproj 文件夹中添加图像 flag.jpg。

（4）在 es.lproj 文件夹中添加图像 flag.jpg。
（5）打开 MainStoryboard.storyboard 文件，对主视图进行设置，效果如图 13.12 所示。

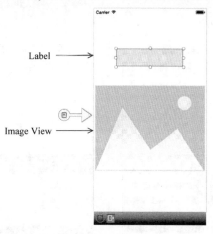

图 13.12　主视图的效果

需要添加的视图以及设置，如表 13-2 所示。

表 13-2　设置主视图

视图	设置
Label1	Name：label Text：（空） Font：System Bold 24 pt Alignment：居中 位置和大小：(66,114, 188, 47)
Image View	Name：iv 位置和大小：(3,214, 314, 240)

（6）打开 13-2ViewController.cs 文件，编写代码，实现根据用户本地化的偏好设置实现加载和显示资源的功能。代码如下：

```
using System;
using System.Drawing;
using MonoTouch.Foundation;
using MonoTouch.UIKit;
namespace Application
{
    public partial class _3_2ViewController : UIViewController
    {
        ……
        #region View lifecycle
        public override void ViewDidLoad ()
        {
            base.ViewDidLoad ();
            label.Text = NSLocale.PreferredLanguages[0];
            //获取当前使用语言
            iv.Image=UIImage.FromFile(NSBundle.MainBundle.PathFor
            Resource ("flag", "jpg"));
        }
        ……
```

```
        #endregion
    }
}
```

运行效果如图 13.13 所示。

语言为英语

语言为西班牙语

图 13.13　运行效果

13.3　区　域　格　式

不同国家和地区所使用的计量单位均有所差异。区域格式就是设置显示各种信息的方式，如货币、日期和时间，这些都是根据设置的不同区域进行显示的。

【示例 13-3】 以下将根据用户的区域格式设置显示日期、时间和货币等内容。具体的操作步骤如下所述。

（1）创建一个 Single View Application 类型的工程。

（2）打开 MainStoryboard.storyboard 文件，对主视图进行设置，效果如图 13.14 所示。

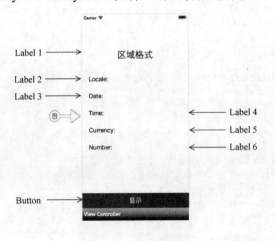

图 13.14　主视图的效果

需要添加的视图以及设置,如表 13-3 所示。

表 13-3 设置主视图

视 图	设 置
Label1	Text:区域格式 Font:System 27 pt Alignment:居中 位置和大小:(0,101, 320, 39)
Label2	Name:lblLocale Text:Locale: Font:System 18 pt 位置和大小:(20,180, 280, 21)
Label3	Name:lblDate Text:Date: Font:System 18 pt 位置和大小:(20,230, 280, 21)
Label4	Name:lblTime Text:lblTime Font:System 18 pt 位置和大小:(20,280, 280, 21)
Label5	Name:lblCurrency Text:Currency: Font:System 18 pt 位置和大小:(20,330, 280, 21)
Label6	Name:lblNumber Text:Number: Font:System 18 pt 位置和大小:(20,380, 280, 21)
Button	Name:btn Title:显示 Font:System 19 pt Text Color:白色 Background:深灰色 位置和大小:(0,525, 320, 43)

(3)打开 13-3ViewController.cs 文件,编写代码,实现日期、时间和货币的显示。代码如下:

```
using System;
using System.Drawing;
using MonoTouch.Foundation;
using MonoTouch.UIKit;
namespace Application
{
    public partial class _3_3ViewController : UIViewController
    {
        ……
        #region View lifecycle
        public override void ViewDidLoad ()
        {
            base.ViewDidLoad ();
```

```
            //货币按钮显示信息
            btn.TouchUpInside += (sender, e) => {
                //显示区域环境
                lblLocale.Text = string.Format("Locale: {0}", NSLocale.
                CurrentLocale. LocaleIdentifier);
                //显示日期
                lblDate.Text = string.Format("Date: {0}", DateTime. Now.
                ToLongDateString());
                //显示时间
                lblTime.Text = string.Format("Time: {0}", DateTime.Now.
                ToLongTimeString());
                lblCurrency.Text = string.Format("Currency: {0:c}", 250);
                //显示货币
                lblNumber.Text = string.Format("Number: {0:n}", 1350);
            //显示数字
            };
    }
    ……
    #endregion
    }
}
```

运行效果如图 13.15 所示。

图 13.15 运行效果

第 14 章 发布应用程序

开发者所制作的每一个应用程序往往都是要盈利的,所以并不是简简单单的制作而已。想要让自己的应用程序盈利,就需要将其发布到 App Store 上。这样,当用户下载购买你的应用程序后,你就会得到相应的报酬。本章将讲解如何发布应用程序。

14.1 申请发布证书

要发布应用程序,首先要申请发布证书。本节将讲解如何申请发布证书。

14.1.1 申请证书

以下是申请发布证书的具体步骤。

(1)在 Safari 的搜索栏中输入网址(https://developer.apple.com/ devcenter/ios/ index.action),然后按下回车键,进入 iOS Dev Center-App Developer 网页。

(2)单击 Log in 按钮,进入 Sign in with your Apple ID-Apple Developer 网页。在此网页中需要开发者输入 App ID 以及密码。然后单击 Sign In 按钮,此时会再次进入 iOS Dev Center-App Developer 网页。

(3)选择 Certificates,Identifiers&Profiles 选项,进入 Certificates,Identifiers &Profiles-App Developer 网页。

(4)选择 Certificates 选项,进入 iOS Certificates-Apple Developer 网页。如果是第一次进入此网页,并且没有申请发布证书会看到如图 14.1 所示的效果。

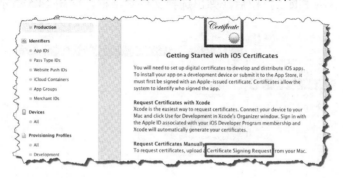

图 14.1 操作步骤 1

(5)在此网页中,选择蓝色的 Certificate Signing Request 字符串,进入 Add-iOS Certificates-AppleDeveloper 网页,如图 14.2 所示。

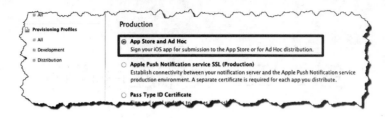

图 14.2　操作步骤 2

（6）选择 App Store and Ad Hoc 单选按钮。单击 Continue 按钮，进入 Request 选项卡的网页中，如图 14.3 所示。

图 14.3　操作步骤 3

（7）单击 Continue 按钮，进入 Generate 选项卡的网页中，如图 14.4 所示。

图 14.4　操作步骤 4

（8）选择 Choose File…按钮后，弹出选择文件对话框，如图 14.5 所示。

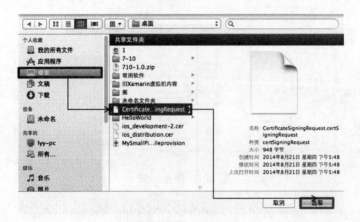

图 14.5　操作步骤 5

(9）选择在桌面的 CertificateSigningRequest.certSigningRequest 文件，此文件就是生成的证书签名申请（在第 1 章中讲解过此文件的申请），单击"选取"按钮。然后单击 Generate 按钮，进入 Download 选项卡的网页中，如图 14.6 所示。

图 14.6　操作步骤 6

（10）单击 Download 按钮，对生成的证书进行下载。下载后的证书名为 ios_distribution.cer。

（11）双击下载的 ios_distribution.cer 证书，将此证书添加到钥匙串中。

14.1.2　申请证书对应的配置文件（Provision File）

以下是申请发布证书对应的配置文件的具体步骤。

（1）如果开发者还处于下载证书的网页，可以选择此网页右侧的 Provisioning Profiles 的 Distribution 选项，进入 iOS Provisioning Profiles (Distribution)-Apple Developer 网页，如图 14.7 所示。

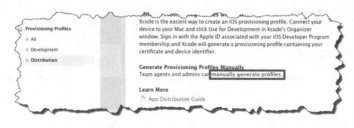

图 14.7　操作步骤 1

（2）选择蓝色的 manually generate profiles 字符串，进入 Add-iOS Provisioning Profile-Apple Developer 网页，如图 14.8 所示。

图 14.8　操作步骤 2

（3）选择 App Store，单击 Continue 按钮，进入 Configure 选项卡的网页中，如图 14.9 所示。

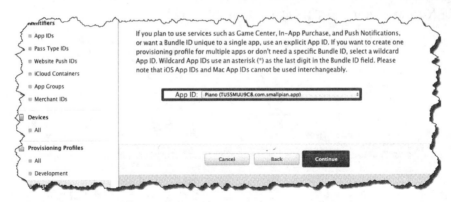

图 14.9　操作步骤 3

（4）选择 App ID，单击 Continue 按钮，进入 Configure 选项卡的选择证书的网页中，如图 14.10 所示。

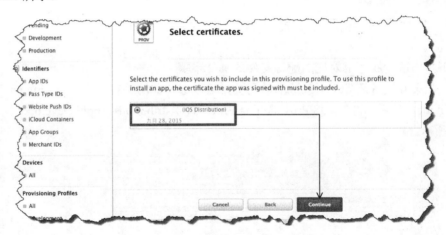

图 14.10　操作步骤 4

（5）选择某一个证书单选按钮，单击 Continue 按钮，进入 Generate 选项卡的网页中，如图 14.11 所示。

图 14.11　操作步骤 5

（6）输入配置的文件名，单击 Generate 按钮，进入 Download 选项卡的网页中，如图 14.12 所示。

图 14.12　操作步骤 6

（7）选择 Download 按钮，对 Provisioning Profiles 进行下载，下载后的文件为 MySmallPiano.mobileprovision。

（8）双击下载的 MySmallPiano.mobileprovision 文件，将此文件添加到 Organizer 的 Provisioning Profiles 中。

14.2　准备提交应用程序

在提交应用程序之前，需要做一些准备工作。否则，你的应用程序是无法提交的。本节将针对这些准备工作进行讲解。

14.2.1　创建应用及基本信息

要提交一个应用程序，首先需要在 iTunes Connect 的网页中对这个应用程序进行创建和填写一些基本信息。以下是它的具体操作步骤。

（1）在 Safari 的搜索栏中输入网址（http://itunesconnect.apple.com），然后按回车键，进入 iTunes Connect 的登录网页，如图 14.13 所示。

图 14.13　操作步骤 1

（2）输入苹果账号和密码后，单击跳转按钮进入服务条款的网页，如图 14.14 所示。

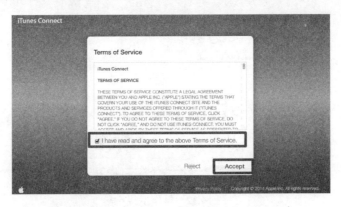

图 14.14　操作步骤 2

（3）选择 I have read and agree to the above Terms of Service 复选框，然后触摸 Accept 按钮，进入 iTunes Connect 的网页，如图 14.15 所示。

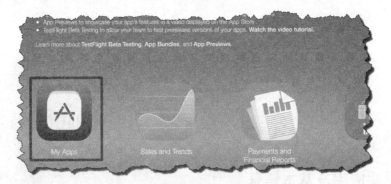

图 14.15　操作步骤 3

（4）选择 My Apps，进入 My Apps 网页，在这里面放置了一些上传的应用程序，如果你是第一次使用，就是空的，如图 14.16 所示。

图 14.16　操作步骤 4

（5）触摸+按钮，弹出下拉菜单，如图 14.17 所示。

第 14 章 发布应用程序

> 注意：如果在你的 My Apps 中存在一些应用程序，那么在触摸+按钮后，出现的下拉菜单如图 14.18 所示。可以选择其中的 New iOS App 创建一个新的 iOS 应用程序。

图 14.17　操作步骤 5　　　　　　　　图 14.18　下拉列表

（6）在弹出的下拉菜单中选择 New iOS App 选项，弹出 New iOS App 对话框，在此对话框中输入相应的内容，如图 14.19 所示。

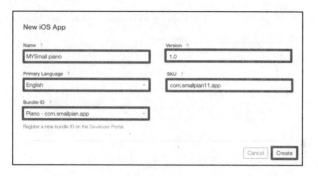

图 14.19　操作步骤 6

> 注意：App Name 必须是 App Store 未使用的，当填入的时候，系统会检查。Bundle ID 中输入应用程序标识符，它是在 iOS 开发中心的配置门户创建 App ID 时生成的，如果在配置门户网站中有就可以在下拉列表中找到。SKU 是应用程序编号，具有唯一性。

（7）单击 Create 按钮，进入此应用的详细信息的网页，如图 14.20 所示。开发者需要在此网页中输入一些信息。

图 14.20　操作步骤 7

· 453 ·

14.2.2 工程的相关设置

一个应用程序，在编程过程中，属性的设置并不影响开发，即使这些属性的设置是错误的。但是在发布时，正确地设置这些属性是很重要的。以下将讲解一些属性的设置。

1. 设置标识符

标识符在开发过程中对我们来说并没有什么影响，但是在发布时非常重要。实际上，我们在第 4 章就已经讲过如何在工程中设置它了。选择"引用"文件夹上方的工程名，选择 iOS Application 选项，在其中找到 Bundle Identifier 对标识符进行设置，如图 14.21 所示。

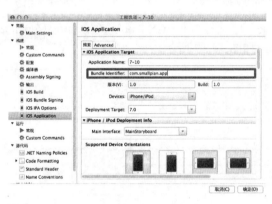

图 14.21　设置标识符

> 注意：这里的标识符是图 14.19 中使用的 Bundle Identifier。

2. 设置图标

每一个应用程序都必须要有图标，否则发布的应用程序是通不过审核的，图标的设置在第 1 章中也有讲过，如图 14.22 所示。

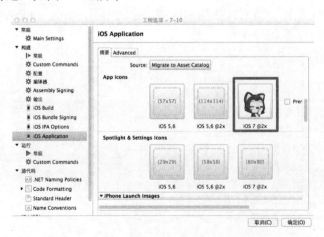

图 14.22　图标的设置

图标设置好后,单击"确定"按钮,退出设置。

3. 设置 iOS Bundle Signing

在运行按钮的右边有一个配置选项,将其设置为 Ad-Hoc,如图 14.23 所示。

图 14.23 操作步骤 1

选择"引用"文件夹上方的工程名,选择 iOS Bundle Signing 选项,将 Identity 设置为申请并下载的发布证书。将 Provisioning profile 设置为申请并下载的配置文件,如图 14.24 所示。

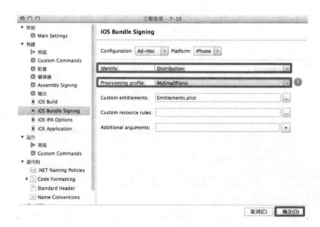

图 14.24 操作步骤 2

14.3 提交应用程序到 App Store 上

本节将讲解提交应用程序到 App Store 上的具体步骤。
(1) 选择"工程"|Zip App Bundle…命令,弹出 Save zipped app bundle 对话框,如图 14.25 所示。

图 14.25 操作步骤 1

(2)单击 Save 按钮,将打包的应用程序放在桌面。

(3)选择 Build|Archive 命令,打开 Archives,如图 14.26 所示。

图 14.26　操作步骤 2

(4)选择 Distribute…按钮,弹出"欢迎使用 Application Loader"对话框,如图 14.27 所示。

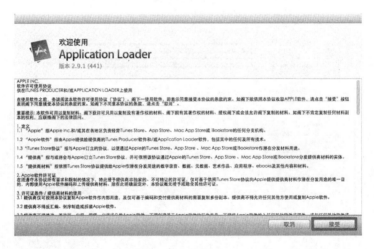

图 14.27　操作步骤 3

注意:从 2010 年年中开始,开发者上传软件必须使用 Application Loader 这个 MAC 机上的应用程序。如果您安装了最新版的 Xcode 开发环境,对于在 4.2 及以上版本,Developer/Applications/Utilities 目录中已经有 Application Loader 程序,无须单独安装。对于 Xcode 4.3 及以后版本,在/Applications/XCode.app/Contents/Applications 目录中可以找到(右键 Xcode 选择 Open Developer Tool 页可以看到 Application Loader)。当开发者第一次使用 Application Loader 程序,会看到如图 14.27、图 14.28 和图 14.29 所示的对话框,如果不是第一次使用就会看到如图 14.30 所示的对话框。

(5)单击"接受"按钮,输入苹果账号和密码,如图 14.28 所示。

第 14 章　发布应用程序

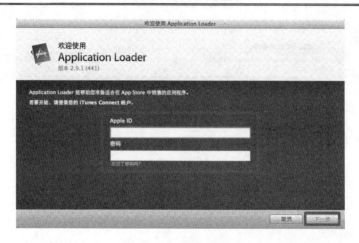

图 14.28　操作步骤 4

（6）单击"下一步"按钮，会看到如图 14.29 所示的对话框。

图 14.29　操作步骤 5

（7）单击"下一步"按钮，会看到如图 14.30 所示的对话框。

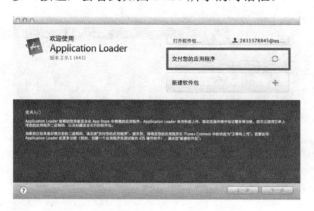

图 14.30　操作步骤 6

（8）单击"交付您的应用程序"按钮，弹出"选择应用程序"对话框，如图 14.31 所示。

第 3 篇　高级应用篇

图 14.31　操作步骤 7

（9）此时会出现你的 iTunes Connect 网页中创建的应用程序，选择在本章中我们创建的 MYSmall piano 应用程序，然后单击"下一步"按钮，弹出"应用程序信息"对话框，如图 14.32 所示。

图 14.32　操作步骤 8

（10）单击"选取…"按钮，弹出选择文件对话框，如图 14.33 所示。

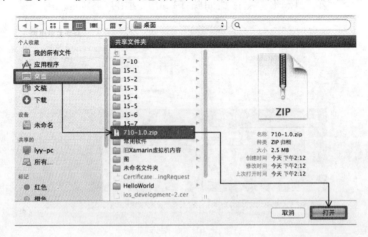

图 14.33　操作步骤 9

（11）选择打包的应用程序，单击"打开"按钮，弹出"正在添加应用程序…"对话框，如图 14.34 所示。

图 14.34　操作步骤 10

（12）单击"发送"按钮，打开"正在添加应用程序…"对话框，如图 14.35 所示。

图 14.35　操作步骤 11

（13）当添加完成后，会出现如图 14.36 所示的界面。

图 14.36　操作步骤 12

（14）单击"下一步"按钮，弹出"谢谢您"对话框，如图 14.37 所示。

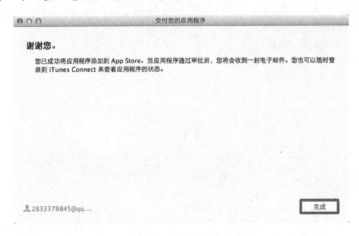

图 14.37　操作步骤 13

此时应用程序就提交到了 App Store 上，这时进行等待审核状态，审核通过后，应用程序就发布成功了。

注意：在添加应用程序时出现如图 14.38 所示的错误，它的解决方式就是在 iOS 开发中心的配置门户中，将创建的 App ID 中的 Associate Domains 选项取消，如图 14.39 所示。

图 14.38　错误

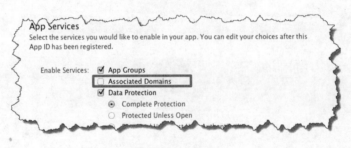

图 14.39　解决

14.4 常见审核不通过的原因

App Store 的审核是很严格的。苹果官方提供了一份详细的审核指南，包括 22 大项、100 多小项的拒绝上线条款，并且条款在不断地增加，此外，还包括一些模棱两可的条例，所以稍微有"闪失"，应用就有可能被拒绝。以下将介绍常见的应用被拒的原因。

1. 功能问题

在开发应用前，开发者一定要对产品进行认真的测试。如果你的应用程序存在崩溃、错误、使用非公开 API、有意提供隐蔽或虚假功能，却又不能明显表示的问题，无疑是被审核小组拒绝的对象。

2. 界面问题

苹果审核指南规定开发者的应用必须遵守《iOS 用户界面指导原则》中解释的所有条款和条件，如果违反这些规则，就会拒绝上线。

3. 商业问题

在发布应用时，首先不可以侵犯苹果公司的商标和版权。也就是在应用中不能出现苹果的图标，不能使用苹果公司现有产品的类似名称为应用命名。

4. 不当内容

一些不合适，不合法的内容，苹果公司也不允许上架，如涉嫌诽谤、侮辱、狭隘内容、打击个人或团体的应用；展示人或动物被杀戮、致残、枪击、针刺或其他伤害的真实图片的应用；描述暴力或虐待儿童的应用；含有韦氏词典中定义的色情素材的应用等。

> 注意：Steve Jobs 非常在意 App Store 的色情内容。他曾说："如果你想要色情内容，那么就用 Android 手机吧"。

5. 其他

除以上这些内容被拒绝外，还有位置、推送通知、iAD 相关的、媒体内容、购买与流通、抓取和聚合、设备损害、暴力等存在的问题也会被拒绝。

> 说明：详细内容请开发者参考 iOS APP 审核规则（https://developer.apple.com/app-store/review/）。

第15章 高级功能

本章将讲解一些有关 iOS 开发的高级内容。例如，在创建的工程中显示分页内容，类似于书一样的导航效果，即卷页效果；使用粒子系统实现礼花的效果；实现自定义过渡动画；在 UI 的元素中添加仿真物理引擎等。

15.1 卷页效果

在第 11 章中卷页的效果已经实现过了。本节将使用另外一种方式实现卷页效果——UIPageViewController。UIPageViewController 在 iOS 5 中被推出。它封装了卷页功能的操作。

【示例 15-1】 以下将使用 UIPageViewController 类实现卷页的效果。具体操作步骤如下所述。

（1）创建一个 Single View Application 类型的工程。
（2）在工程中创建一个文件夹，命名为 Images。
（3）添加图像 1.jpg、2.jpg 和 3.jpg 到创建工程的 Images 文件夹中。
（4）在创建的工程中添加一个类型为 iPhone View Controller 的文件，命名为 Page。
（5）打开 Page.xib 文件，对主视图进行设置，效果如图 15.1 所示。

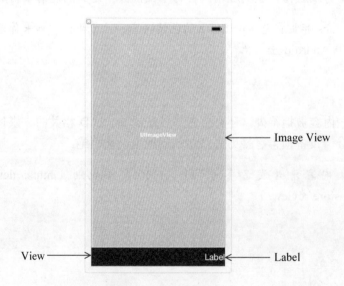

图 15.1　主视图的效果

需要添加的视图以及设置，如表 15-1 所示。

表 15-1 设置主视图

视　　图	设　　置
Image View	位置和大小：(0,0,320,525) 为图像视图声明插座变量 imgView
Label	Color：白色 Font：System 20.0 Alignment：居中 位置和大小：(0,525,320,43) 为按钮声明插座变量 lblPageNumber
View	Background：深灰色 位置和大小：(0,525,320,43)

（6）回到 Xamarin Studio，打开 Page.cs 文件，编写代码，实现对 Page 主视图的设置。代码如下：

```
using System;
using System.Drawing;
using MonoTouch.Foundation;
using MonoTouch.UIKit;
namespace Application
{
    public partial class Page : UIViewController
    {
        //Page 视图控制器的构造方法
        public Page (int pageIndex) : base ("Page", null)
        {
            this.PageIndex = pageIndex;
        }
        //设置 PageIndex 属性
        public int PageIndex
        {
            get;
            private set;
        }
        ……                          //这里省略了视图控制器的析构方法
        public override void ViewDidLoad ()
        {
            base.ViewDidLoad ();
            imgView.Image = UIImage.FromFile(string.Format ("Images/{0}.jpg", this. PageIndex + 1));
            lblPageNumber.Text = string.Format("Page {0}", this.PageIndex + 1);
        }
    }
}
```

（7）打开 15-1ViewController.cs 文件，编写代码，实现卷页动画效果。代码如下：

```
using System;
using System.Drawing;
using MonoTouch.Foundation;
using MonoTouch.UIKit;
namespace Application
{
```

```csharp
public partial class _5_1ViewController : UIViewController
{
    private UIPageViewController pageViewController;
    private int pageCount = 3;
    ……                                    //这里省略了视图控制器的构造方法和析构方法
    #region View lifecycle
    public override void ViewDidLoad ()
    {
        base.ViewDidLoad ();
        Page firstPage = new Page(0);                    //得到第一页
        //实例化对象
        this.pageViewController = new UIPageViewController (UIPage
        ViewController Transition Style. PageCurl, UIPageView
        Controller Navigation Orientation.Horizontal, UIPageView Con
        troller Spine Location. Min);
        //设置首页
        this.pageViewController.SetViewControllers(new UIViewCon
        troller [] { firstPage }, UIPageViewController Navigation 
        Direction. Forward,false, s => { });
        //下一页
        this.pageViewController.GetNextViewController = this.Get Next
        View Controller;
        //上一页
        this.pageViewController.GetPreviousViewController = this.Get
        Previous ViewController;
        this.pageViewController.View.Frame = this.View.Bounds;
        //设置视图的框架
        this.View.AddSubview(this.pageViewController.View);
        //为主视图添加视图
    }
    //返回下一个ViewController对象
    private UIViewController GetNextViewController(UIPageView Con
    troller pageController, UIViewController referenceViewController)
    {
        Page currentPageController = referenceViewController as Page;
        //判断当前的页的索引是否为最后一页
        if (currentPageController.PageIndex >= (this.pageCount - 1))
        {
            return null;
        } else
        {
            int nextPageIndex = currentPageController.PageIndex + 1;
            return new Page(nextPageIndex);
        }
    }
    //返回上一个ViewController对象
    private UIViewController GetPreviousViewController (UIPageView
    Controller pageController, UIViewController referenceView Controller)
    {
        Page currentPageController = referenceViewController as Page;
        //判断当前的页的索引是否为首页
        if (currentPageController.PageIndex <= 0)
        {
            return null;
        } else
        {
            int previousPageIndex =currentPageController.PageIndex- 1;
            return new Page(previousPageIndex);
        }
```

```
        }
        ……                    //这里省略了视图加载和卸载前后的一些方法
        #endregion
    }
}
```

运行效果如图 15.2 所示。

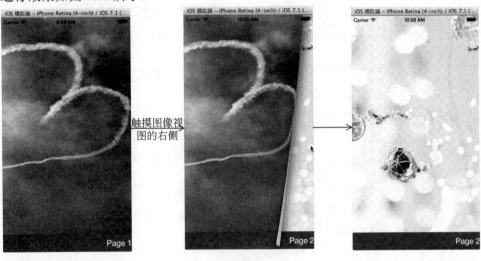

图 15.2　运行效果

15.2　粒子系统

CAEmitterLayer 提供了一个基于 Core Animation 的粒子发射系统，使用它可以实现各种各样的动画效果。

【示例 15-2】以下将使用 CAEmitterLayer 类实现一个礼花的效果。具体的操作步骤如下所述。

（1）创建一个 Single View Application 类型的工程。
（2）添加图像 1.png 和 2.png 到创建工程的 Resources 文件夹中。
（3）打开 15-2ViewController.cs 文件，编写代码，实现礼花效果。代码如下：

```
using System;
using System.Drawing;
using MonoTouch.Foundation;
using MonoTouch.UIKit;
using MonoTouch.CoreAnimation;
namespace Application
{
    public partial class _5_2ViewController : UIViewController
    {
        ……                    //这里省略了视图控制器的构造方法和析构方法
        #region View lifecycle
        public override void ViewDidLoad ()
        {
            base.ViewDidLoad ();
```

```csharp
CAEmitterLayer fireworksEmitter = new CAEmitterLayer();
RectangleF viewBounds = this.View.Layer.Bounds;
//设置发射位置
fireworksEmitter.Position = new PointF (viewBounds.Size.Width
/ 2, viewBounds.Size.Height);
fireworksEmitter.Size = new SizeF (viewBounds.Size.Width / 2,
0.0f);                                      //发射源的尺寸大
fireworksEmitter.Mode = CAEmitterLayer.ModeOutline;
                                            //发射模式
fireworksEmitter.Shape = CAEmitterLayer.ShapeLine;
                                            //发射形状
fireworksEmitter.RenderMode = CAEmitterLayer.RenderAdditive;
                                            //渲染模式
CAEmitterCell rocket = CAEmitterCell.EmitterCell();
rocket.BirthRate = 1.0f;             //粒子参数的速度乘数因子
rocket.EmissionRange = 0.25f *(float) Math.PI;   //周围发射角度
rocket.Velocity = 380;               //速度
rocket.VelocityRange = 100;          //速度范围
rocket.AccelerationY = 75;           //粒子 y 方向的加速度分量
rocket.LifeTime = 1.02f;
                                     //粒子展现的图片
rocket.Contents = UIImage.FromFile ("1.png").CGImage;
rocket.Scale = 0.2f;                 //缩放
rocket.Color = UIColor.Red.CGColor;//粒子的颜色
rocket.GreenRange = 1.0f;            //一个粒子的绿颜色能改变的范围
rocket.RedRange = 1.0f;              //一个粒子的红颜色能改变的范围
rocket.BlueRange = 1.0f;             //一个粒子的蓝颜色能改变的范围
rocket.SpinRange = (float)Math.PI;  //子旋转角度范围
CAEmitterCell burst = CAEmitterCell.EmitterCell();
burst.BirthRate = 1.0f;
burst.Velocity = 0;
burst.Scale = 2.5f;
burst.RedSpeed = -1.5f;
burst.BlueSpeed = 1.5f;
burst.GreenSpeed = 1.0f;
burst.LifeTime = 0.35f;              //生命周期范围
CAEmitterCell spark =CAEmitterCell.EmitterCell();
spark.BirthRate = 400;
spark.Velocity = 125;
spark.EmissionRange = 2 * (float)Math.PI;
spark.AccelerationY = 75;
spark.LifeTime = 3;
spark.Contents = UIImage.FromFile ("2.png").CGImage;
spark.ScaleSpeed = -0.2f;
spark.GreenSpeed = -0.1f;
spark.RedSpeed = 0.4f;
spark.BlueSpeed = -0.1f;
spark.AlphaSpeed = -0.25f;
spark.Spin = 2 * (float)Math.PI;
spark.SpinRange = 2 * (float)Math.PI;
CAEmitterCell[] r = { rocket };
CAEmitterCell[] b = { burst };
CAEmitterCell[] s = {spark };
fireworksEmitter.Cells = r;
rocket.Cells = b;
burst.Cells = s;
this.View.Layer.AddSublayer (fireworksEmitter);
}
```

第 15 章　高级功能

```
      ……             //这里省略了视图加载和卸载前后的一些方法
      #endregion
    }
}
```

运行效果如图 15.3 所示。

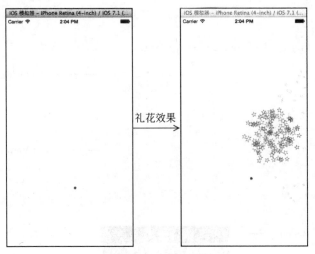

图 15.3　运行效果

15.3　内 容 共 享

在很多的应用程序中都有内容共享功能，如图 15.4 所示。

图 15.4　内容共享

要实现内容共享可以通过使用 UIActivityViewController 控制器，它可以为应用程序提供各种服务，例如发送短消息和邮件、复制内容到剪贴板，以及发布消息到 Twitter、Facebook

及微博。本节将讲解内容共享的实现。

【示例 15-3】 以下将添加内容共享功能到创建的应用程序中。具体的操作步骤如下所述。

（1）创建一个 Single View Application 类型的工程。

（2）打开 MainStoryboard.storyboard 文件，对主视图进行设置，效果如图 15.5 所示。

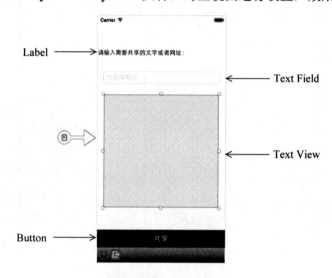

图 15.5　主视图的效果

需要添加的视图以及设置，如表 15-2 所示。

表 15-2　设置主视图

视图	设置
Label	Text：请输入需要共享的文字或者网址： Font：System 15 pt 位置和大小：(0,72,292,37)
Text Field	Name：tf Text：（空） Font：System 16 pt Placeholder：内容或网址 位置和大小：(14,137,292,30)
Text View	Name：tv Text：（空） Font：System 16 pt 取消 Editable 复选框
Button	Name：btn Title：共享 Font：System 18.0 Text Color：白色 Background：黑色 位置和大小：(0,525,320,43)

（3）打开 15-3ViewController.cs 文件，编写代码，实现触摸按钮，弹出 UIActivityViewController 控制器。代码如下：

```
using System;
using System.Drawing;
using MonoTouch.Foundation;
using MonoTouch.UIKit;
namespace Application
{
    public partial class _5_3ViewController : UIViewController
    {
        private UIActivityViewController shareController;
        ……                              //这里省略了视图控制器的构造方法和析构方法
        #region View lifecycle
        public override void ViewDidLoad ()
        {
            base.ViewDidLoad ();
            //触摸 return 键，关闭键盘
            tf.ShouldReturn=delegate{
                tf.ResignFirstResponder();
                tv.Text=tf.Text;
                return true;
            };
            //触摸按钮，打开 UIActivityViewController 控制器
            btn.TouchUpInside += async (sender, e) => {
                NSString contents = new NSString(tv.Text);
                //实例化 UIActivityViewController 对象
                this.shareController = new UIActivityViewController(new
                NSObject[] {
                    contents
                } , null);
                //当 UIActivityViewController 控制器退出后执行的方法
                this.shareController.CompletionHandler = this.Activity
                Completed;
                await this.PresentViewControllerAsync(this.shareCon
                troller, true);
            } ;
        }
        //实现警告视图的弹出
        private void ActivityCompleted(NSString activityType, bool
        completed)
        {
            //判断布尔值是否共享
            if (completed == true)
            {
                UIAlertView alert = new UIAlertView ();
                alert.Title = "已完成共享";
                alert.AddButton ("Cancel");
                alert.Show ();
            } else
            {
                UIAlertView alert = new UIAlertView ();
                alert.Title = "由于某些原因，共享失败";
                alert.AddButton ("Cancel");
                alert.Show ();
            }
        }
        ……                              //这里省略了视图加载和卸载前后的一些方法
        #endregion
    }
}
```

运行效果如图 15.6 所示。

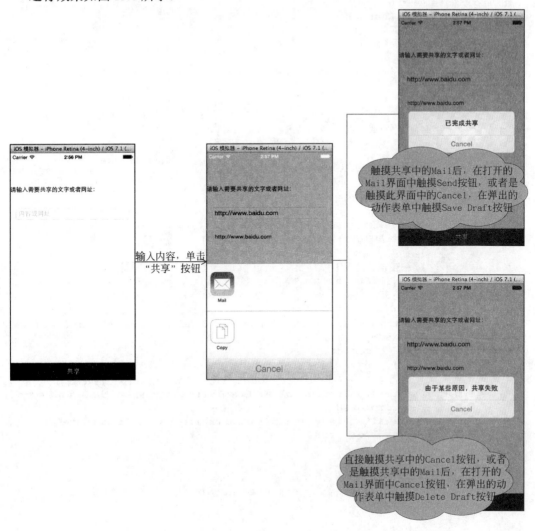

图 15.6　运行效果

15.4　动作表单

UIActionSheet（动作表单）提供了从屏幕底部向上滚动的菜单，同 UIAlertView 一样，它也是一个模态对话框，如图 15.7 所示。本节将讲解动作表单的使用。

【示例 15-4】　以下将在自制的应用程序中使用动作表单，让其实现背景的选择。具体的操作步骤如下所述。

（1）创建一个 Single View Application 类型的工程。

（2）添加图像 1.jpg、2.png 和 3.png 到创建工程的 Resources 文件夹中。

（3）打开 MainStoryboard.storyboard 文件，对主视图进行设置，效果如图 15.8 所示。

第 15 章 高级功能

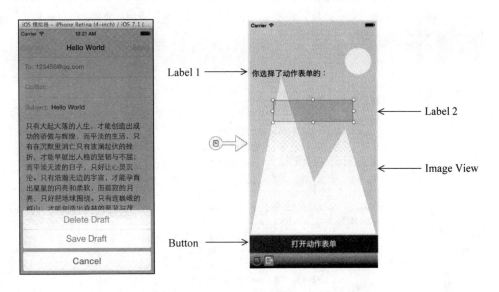

图 15.7 动作表单　　　　图 15.8 主视图的效果

需要添加的视图以及设置，如表 15-3 所示。

表 15-3 设置主视图

视　　图	设　　置
Label1	Text：你选择了动作表单的: Font：System 19 pt 位置和大小：(8,118,210,21)
Label2	Name：label Text：（空） Font：System Bold 25 pt Alignment：居中 位置和大小：(62,196,197,50)
Button	Name：btn Title：打开动作表单 Font：System 19.0 Text Color：白色 Background：深灰色 位置和大小：(0,525,320,43)
Image View	Name：iv 位置和大小：(0,0,320,525)

（4）打开 15-4ViewController.cs 文件，编写代码，实现触摸按钮，弹出动作表单的功能。代码如下：

```
using System;
using System.Drawing;
using MonoTouch.Foundation;
using MonoTouch.UIKit;
namespace Application
{
    public partial class _5_4ViewController : UIViewController
    {
```

• 471 •

```csharp
......                           //这里省略了视图控制器的构造方法和析构方法
#region View lifecycle
public override void ViewDidLoad ()
{
    base.ViewDidLoad ();
    btn.TouchUpInside += (sender, e) => {
        UIActionSheet sheet=new UIActionSheet();
        sheet.Title="选择背景";
        sheet.AddButton("风景壁纸");
        sheet.AddButton("人物壁纸");
        sheet.AddButton("卡通壁纸");
        sheet.AddButton("默认");
        sheet.AddButton("Cancel");
        sheet.ShowInView(this.View);                     //显示动作表单
        //响应动作表单
        sheet.Dismissed += (sender1, e1) =>{
            if (e1.ButtonIndex == 0)
            {
                label.Text="风景壁纸";
                iv.Image=UIImage.FromFile("1.jpg");
            } else if (e1.ButtonIndex == 1)
            {
                label.Text="人物壁纸";
                iv.Image=UIImage.FromFile("2.png");
            } else if (e1.ButtonIndex == 2)
            {
                label.Text="卡通壁纸";
                iv.Image=UIImage.FromFile("3.png");
            } else if (e1.ButtonIndex == 3)
            {
                label.Text="默认";
                iv.Image=UIImage.FromFile("");
            } else if (e1.ButtonIndex == 4)
            {
                UIAlertView alert=new UIAlertView();
                alert.Title="退出动作表单" ;
                alert.AddButton("Cancel");
                alert.Show();
            }
        };
    };
}
......                           //这里省略了视图加载和卸载前后的一些方法
#endregion
```

运行效果如图 15.9 所示。

15.5　实现自定义过渡动画

在第 11 章中所讲的过渡动画都是 iOS 所提供的。本节将讲解自定义的过渡动画。

【示例 15-5】 以下将实现一个自定义的过渡动画效果。具体的操作步骤如下所述。

（1）创建一个空类型的工程。

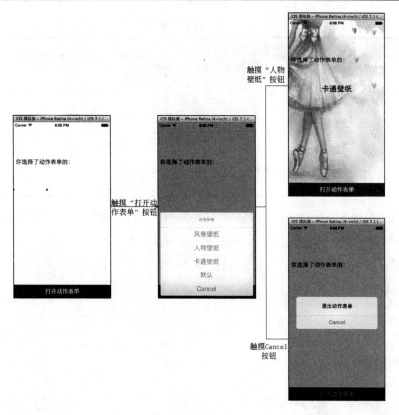

图 15.9　运行效果

（2）添加图像 1.jpg 到创建工程的 Resources 文件夹中。

（3）在创建的工程中添加一个类型为 iPhone View Controller 的文件，命名为 Custom Transition AppViewController。

（4）打开 CustomTransitionAppViewController.xib 文件，对主视图进行设置，效果如图 15.10 所示。

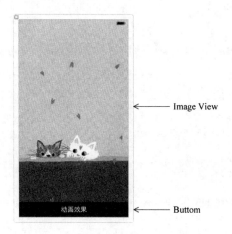

图 15.10　主视图的效果

需要添加的视图以及设置，如表 15-4 所示。

表 15-4 设置主视图

视图	设置
Image View	Image: 1.jpg 位置和大小: (0,0,320,568)
Button	Title: 动画效果 Font: System 19.0 Text Color: 白色 Background: 深灰色 位置和大小: (0,525,320,43) 为按钮声明插座变量 btnPresent

（5）打开 CustomTransitionAppViewController.cs 文件，编写代码，实现自定义的动画效果。代码如下：

```
using System;
using System.Drawing;
using MonoTouch.Foundation;
using MonoTouch.UIKit;
namespace Application
{
    public partial class CustomTransitionAppViewController : UIViewController
    {
        ……
        public override void ViewDidLoad ()
        {
            base.ViewDidLoad ();
            this.btnPresent.TouchUpInside += async (sender, e) => {
                ModalController modalController = new ModalController();
                //设置显示风格
                modalController.ModalPresentationStyle = UIModalPresentationStyle.Custom;
                modalController.TransitioningDelegate = new MyTransitionDelegate();     //设置委托
                await this.PresentViewControllerAsync(modalController, true);
            } ;
        }
        //创建 UIViewControllerAnimatedTransitioning 类的子类 MyTransitionAnimator
        public class MyTransitionAnimator : UIViewControllerAnimatedTransitioning
        {
            //设置属性
            public bool IsPresenting
            {
                get;
                set;
            }
            //过渡动画的持续时间
            public override double TransitionDuration (IUIViewControllerContextTransitioning transitionContext)
            {
                return 1;
            }
            //自定义过渡动画
```

第15章 高级功能

```csharp
public override void AnimateTransition (IUIViewController
Context Transitioning transitionContext)
{
    if (this.IsPresenting)
    {
        //创建实现过渡动画的容器视图
        UIView containerView = transition Context. Container
        View;
        //获取切出的视图控制器
        UIViewController toViewController = transitionContext.
        GetViewControllerForKey (UITransitionContext. ToView
        ControllerKey);
        containerView.AddSubview(toViewController.View);
        //为容器视图添加视图
        RectangleF frame = toViewController.View.Frame;
        toViewController.View.Frame = RectangleF.Empty;
        //动画效果
        UIView.Animate(this.TransitionDuration(transition
        Context),
            () => toViewController.View.Frame = new RectangleF
            (0f, 0f, frame.Width, frame.Height ),
            () => transitionContext.CompleteTransition
            (true));
    } else
    {
        //获取切入的视图控制器
        UIViewController fromViewController = transition
        Context.GetViewControllerForKey (UITransitionContext.
        FromView ControllerKey);
        RectangleF frame = fromViewController.View.Frame;
        frame = RectangleF.Empty;
        //动画效果
        UIView.Animate(this.TransitionDuration(transition
        Context),
            () => fromViewController.View.Frame = frame,
            () => transitionContext.CompleteTransition
            (true));
    }
}
}
//创建 UIViewControllerTransitioningDelegate 类的子类 MyTransition
Delegate
public class MyTransitionDelegate : UIViewController
TransitioningDelegate
{
    private MyTransitionAnimator animator;
    //重写显示视图控制器时的动画
    public override IUIViewControllerAnimatedTransitioning
    PresentingController (UIViewController presented, UIView
    Controller presenting, UIViewController source)
    {
        this.animator = new MyTransitionAnimator();
        this.animator.IsPresenting = true;
        return this.animator;
    }
    //重写退出视图控制器时的动画
    public override IUIViewControllerAnimatedTransitioning
    GetAnimationControllerForDismissedController (UIView
    Controller dismissed)
    {
```

```
                this.animator.IsPresenting = false;
                return this.animator;
            }
        }
    }
}
```

（6）在创建的工程中添加一个类型为 iPhone View Controller 的文件，命名为 ModalController。

（7）打开 ModalController.xib 文件，将主视图的背景设置为浅蓝色。

（8）打开 ModalController.cs 文件，编写代码，实现退出视图控制器。代码如下：

```
using System;
using System.Drawing;
using MonoTouch.Foundation;
using MonoTouch.UIKit;
namespace Application
{
    public partial class ModalController : UIViewController
    {
        ……
        public override void ViewDidLoad ()
        {
            base.ViewDidLoad ();
            UIButton btnDismiss = new UIButton ();
            btnDismiss.Frame = new RectangleF (0, 0, 320, 568);
            this.View.AddSubview (btnDismiss);
            //触摸按钮，退出此视图控制器
            btnDismiss.TouchUpInside +=
                async (sender, e) => await this.DismissViewControllerAsync
                (true);
        }
    }
}
```

（9）打开 AppDelegate.cs 文件，编写代码，实现使用控制器加载视图的功能。代码如下：

```
using System;
using System.Collections.Generic;
using System.Linq;
using MonoTouch.Foundation;
using MonoTouch.UIKit;
namespace Application
{
    [Register ("AppDelegate")]
    public partial class AppDelegate : UIApplicationDelegate
    {
        UIWindow window;
        CustomTransitionAppViewController viewController;
        public override bool FinishedLaunching (UIApplication app,
        NSDictionary options)
        {
            window = new UIWindow (UIScreen.MainScreen.Bounds);
            viewController = new CustomTransitionAppViewController ();
            window.RootViewController = viewController;
            window.MakeKeyAndVisible ();
            return true;
        }
```

        }
}

运行效果如图 15.11 所示。

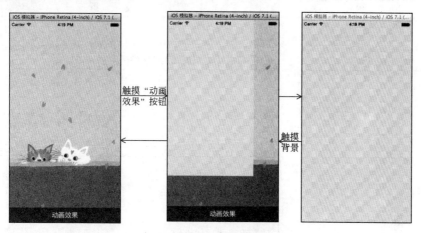

图 15.11　运行效果

15.6　在 UI 元素中使用物理引擎

UIDynamicAnimator 类可以实现物理仿真，它是一个仿真物理引擎。所谓物理仿真就是模拟现实中的物理现象，如重力、碰撞和弹性等现象。本节将讲解此类的使用。

【示例 15-6】以下将实现当触摸 Drop 按钮后，蘑菇就会自由下落，遇到边界或者其他物体时就会出现碰撞行为。触摸 Reset 按钮后，蘑菇就会回到原先的位置。具体的操作步骤如下所述。

（1）创建一个 Single View Application 类型的工程。
（2）添加图像 1.png 到创建工程的 Resources 文件夹中。
（3）打开 MainStoryboard.storyboard 文件，对主视图进行设置，效果如图 15.12 所示。

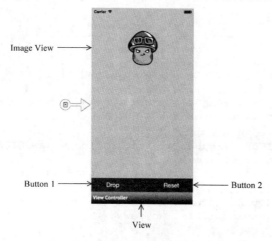

图 15.12　主视图的效果

需要添加的视图以及设置，如表 15-5 所示。

表 15-5 设置主视图

视图	设置
Image View	Name：imgView Image：1.png 位置和大小：(106,52,109,136)
Button1	Name：dbtn Title：Drop Font：System 19.0 Text Color：白色 位置和大小：(20,6,96,30)
Button2	Name：rbtn Title：Reset Font：System 19.0 Text Color：白色 位置和大小：(204,6,96,30)

（4）打开 15-6ViewController.cs 文件，编写代码，实现当触摸 Drop 按钮后，蘑菇就会自由下落，遇到边界或者其他物体时就会出现碰撞行为。触摸 Reset 按钮后，蘑菇就会回到原先的位置。代码如下：

```
using System;
using System.Drawing;
using MonoTouch.Foundation;
using MonoTouch.UIKit;
namespace Application
{
    public partial class _5_6ViewController : UIViewController
    {
        private RectangleF imageRect;
        private UIDynamicAnimator animator;
        ……              //这里省略了视图控制器的构造方法和析构方法
        #region View lifecycle
        public override void ViewDidLoad ()
        {
            base.ViewDidLoad ();
            this.View.InsertSubviewBelow(this.imgView, this.rbtn);
            this.animator = new UIDynamicAnimator(this.View);

            this.dbtn.TouchUpInside += (sender, e) => {
                //实例化重力行为对象
                UIGravityBehavior gravity = new UIGravityBehavior
                (this.imgView);
                //实例化碰撞行为对象
                UICollisionBehavior collision = new UICollisionBehavior
                (this.imgView);
                collision.TranslatesReferenceBoundsIntoBoundary = true;
                //实例化动力行为对象
                UIDynamicItemBehavior dynBehavior = new UIDynamicItem
                Behavior (this.imgView);
                dynBehavior.Density = 1f;
                //相对密度
                dynBehavior.Elasticity = 0.7f;
```

```
            //弹性系数
            dynBehavior.Friction = 1f;                      //摩擦系数
            //添加动力行为
            this.animator.AddBehaviors(gravity, collision, dynBehavior);
        };
        //触摸按钮,让图像回到原先的位置
        this.rbtn.TouchUpInside += (sender, e) => {
            this.animator.RemoveAllBehaviors();
            //删除所有的行为
            imageRect=new RectangleF(106,53,109,136);
            this.imgView.Frame = this.imageRect;
        };
    }
    ……                          //这里省略了视图加载和卸载前后的一些方法
    #endregion
}
```

运行效果如图 15.13 所示。

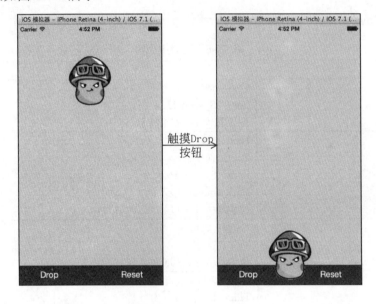

图 15.13　运行效果

15.7　实现文本到语言的功能

AVSpeechSynthesizer 可以实现 Siri 上集成的语言混合功能。本节将讲解此类的使用。

【示例 15-7】 以下将使用 AVSpeechSynthesizer 类实现文本到语言的功能。具体的操作步骤如下所述。

（1）创建一个 Single View Application 类型的工程。

（2）打开 MainStoryboard.storyboard 文件，对主视图进行设置，效果如图 15.14 所示。

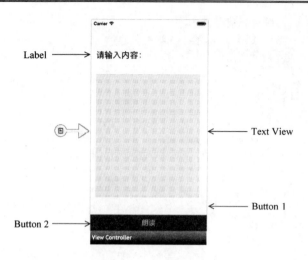

图 15.14 主视图的效果

需要添加的视图以及设置如表 15-6 所示。

表 15-6 设置主视图

视 图	设 置
Label	Text：请输入内容： Font：System 23 pt 位置和大小：(14,72,200,45)
Text View	Name：tv Text：（空） Font：System 16.0 Background：浅灰色 位置和大小：(14,146,286,332)
Button1	Name：cbtn Title：（空） 位置和大小：(0,0,320,568)
Button2	Name：sbtn Title：朗读 Font：System 19.0 Text Color：白色 Background：深灰色 位置和大小：(0,525,320,43)

（3）打开 15-7ViewController.cs 文件，编写代码，实现对输入文本的朗读。代码如下：

```
using System;
using System.Drawing;
using MonoTouch.Foundation;
using MonoTouch.UIKit;
using MonoTouch.AVFoundation;
namespace Application
{
    public partial class _5_7ViewController : UIViewController
    {
        ……                      //这里省略了视图控制器的构造方法和析构方法
```

第 15 章 高级功能

```
        #region View lifecycle
        public override void ViewDidLoad ()
        {
            base.ViewDidLoad ();
            //触摸按钮，关闭键盘
            cbtn.TouchUpInside += (sender, e) => {
                tv.ResignFirstResponder();
            } ;
            //触摸按钮，发音
            sbtn.TouchUpInside += (sender, e) => {
                AVSpeechSynthesizer synth = new AVSpeechSynthesizer();
                AVSpeechUtterance utterance = new AVSpeechUtterance
                (tv.Text);
                utterance.Rate = 0.3f;                  //设置语速快慢
                //对讲话的语音进行设置
                utterance.Voice = AVSpeechSynthesisVoice. FromLanguage
                ("en-US");
                synth.SpeakUtterance(utterance);
                //创建的一段讲话放到语音合成器中，形成音频
            } ;
        }
        ……                       //这里省略了视图加载和卸载前后的一些方法
        #endregion
    }
}
```

运行效果如图 15.15 所示。

图 15.15　运行效果